Studies in Systems, Decision and Control

Volume 663

The series "Studies in Systems, Decision and Control" (SSDC) covers both new developments and advances, as well as the state of the art, in the various areas of broadly perceived systems, decision making and control–quickly, up to date and with a high quality. The intent is to cover the theory, applications, and perspectives on the state of the art and future developments relevant to systems, decision making, control, complex processes and related areas, as embedded in the fields of engineering, computer science, physics, economics, social and life sciences, as well as the paradigms and methodologies behind them. The series contains monographs, textbooks, lecture notes and edited volumes in systems, decision making and control spanning the areas of Cyber-Physical Systems, Autonomous Systems, Sensor Networks, Control Systems, Energy Systems, Automotive Systems, Biological Systems, Vehicular Networking and Connected Vehicles, Aerospace Systems, Automation, Manufacturing, Smart Grids, Nonlinear Systems, Power Systems, Robotics, Social Systems, Economic Systems and other. Of particular value to both the contributors and the readership are the short publication timeframe and the world-wide distribution and exposure which enable both a wide and rapid dissemination of research output.

Indexed by SCOPUS, DBLP, WTI Frankfurt eG, zbMATH, SCImago.

All books published in the series are submitted for consideration in Web of Science.

Fadele Ayotunde Alaba · Uhuegho Kole Osaretin · Alvaro Rocha

Fundamentals of Qualitative Research

Springer

Fadele Ayotunde Alaba
Department of Computer Science
Federal University of Education, Zaria
Zaria, Nigeria

Uhuegho Kole Osaretin
Nigerian College of Aviation Technology
Zaria, Nigeria

Alvaro Rocha
ISEG Lisbon School of Economics
and Management
University of Lisbon
Lisbon, Portugal

ISSN 2198-4182 ISSN 2198-4190 (electronic)
Studies in Systems, Decision and Control
ISBN 978-3-032-18426-9 ISBN 978-3-032-18427-6 (eBook)
https://doi.org/10.1007/978-3-032-18427-6

This Springer imprint is published by the registered company Springer Nature Switzerland AG
The registered company address is: Gewerbestrasse 11, 6330 Cham, Switzerland

Introduction

Purpose and Scope of the Book

The primary purpose of this book is to provide readers with a clear and comprehensive understanding of research methodologies, with particular emphasis on the application of qualitative research. It aims to bridge the gap between theory and practice by offering conceptual explanations alongside practical illustrations. The scope extends to students, early-career researchers, and practitioners who require methodological knowledge to design, conduct, and interpret studies effectively. By addressing foundational concepts as well as contemporary applications, the book ensures that readers develop both technical skills and critical thinking abilities essential for research in education, health, and social sciences.

Structure of the Book

The book is organised in a logical sequence to guide readers from fundamental principles to advanced applications. It begins with an introduction to the philosophy of research, followed by chapters on qualitative, quantitative, and mixed-methods designs. Subsequent sections address data collection techniques, sampling strategies, ethical considerations, and statistical tools. Each chapter is designed to balance theoretical discussion with real-world case studies, ensuring accessibility for novices while still providing depth for more experienced readers. This structure allows for progressive learning while enabling selective reading based on individual research needs.

Significance of Qualitative Research

Qualitative research is significant because it offers systematic, objective, and replicable ways to investigate phenomena. Through numerical analysis, researchers can measure variables, test hypotheses, and establish relationships among constructs. This approach enhances the precision and generalisability of findings, making it indispensable in fields that demand evidence-based decision-making, such as healthcare, education, business, and public policy. Qualitative research not only supports predictive modelling but also enables comparisons across contexts and populations. Its significance lies in its ability to transform abstract concepts into measurable indicators, thereby advancing scientific knowledge and practical interventions.

The Researcher's Role

The role of the researcher in qualitative inquiry extends beyond data collection and analysis. Researchers must act as designers, ensuring that their studies are methodologically rigorous and ethically sound. They are responsible for formulating research questions, selecting appropriate instruments, and maintaining objectivity throughout the process. At the same time, researchers must acknowledge their positionality, recognising that even qualitative investigations are influenced by human choices in design, interpretation, and reporting. Upholding ethical standards, such as informed consent, confidentiality, and transparency, is integral to the researcher's role. By combining technical competence with ethical responsibility, researchers enhance the trustworthiness and impact of their work.

Conclusion

In summary, this book seeks to equip readers with the knowledge and tools to engage confidently with research processes. By outlining its purpose, structure, and scope, the text establishes a foundation for understanding both the philosophy and practice of research. The significance of qualitative inquiry is highlighted as a key contributor to evidence-based knowledge, while the role of the researcher is underscored as central to methodological integrity. Together, these sections set the stage for deeper exploration in subsequent chapters, reinforcing the book's commitment to fostering critical, ethical, and effective research practice.

Contents

Chapter 1
Introduction to the Key Features of Qualitative Research

1.1 Introduction to Qualitative Research

Qualitative research is a vital scientific methodology aimed at understanding human experiences, attitudes, behaviours, and cultural contexts. Unlike qualitative research, which relies on numerical data, qualitative research focuses on narrative data gathered through interviews, observations, and documents. Its primary goal is to explore and interpret the complexities of human experiences rather than to generalize findings statistically. The philosophical foundations of qualitative research are rooted in interpretivist and constructivist epistemologies. Interpretivism emphasizes understanding the subjective meanings individuals assign to their experiences, while constructivism highlights the co-creation of knowledge between researchers and participants. These perspectives allow qualitative research to delve into how individuals perceive and make sense of their social worlds. Several central characteristics distinguish qualitative research from quantitative methods. Firstly, qualitative research is conducted in naturalistic settings, enabling researchers to observe behaviours and interactions in real-life contexts. This approach emphasizes the importance of meaning and understanding, as researchers seek to explore the "why" and "how" behind people's actions and thoughts. The researcher acts as an instrument in the research process, engaging directly with participants and requiring a high level of reflexivity to acknowledge personal biases [1].

Qualitative research employs an inductive approach to data analysis, allowing theories and patterns to emerge from the data rather than imposing preconceived notions. This flexibility in research design enables researchers to adapt their focus based on new insights gained during data collection. Additionally, qualitative research takes a holistic perspective, considering the interconnectedness of various elements and exploring cultural, social, and historical influences on experiences. Rich, thick descriptions are a hallmark of qualitative research, providing detailed accounts that contextualize participant experiences. Data collection methods in qualitative research are diverse and tailored to capture meaningful insights. In-depth

F. A. Alabá et al., *Fundamentals of Qualitative Research*, Studies in Systems, Decision and Control 663, https://doi.org/10.1007/978-3-032-18427-6_1

interviews, focus groups, participant observation, document analysis, and narrative research are common methods used to gather qualitative data. Each method serves to uncover different dimensions of human experience, allowing researchers to explore thoughts, emotions, and social dynamics [2].

Sampling in qualitative research typically employs non-probability techniques, focusing on depth rather than breadth. Purposive sampling selects participants based on specific characteristics relevant to the research question, while snowball sampling uses referrals to identify additional participants. Theoretical sampling is guided by emerging findings, particularly in grounded theory research. Sample sizes are generally small, with an emphasis on reaching saturation—the point at which no new themes emerge. Data analysis in qualitative research is an iterative and interpretive process aimed at uncovering patterns and themes. Common methods include thematic analysis, content analysis, grounded theory, and narrative analysis. Researchers engage closely with the data, coding it into categories and constructing narratives that explain findings within their contextual frameworks [3].

Ensuring trustworthiness in qualitative research is crucial, as it does not rely on statistical measures of validity. Trustworthiness encompasses credibility, transferability, dependability, and confirmability. Researchers achieve credibility through member checking and triangulation, while transferability is enhanced by providing detailed contextual descriptions. Dependability is supported by clear documentation of procedures, and confirmability is ensured through reflexivity and transparency. Ethical considerations are fundamental in qualitative research. Key principles include informed consent, confidentiality, the right to withdraw, and the minimization of harm. Researchers must navigate ethical challenges, particularly when working with vulnerable populations or sensitive topics, ensuring that participants' rights and well-being are prioritized. Qualitative research has both strengths and limitations. Its strengths include offering deep insights into complex social phenomena, capturing participants' perspectives, and allowing for flexible and adaptive research designs. It is particularly useful in understudied or emerging areas. However, limitations include the potential lack of generalizability, time-consuming analysis, the risk of researcher bias, and challenges in replication [4].

Thus, qualitative research is an essential approach for understanding human behaviour, beliefs, and social processes within their contexts. By focusing on meaning, subjectivity, and complexity, qualitative research provides insights that are often inaccessible through numerical analysis alone. This chapter has outlined the foundational elements of qualitative research, including its philosophical roots, methodological practices, and ethical considerations. As society faces increasingly complex social challenges, qualitative research remains a powerful tool for generating empathetic, culturally aware, and actionable knowledge [5].

The objective of the sampling method known as probability sampling is to eliminate bias by ensuring that every member of the population has an equal chance of being picked. This method is based on randomisation. Drawing samples by techniques such as lottery methods or computerised random number generation ensures that it is completely random, which in turn ensures that it is considered objective. Nevertheless, non-probability sampling entails making a purposeful selection based on criteria, availability, or suggestion, which introduces subjectivity into the collection process. Through the use of random selection, bias is reduced, which ultimately leads to more accurate outcomes. On the other hand, non-probability sampling presents a greater exposure to the possibility of bias, which has the potential to skew results and diminish trustworthiness. The use of probability sampling helps to guarantee that the sample is representative of the larger group by enabling it to accurately reflect the variety and distribution of the larger group. The capacity of non-probability sampling to correctly reflect the whole population is hindered by the fact that it is less representative than probability sampling since it is selected based on certain characteristics or convenience [6].

The capability of generalisability is accomplished by the use of probability sampling, which enables the results to be confidently generalised to the full population. The non-probability sampling method, on the other hand, may not be as accurate as probability sampling since the sample is not picked at random. The following are some examples of approaches that fall under the category of probability sampling: simple random sampling, stratified sampling, systematic sampling, cluster sampling, intentional sampling, convenience sampling, snowball sampling, and quanta sampling. When making statistical judgements about a population, such as in the case of national surveys, elections, or large-scale health studies, probability sampling is the method that is most often used. Non-probability sampling is commonly used when exploring new areas of research, conducting qualitative studies, or when the population is difficult to access, such as marginalized groups or secretive organizations.

The sample size requirements for probability sampling and non-probability sampling are different from one another. In order to ensure statistical validity and accuracy, probability sampling necessitates the use of larger sample sizes. On the other hand, non-probability sampling may be used with lower sample sizes, especially in qualitative research. It may be concluded that probability sampling is appropriate for statistical analysis, but non-probability sampling is the most acceptable method for qualitative research [7].

Table 1.1 present the comparison of probability and non-probability sampling techniques.

1.2 Key Characteristics of Qualitative Research

Qualitative research is a methodological technique focused on understanding the meaning that people or groups give to a social or human situation. Unlike qualitative research, which aims to measure and quantify events using numerical data, qualitative research examines the richness and depth of human experience via open-ended inquiry [8] (Fig. 1.1).

The following is a list and diagram of the primary qualities that establish qualitative research:

Table 1.1 Comparing probability and non-probability sampling techniques

Aspect	Probability sampling	Non-probability sampling
Definition	Every member of the population has a known, non-zero chance of being selected	Not every member has a known or equal chance of being selected
Randomness	Based on random selection	Based on subjective judgment or convenience
Bias	Less prone to selection bias	Higher risk of selection bias
Representativeness	More representative of the population	Less representative; focuses on specific subgroups
Generalizability	Findings can be generalized to the entire population	Limited generalizability
Examples	Simple random sampling, Stratified sampling, Systematic sampling, Cluster sampling	Purposive sampling, Convenience sampling, Snowball sampling, Quota sampling
Cost and time	Often more time-consuming and expensive	Usually quicker and more cost-effective
Use case	When population parameters are known and statistical inference is needed	When exploring in-depth information or working with hard-to-reach populations
Sample size requirements	Often requires a larger sample size for accuracy	Can be effective with smaller sample sizes
Suitability	Best for qualitative research	Best for qualitative research

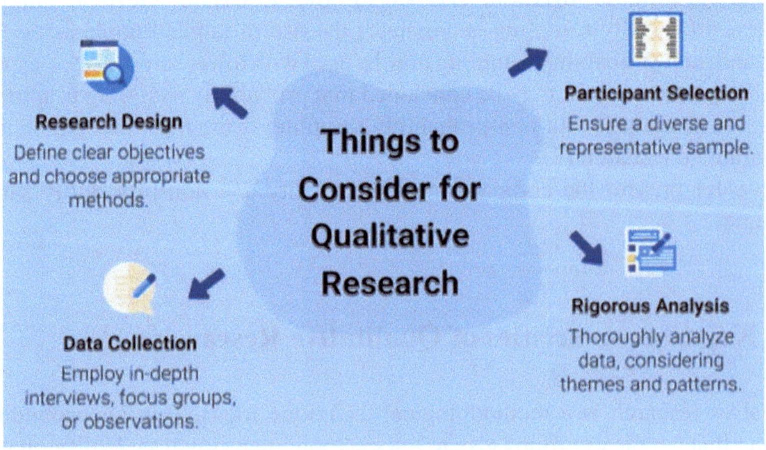

Fig. 1.1 Key characteristics of qualitative research [9]

1. **Naturalistic Setting**

The naturalistic setting of qualitative research is one of its core characteristics. This setting emphasizes the value of examining human behaviour and experiences in the circumstances in which they occur in their natural environments. The goal of this technique is to get an understanding of events in their natural state, without the intrusion of artificial modification or external control. Qualitative researchers aim to get a thorough and nuanced grasp of the topics they are examining by immersing themselves in surroundings that are representative of the actual world. When compared to qualitative research, which often entails changing variables in controlled settings to show cause-and-effect linkages, this stands in sharp contrast to the traditional approach. One of the most important aspects of qualitative research is naturalistic inquiry, which focuses on the investigation of people or groups within their daily contexts, such as the places where they live, work, socialise, and interact with one another. The researchers who use this strategy do not make any changes to the circumstances or introduce any experimental treatments. In its place, kids engage in activities such as observation, attentive listening, and interpretation of things as they occur in their natural state. In the course of the research process, the environment in which the investigation is carried out becomes an essential component, which in turn has a substantial impact on the data that is gathered. For instance, a qualitative researcher would visit the classroom, watch behaviours, listen to interactions, and take field notes to conduct a study that investigates the dynamics of the classroom among primary school instructors and pupils. This would be done without altering the conventional procedures or structures that are already in place. Instead of trying to exert control over the environment, the goal is to gain knowledge from it [10].

2. **Researcher as Key Instrument**

Qualitative research differentiates itself from qualitative approaches by positioning the researcher as the focal point of the investigative process. In contrast to qualitative methodologies that use structured instruments like surveys or experiments, qualitative research prioritises the researcher as the primary instrument for data gathering, interpretation, and analysis. This distinctive perspective underscores the need of human engagement, reflexivity, and contextual awareness in producing reliable qualitative information. Qualitative research fundamentally aims to comprehend human experiences, behaviours, cultures, and social realities. This comprehension cannot be attained via remote observation or impersonal instruments; it requires intimate connection, empathy, and active listening, all enabled by the researcher's participation. Qualitative researchers use techniques such as in-depth interviews, participant observation, focus groups, and ethnographic immersion, which rely significantly on their capacity to engage meaningfully with individuals and understand answers within the framework of their lived experiences [11].

The researcher's work encompasses more than simple data collecting; they are also tasked with evaluating intricate and unstructured qualitative data. This interpretation is intrinsically subjective, shaped by the researcher's theoretical framework, cultural values, and personal experiences. This subjectivity, however vulnerable to scrutiny, allows qualitative research to investigate depth, context, and complexity sometimes neglected by qualitative approaches. Various researchers may interpret identical data differently, reflecting their distinct views, a hallmark of qualitative investigation. Reflexivity is essential in qualitative research, since it entails a critical examination of how the researcher's identity, values, and emotions affect their engagement with participants and the interpretation of findings. A reflective researcher recognises their positionality and the influence of issues like as gender, racism, and power dynamics on their comprehension. Methods like as maintaining a reflective diary and participating in peer debriefing improve self-awareness, so bolstering the study's credibility and ethical integrity [12].

Designating the researcher as the primary instrument in qualitative research has several benefits, such as adaptability and reactivity to emergent themes, contextual awareness of social and cultural factors, and the capacity to cultivate empathy and connection with participants. These traits increase the profundity and sincerity of replies, especially when discussing delicate subjects. Moreover, researchers may include many cues—verbal, non-verbal, environmental, and emotional—into their analyses, resulting in a more comprehensive comprehension of the topic. This pivotal function also poses obstacles. The potential for prejudice and subjectivity in interpretation requires stringent reflexivity and transparency. Researchers may encounter emotional challenges when dealing with sensitive narratives, and ethical difficulties may emerge from ambiguous limits in participant interactions. The subjective aspect of qualitative research might hinder replicability, making it difficult for other researchers to draw identical conclusions. Qualitative researchers have substantial ethical obligations, including securing informed consent, maintaining anonymity and confidentiality, safeguarding participants from abuse, and addressing power imbalances to guarantee a genuine portrayal of voices. Ongoing ethical scrutiny is crucial, especially when researchers are intimately involved in the lives and narratives of participants. In summary, the concept of the researcher as the primary instrument is essential to qualitative research. This job incorporates a human dimension to research that mechanical instruments cannot emulate, facilitating flexibility, depth, empathy, and contextual complexity. Nevertheless, this authority entails the need for reflexivity, ethical behaviour, and openness. By recognising both the advantages and constraints of their work, qualitative researchers may enhance information that is rigorous, significant, empathetic, and transformational [13].

3. **Multiple Sources of Data**

Qualitative research is a methodological approach that relies on multiple sources of data to capture the richness and complexity of human experiences. This multi-method approach enhances the depth, context, and credibility of findings by allowing researchers to explore the phenomenon from multiple perspectives and layers. The use of multiple sources, such as interviews, focus groups, observations, field notes, documents, photos, and audio-visual materials, is not only a methodological strength but also a critical element of qualitative rigor. Understanding the value of multiple data sources is essential for qualitative researchers to capture different dimensions of a phenomenon, validate findings through triangulation, enrich the contextual understanding, mitigate biases or limitations inherent in any single method, and reflect the voices of diverse participants or stakeholders. This multidimensional approach supports the goal of holistic inquiry, where the focus is not just on what is said or observed, but also on how it is expressed, when, and in what context [14].

Qualitative research commonly integrates several types of data sources, each offering unique contributions to the richness of the study. Interviews are the most widely used method in qualitative research, providing deep, personal insights into participants' beliefs, feelings, and lived experiences. Focus groups involve discussions among small groups of participants, guided by a facilitator, revealing social dynamics, collective opinions, and group norms. Observations allow researchers to witness behaviours, events, and interactions as they occur in real time, uncovering the difference between what people say they do and what they do. Field notes are written accounts of what researchers observe and experience in the research setting, often including descriptions, impressions, analytical reflections, and emerging themes. Documents and texts offer historical and institutional insights, while photographs and visual materials provide non-verbal data that complement spoken or written accounts. Audio-visual recordings are valuable for capturing the tone, pauses, gestures, and expressions that accompany spoken language. Triangulation, a strategy for increasing the credibility and validity of qualitative research, involves cross-verifying information obtained from different methods or perspectives. By comparing data across these sources, researchers can identify convergences, contradictions, and nuances that would otherwise go unnoticed [15].

However, using multiple data sources presents challenges such as complexity in data management, time and resource demands, interpretative complexity, and ethical considerations. Despite these challenges, the benefits of a comprehensive, nuanced understanding often outweigh the difficulties. The use of multiple sources of data strengthens qualitative research in several key ways: depth. By capturing diverse aspects of the phenomenon, the study becomes more complete and contextually grounded. Contextual richness: Multiple data sources offer insights into environmental, social, and emotional factors that influence participants. Validation: When findings are supported by different data types, they gain credibility. Thus, the integration of multiple sources of data is a cornerstone of qualitative research, rooted in the commitment to depth, authenticity, and trustworthiness. Through interviews, observations, documents, and visual materials, researchers can explore phenomena from varied angles, making their findings more comprehensive and credible. Triangulation not only reinforces the reliability of conclusions but also enriches the narrative with diverse voices and complex insights [16].

4. **Emergent Design**

The emergent design is a hallmark of qualitative research; it permits the study's framework to change in response to the emergence of new ideas and patterns. Researchers may tailor their research objectives, data collecting methods, sample procedures, and theoretical frameworks to fit the study's evolving needs because of this adaptability. In addition to being an absolute must-have, emergent design is a philosophical and methodological tenet of transparency, inquiry, and adaptability. Unexpected ideas are captured, authenticity is enhanced, relevance is improved, and theory development via approaches like grounded theory is supported. Research topics, data collecting methodologies, sample tactics, and theoretical lenses may all undergo revision when a qualitative study undergoes emergent design. Depth and richness, participant-centeredness, response to context, originality, and invention are some of the advantages of using an emergent design in qualitative research. Nevertheless, there are obstacles to overcome, such as problems with planning, demands on time and resources, data overload, and the need for researcher discipline. Researchers must keep transparent and coherent documentation in the form of notes, journals, and audit trails in order to tackle these issues.

Qualitative research methods such as action research, phenomenological investigations, and ethnographic research all use emergent designs. When doing ethnographic research, one must first observe how societal norms impact decision-making before moving the emphasis from agricultural methods to gender roles. In phenomenological research, early participant input is used to examine coping mechanisms rather than clinical symptoms in order to better understand patients' experiences with chronic disease. During action research, new concerns are identified during implementation and the study's path is co-defined. Emergent design also places a premium on ethical concerns. Ethical reflexivity is essential for managing the dynamic nature of emergent design while upholding participants' rights and dignity. Researchers must maintain ongoing informed consent, keep participants updated about any major changes in the research focus, ethically review new data collection methods or participant groups, report design changes transparently, and keep informed consent [17].

Finally, qualitative research is characterised by its open, responsive, and exploratory character, which is reflected in emergent design. Instead of being limited by strict methodological guidelines, it allows researchers to follow the data wherever it may take them. The findings are significant, culturally grounded, and often surprising because of this flexibility. Qualitative researchers have a better grasp of complicated social realities when they are open to change and ambiguity. There are certain planning and analytical hurdles to go over, but the payoff is research that is richer and more genuine because of the flexibility. Exploring the richness, complexity, and unpredictability of human existence is, at its core, affirmed by emergent design.

5. **Inductive Data Analysis**

One of the most basic analytical approaches in qualitative research is known as inductive data analysis. This method enables the spontaneous development of patterns, themes, and categories from the data. This bottom-up method guarantees that results continue to be anchored in the experiences of participants and real-world situations, honouring the complexity and unpredictability of human behaviour and social relationships. When using inductive reasoning, one begins with raw data, such as interview transcripts, observation notes, or papers, and then works their way towards more general patterns, ideas, or hypotheses. Inductive reasoning proceeds from the particular to the general.

Researchers engage with data in a manner that is both methodical and flexible, seeking to identify repeating ideas, phrases, behaviours, or observations that may hint at underlying patterns. This kind of analysis is known as inductive analysis. The next step is to organise these patterns into topics or categories, which serve as the basis for developing new theoretical discoveries. By way of illustration, a researcher who is investigating the experiences of online learning may discover themes such as "isolation," "self-regulation," and "digital fatigue" by naturally immersing themselves in the data. Data familiarisation, initial coding, discovering patterns and themes, continual comparison, and theoretical integration are the usual processes involved in the process of inductive data analysis. These steps are often interrelated with one another within the process. Making sense of the data in a manner that contributes to a better understanding of the phenomenon that is being investigated is the objective of this endeavour. There are several benefits associated with inductive analysis, such as the creation of theory, flexibility, richness and depth, and ideas that are prompted by facts. The findings are based on the real experiences of the participants, which provides an increased level of authenticity and credibility to the study undertaken. Instead of being bound by pre-existing hypotheses, inductive analysis gives researchers the ability to respond to new evidence as it becomes available. In addition to this, it makes it possible to discover intricate and context-specific patterns that qualitative tools may not be able to capture. In the grand scheme of things, inductive analysis is an essential instrument in qualitative research since it encourages investigation, openness, and contextual awareness [18].

Inductive analysis is a powerful method used in qualitative research to uncover patterns, generate themes, and develop theories from participants' lived realities. However, it presents challenges such as subjectivity, time-consuming nature, data overload, and credibility concerns. Researchers must work diligently to demonstrate transparency and trustworthiness using techniques like member checking, peer review, and audit trails.

Tools and techniques used in inductive analysis include NVivo, ATLAS.ti, MAXQDA, and Dedoose. These tools support coding, theme development, and visualization, helping to organize large volumes of data systematically while maintaining a qualitative approach. Inductive data analysis is central to many qualitative methodologies, such as grounded theory, thematic analysis, phenomenology, and ethnography. Ethical considerations are essential during inductive analysis, as researchers must acknowledge their biases, ensure participant voices are accurately represented, and ground themes in data. Transparency in decision-making during coding and analysis strengthens the trustworthiness and credibility of the research. Hence, inductive data analysis is at the heart of qualitative research, providing a powerful means of uncovering patterns, generating themes, and developing theories from participants' lived realities. While complex, subjective, and labour-intensive, its flexibility and openness make it ideally suited to the exploratory nature of qualitative inquiry. When conducted with reflexivity, transparency, and methodological rigor, inductive analysis offers meaningful and transformative insights [19] (Table 1.2).

Table 1.2 Inductive analysis versus deductive analysis

Aspect	Inductive analysis	Deductive analysis
Approach	Bottom-up: builds theory from data	Top-down: tests the theory using data
Purpose	Discover patterns and generate new ideas	Confirm or refute existing hypotheses
Process	Flexible, iterative, exploratory	Structured, pre-defined, confirmatory
Application	Qualitative research	Qualitative research
Examples of use	Grounded theory, thematic analysis, and ethnography	Experiments, surveys, and quasi-experimental designs

6. Focus on Meaning

Understanding the meaning that people ascribe to their lives, experiences, and relationships is the goal of qualitative research, which is a way of doing research. Qualitative research, on the other hand, is often centred on the measurement of variables or the statistical testing of hypotheses rather than qualitative research. The methodology of qualitative research emphasizes the ways in which individuals make sense of their surroundings. This includes the ideas, feelings, beliefs, cultural norms, and personal values that influence their perceptions and direct their behaviours. Researchers can unearth insights that go beyond surface-level data by interacting intimately with participants and investigating their viewpoints. This allows them to uncover the deeper levels of human knowledge and motivation. After gaining knowledge of the significance of meaning in qualitative research, researchers can capture the breadth of human experience, comprehend the implications of contextual factors, magnify the voices of participants, foster empathy and comprehension, and provide support for the creation of theories. The investigation of meaning may be accomplished using a variety of methods, such as in-depth interviews, focus groups, observations, document and text analysis, narrative tales, and life histories [20].

Within the realm of qualitative research, interpretation and meaning-making are not only mechanical processes; rather, they are acts that are creative, iterative, and reflective. Transcripts, notes, or writings are analyzed by researchers to find patterns, metaphors, emotions, and themes that reflect how participants view the world around them. When going through this process, it is common to go through numerous rounds of reading, coding, comparing, and re-contextualising, all to capture the depth and complexity of meaning. The fields of health research, educational research, workplace studies, and cultural research are all examples of studies that are focused on meaning specifically. Educational research focuses on what “learning” or “success” means to students from diverse backgrounds, discovering assumptions and disparities that impact engagement and accomplishment. Health research investigates the meaning of sickness from the standpoint of a patient, whereas clinical research investigates the meaning of illness from the perspective of a patient. Research conducted in the workplace seeks to understand how workers view leadership, job satisfaction, or burnout by gaining knowledge of the significance of these events in their day-to-day life. The purpose of cultural research studies is to get an understanding of the common meanings that bind societies together via the study of rituals, symbols, or traditions.

This focus on meaning, however, also carries with it a number of obstacles, including subjectivity, interpretive complexity, believability, and ethical issues respectively. The fact that meanings are subjective and may change not only from person to person but also over time and in different contexts makes it challenging to standardize and differentiate between them. In order to be credible, researchers need to be transparent about how they arrived at their results, whereas interpretative intricacy requires expertise, sensitivity, and methodological rigor. Concerns pertaining to ethics include safeguarding the psychological well-being of participants and accurately conveying the significance of their contributions. As a result, the emphasis on meaning is a fundamental component of qualitative research. This shifts the attention away from just observing what people do and toward comprehending the reasons behind their actions, the emotions they experience, and the significance of the actions to them. Research that is qualitative is able to capture the richness, complexity, and depth of human existence in ways that statistics alone are unable to represent. This is because qualitative research emphasizes the subjective viewpoints of people. By serving as a critical contrast in a society that is becoming more dominated by statistics and analytics, the qualitative method works to remind us that knowledge is not only about quantity but also about stories, insights, emotions, and meaning. The purpose of qualitative research is to shed light on how individuals make meaning of their lives. This kind of study not only adds to academic knowledge, but also to empathy, policy, and social change [21].

7. **Participant Perspective (Emic View)**

Understanding the participant's point of view, also known as the emic viewpoint, is a basic premise that is emphasized in qualitative research. This viewpoint, which is derived from the field of anthropology, describes the knowledge of cultural and social experiences that comes from being on the inside. The objective of qualitative researchers is to discover how people themselves understand, define, and create a sense of their lives. This is in contrast to the etic approach, which examines human behaviour via an external lens. This participant-centred method differentiates qualitative inquiry from many kinds of qualitative research by placing a higher value on lived experience as opposed to imposed theory, subjectivity as opposed to objectivity, and conversation as opposed to measurement. Within the realm of qualitative research, the emic viewpoint contributes to an increase in authenticity by ensuring that the voices of the participants are heard in their own words, without being filtered via the interpretations of the researcher. This ultimately results in outcomes that are more based in reality and represent the intricacies of the human experience reflected in the findings. In addition to this, it takes into account the cultural and social circumstances, which enables researchers to comprehend these situations in the same way that outsiders do, so preventing cultural bias or distortion.

It is helpful to centre the viewpoint of the participant because it helps to balance the power dynamics between the researcher and the participant. It also encourages conversation, mutual respect, and informed permission, which helps to reinforce ethical norms throughout the research process. Researchers can unearth previously unknown or unexpected aspects of the phenomenon that is being investigated if they pay close attention to the opinions of the participants. In-depth interviews, participant observation, narrative and life story techniques, focus groups, participatory and community-based methods, and other experimental approaches are some of the methodological approaches that may be used to capture the emic perspective. The emic approach, on the other hand, is fraught with difficulties such as subjectivity, translation and interpretation, researcher bias, and ethical conundrums. For researchers to handle these issues, they need to engage in reflexivity, make use of member checking, and preserve transparency in the process by which interpretations are produced.

Investigating the traditional methods of birthing that are used in a community located in rural Africa is one example of this. An etic approach might evaluate these practices in comparison to Western medical standards and conclude that they are unsafe. On the other hand, an emic perspective would seek to understand what these practices mean to the women who use them, including their spiritual significance, social importance, and perceived effectiveness. Through the use of an emic perspective, it is possible to uncover profound cultural values of motherhood, community, and ancestral connection that cannot be conveyed by biological models alone.

Skilled qualitative researchers need to be able to traverse both views to attain the emic-etic balance. They must respect the insider's viewpoint while simultaneously relying on their own knowledge and theoretical tools to analyze and convey the results. This equilibrium guarantees a detailed description as well as an in-depth study. As a result, the participant perspective, also known as the emic view, is an essential component of qualitative research. This viewpoint affirms the notion that individuals are the foremost authorities on their own lives and that research should aim to comprehend the world in the same manner in which they see and relate to it. In order to capture the emic viewpoint, researchers need to demonstrate humility, empathy, and deep listening skills. This requires them to put aside any preconceived beliefs they may have and connect with participants in ways that are relevant and respectful [22].

8. **Holistic Account**

Qualitative research is a methodological technique that focuses on offering a complete account of the topic that is being studied. This approach takes into consideration the social, cultural, historical, political, emotional, and environmental circumstances that impact human experiences. The concept that reality is complicated, interdependent, and reliant on context is reflected in this all-encompassing method for understanding reality. Through the adoption of a holistic approach, qualitative research extends beyond the "what" to investigate the "how" and "why," therefore capturing the richness and complexity of real-life experiences. This ensures that the results are not only accurate in theory but also significant in the participants' daily lives. The term "holistic account" refers to the endeavor of depicting a phenomenon in all of its complete complexity, taking into consideration the fact that absolutely no element of human existence takes place in isolation. It gives researchers the ability to investigate numerous levels of impact, which acknowledges the dynamic and varied character of actual life. Through the process of placing results within the right context, a holistic account improves contextual knowledge, which in turn increases the interpretive value of the findings. It demonstrates empathy and respect for the complexity of their participants' lives by taking into consideration the larger context in which individuals live. This respects the lived reality of the participants [23].

Qualitative researchers use a variety of methods to guarantee that they can provide a comprehensive account. These methods include extended engagement, diverse data sources, reflexivity, dense description, and theoretical sensitivity. For instance, a qualitative study that investigates the experiences of refugee children attending public schools would investigate the child's migration journey and history of traumatic experiences, as well as the child's family and home life, peer relationships and experiences of inclusion or discrimination, the perceptions of teachers and the policies of the school, the child's cultural identity and sense of belonging, as well as the socio-political environment and the attitudes of the community toward refugees. Developing a comprehensive account is not without its problems, which include the rigorous use of time and resources, an overwhelming amount of data, the careful balance of depth and breadth, and the complexity of ethical considerations. On the other hand, the advantages of generating a comprehensive and significant account of human existence much exceed the challenges, particularly when the objective is to comprehend rather than to forecast, and to interpret rather than to measure.

Researchers, practitioners, and policymakers can understand intersecting factors that influence behaviours, context-sensitive design interventions, recognize and challenge structural inequalities, and promote empathy, inclusion, and social justice when they have access to a holistic account. This enhances the trustworthiness, transferability, and relevance of qualitative findings. As a conclusion, the holistic account is more than simply a methodological option in qualitative research; rather, it is a philosophical commitment to comprehending individuals in all of their complexities. Researchers can generate interpretations that are not only plausible but also profoundly significant when they place human experiences within wider social, cultural, historical, and emotional contexts. Being able to perceive the whole picture, which includes the capacity to listen, observe, and evaluate with depth and sensitivity, is a scholastic virtue as well as a social imperative in today's society, which is becoming more complicated and diverse. One narrative at a time, qualitative research continues to shed light on the rich fabric that is human existence by using the lens of a holistic account [24].

9. **Rich, Thick Description**

It is well known that qualitative research places a strong focus on rich, dense description. This refers to the precise and vivid representation of the experiences, locations, interactions, and feelings of the participants in such a manner that the readers feel as if they are there inside the scenario. Using this method, researchers are able to express the nuances of real-life events, and readers are able to connect with the data in a way that is empathic. This strategy adds depth, context, and significance. Clifford Geertz, an anthropologist, is credited with popularizing the idea. Geertz brought attention to the significance of capturing the many layers of meaning that are inherent in cultural practices. Within the realm of qualitative research, a description that is both rich and thick serves several important functions, including the following: contextual comprehension, emotional and experiential depth, transferability, credibility and trustworthiness, and analytical richness. It requires paying attention to a number of different aspects, including the physical surroundings, the actions and behaviour of the participants, verbal and non-verbal communication, the cultural and social background, as well as feelings and meanings from the participants. Through the process of weaving these components together, the researcher can develop a multi-layered narrative that accurately depicts the variety of events that occur in the actual world.

Several techniques are used by researchers to provide extensive descriptions. These techniques include extended periods spent in the field, detailed field notes, in-depth interviews, audio-visual materials, and reflexivity. To get a profound acquaintance with the environment and behaviour, prolonged interaction is necessary. This enables more comprehensive observations to be made. Detailed field notes are used to document sensory details, quotations, comments, and emergent interpretations. On the other hand, in-depth interviews allow participants to tell tales, describe events, and explain the implications of those experiences. The use of audio-visual resources allows for the revisiting of minute nuances that contribute to the depth of description and may be revisited during analysis. If the results of the reflexivity are given openly and honestly, then the account will be more comprehensive. Producing detailed and detailed descriptions, on the other hand, involves several obstacles, including the amount of data, the balancing act between depth and focus, and subjectivity. It is possible to overload readers with an excessive amount of detail that lacks evident relevance, and there is always the possibility of prejudice or distortion. Reporting that is both reflexive and ethical is required to guarantee accuracy while also capturing depth. When executed with finesse, rich description transforms into a potent instrument for narrative and analysis that brings research to life.

An individual conducting research on religious rites might describe a courtyard that was filled with the aroma of burning incense and the rhythmic beat of hand drums throughout the investigation. The shrine was surrounded by women dressed in colourful saris who subtly chanted while the elderly sat on woven mats with their legs crossed and their eyes closed in prayer. Children spent their time playing on the periphery, giggling and imitating the rituals with exaggerated movements. The environment was characterised by a mixture of reverence, community, and celebration; it was a ritual that was deeply ingrained in the culture of everyday life. In conclusion, rich and detailed description is an essential component of qualitative research. It enables researchers to communicate the fullness and complexity of human experience in a manner that is meaningful, contextual, and emotionally resonant. Rich description not only adds depth to analysis but also contributes to the credibility, transferability, and impact of qualitative research. This is accomplished by emphasizing details, context, and the voices of participants. Simply put, it brings the data to life and serves as a reminder that there is a story hidden behind each and every theme or category, one that is abundant in texture, meaning, and humanity [25].

10. **Reflexivity**

The researcher must maintain a state of self-awareness and self-examination during the whole process of doing qualitative research. This is an essential component of the research method known as reflexivity. This method takes into account the impact that the researcher's preconceived notions, history, identity, beliefs, and emotional reactions have on the process of data collection, analysis, and interpretation. It is not the purpose of reflexivity to eradicate subjectivity, which is an intrinsic and often useful component of qualitative inquiry; rather, the objective of reflexivity is to make the research process more transparent, ethical, and believable. Reflexivity may be understood on a number of different levels, including personal reflexivity, epistemological reflexivity, and methodological reflexivity. Transparency is shown by researchers when they acknowledge the impact of the researcher, which helps readers believe that the results were not skewed by any concealed bias. In order to maintain ethical integrity, researchers should practice reflexivity, which encourages them to respect the voices of participants and limit their own contribution. Additionally, it enhances analytical rigor, which ultimately results in findings that are more careful and balanced.

The practice of reflexivity is not a one-time activity; rather, it is an ongoing process that starts before the research and continues all the way through the writing and distribution of the findings. Reflexive journaling, positionality statements, peer debriefing, member checking, and triangulation are all common tactics that are used in the practice of reflexivity. Examples of reflexivity in practice include a qualitative study on religious identity that was carried out by an atheist researcher. In this study, the researcher was required to acknowledge their own personal beliefs, reflect on their emotional reactions during interviews, consider how their identity influences the trust or openness of participants, and remain vigilant to avoid projecting scepticism or judgment onto the narratives of the participants. Another example may be a male researcher conducting interviews with female participants regarding their experiences of gender discrimination. In this scenario, reflexivity would need a continuous awareness of power dynamics, language usage, and body language, as well as a critical assessment of how gendered assumptions could impact the study.

The emotional strain, the time-consuming nature of keeping journals, the potential of overemphasis, and the subjective character of reflexivity are all challenges that are associated with exercising reflexivity. On the other hand, these difficulties highlight the significance of reflexivity in research, which is essential because research is notoriously difficult and subjective. The use of first-person writing, transparent decision-making, and introspective commentary are common examples of types of writing that exhibit reflexivity in final reports or publications. When there is transparency, the quality of the study is improved, and readers are invited to participate in the analysis process. This allows them to assess the honesty of the work. In conclusion, reflexivity is an essential component of qualitative research, since it guarantees that the study is truthful, trustworthy, and morally sound. It recognizes that the researcher is not only a bystander but rather an active participant in the process of gathering information and constructing knowledge. The study is more deliberate, nuanced, and trustworthy as a result of reflexive practice, which does this by transforming possible bias into an advantage via critical self-awareness, honest reporting, and ethical sensitivity. Engaging with research in a manner that is responsible, respectful, and real is made possible via the use of reflexivity in a world that is filled with a wide variety of human experiences and complicated societal challenges [26].

11. **Non-statistical Analysis**

Qualitative and qualitative research paradigms are often the two that emerge as the most prominent in the field of academic research. When it comes to testing hypotheses and measuring variables, qualitative research largely focuses on numerical data and statistical analysis. On the other hand, qualitative research takes a different approach. The investigation of human experiences, perceptions, and meanings has a prominent position in this approach. Interview transcripts, field notes, open-ended survey answers, diary entries, pictures, and videos are some examples of the types of records that are usually used to collect qualitative data. Academics can investigate intricate social phenomena that are difficult to quantify due to the richness of context and depth that this sort of data provides. There is a possibility that thematic analysis is the non-statistical form of analysis that is used the most often in qualitative research. Identification, analysis, and reporting of recurring themes or patterns within the data are all part of the process. There are normally six stages involved in the process of thematic analysis. These stages include becoming acquainted with the data, creating initial codes, looking for themes, evaluating themes, defining and labelling themes, and publishing the report. This approach is particularly useful because it is adaptable and may be used for a broad variety of research subjects and data formats.

Another method that does not rely on statistics is called content analysis, and it includes classifying verbal or behavioural data in order to summarize, tabulate, and categorize the content's information. There are two distinct forms of content analysis: manifest content analysis, which focuses on the surface meaning of a text (for example, counting the number of times a word occurs), and latent content analysis, which is more concerned with deciphering the meaning that lies behind the surface. The focus of content analysis in qualitative research is still on interpretation rather than numerical significance, in contrast to the statistical frequency analysis performed in qualitative research. The study of narrative analysis focuses on the manner in which tales are communicated and the manner in which humans get meaning from their experiences via the use of storytelling. Researchers that do case studies, personal interviews, and life histories will find it very helpful in their investigations. In addition to analyzing the story's content, narrative analysis also investigates the structure, context, and purpose of the narrative. Research that makes use of narrative analysis often investigates the ways in which people construct their narratives, the elements that are included or removed, and the larger societal or cultural narratives that have an impact on personal accounts.

Language as a type of social activity is the primary emphasis of discourse analysis methodology. For the sake of constructing meaning, asserting authority, and establishing social norms, it investigates the ways in which language is utilized in texts and interactions. Discourse analysis, in contrast to theme and content analysis, which often concentrate on the substance of what is said, studies how it is expressed as well as the social effects that it has. Even though all of these techniques are classified as non-statistical analysis, they are distinct from one another in terms of their emphasis and application. When it comes to recognizing broad patterns throughout a data collection, thematic analysis is the optimum method. A further level of classification is added by content analysis, which also has the potential to incorporate frequency without being statistical. The individual's story is the focus of narrative analysis, which investigates the form and meaning of the narrative. The field of discourse analysis pays attention to language in the context of its social setting, with a particular emphasis on power relations and societal assumptions. In conclusion, non-statistical analysis is an essential component of qualitative research since it facilitates the investigation of human behaviour, language, and social phenomena via the use of vital instruments. This allows academics to dive deeper into the complexity of meaning-making, identity, and interaction via the use of approaches such as theme analysis, content analysis, narrative analysis, and discourse analysis. It is expected that non-statistical approaches will continue to play a vital role in capturing the variety of human cognition and activity as research issues continue to change in a world that is becoming more complicated [27].

12. **Subjectivity and Interpretation**

Rather than relying on numerical measurement and statistical analysis, qualitative research takes a different approach to the development of information. Instead, it emphasises meaning, context, and the intricacies of human experiences. It is characterized by subjectivity, which is not considered a weakness or prejudice but rather as a vital component of the interpretative process. Subjectivity is a distinct feature of this. By using this method, subjectivity is acknowledged and embraced, and it is acknowledged that all information is influenced by the viewpoints, experiences, and settings of both the researcher and the participants. In the context of qualitative research, the term "subjectivity" refers to the impact that one's own personal experiences, beliefs, values, and emotions have on the research process. When doing qualitative research, the researcher is not only a bystander but rather serves as an active participant in the process of data gathering and interpretation. The method in which data is acquired, evaluated, and presented is impacted by several factors, including the researcher's worldview, cultural background, theoretical perspective, and interpersonal dynamics. This subjectivity is seen as a resource that has the potential to significantly enhance comprehension by incorporating the individual's points of view into the process of interpretation. Researchers can develop insights that are rooted in both their realities and the realities of the participants if they bring their viewpoints into the interpretative process.

Qualitative research is fundamentally interpretative, to gain an understanding of how people build meaning in their lives and how these meanings are rooted in the cultural, social, and historical settings in which they are presented. The process of interpretation is not a straightforward one; rather, it requires absorption in the data, contemplation, coding, and the formation of themes that represent the core of the experiences that the participants had. When we say that the process is context-bound, we imply that the results are tightly connected to the particular environment, time, and social circumstances that were present during the study that was carried out. Within the realm of qualitative research, the researcher is an essential component in the process of meaning interpretation. Through their interactions with participants, they can mould the data that is created, and their interpretation is what directs the transformation of raw data into codes, categories, and themes, respectively. Because of this, qualitative research is considered to be a kind of knowledge co-construction, in which both the researcher and the participant contribute to the overall comprehension of the topic being considered.

Reflexivity is a methodological tool that is used by qualitative researchers to recognize and critically reflect on their own subjectivity. It entails understanding how one's position could impact the research process, and it improves the credibility and transparency of qualitative research by making the researcher's role obvious. In addition, it helps establish the researcher's credibility. For the purpose of encouraging honesty and self-awareness, researchers can use a reflexive journal to document their thoughts, feelings, and decisions throughout the process of conducting research. Credibility, transferability, dependability, and confirmability are some of the criteria that are utilized in qualitative research in order to establish trustworthiness and rigor. A number of methods, including member checking, triangulation, and peer debriefing, are used to guarantee that interpretations are correct and accurately represent the experiences of the participants, rather than only reflecting the researcher's preconceived notions.

A research approach that is more empathic, nuanced, and human-centred can be achieved by embracing subjectivity within the field. Richness of data, authenticity, flexibility, and cultural sensitivity are all benefits that are fostered by it. The use of methodological rigor and transparency, on the other hand, can be used to address challenges such as researcher bias and inconsistent findings. To summarise, subjectivity and interpretation are essential components of qualitative research, as they contribute to the enhancement of the investigation by adding depth, empathy, and contextual relevance. Qualitative researchers can develop credible and meaningful results that connect with the lived experiences of the people they investigate if they engage reflexively with their data and participants [28].

13. **Transferability over Generalizability**

When it comes to comprehending complicated phenomena inside particular settings and from the points of view of those who are dealing with them, qualitative research is a distinctive method. Transferability is the foundation upon which it is built. Transferability is a term that describes the degree to which the results of a research may be applied or "transferred" to other contexts, circumstances, or populations who have comparable features. This book examines the relevance of transferability in qualitative research, compares it to generalizability in qualitative research, and analyzes the tactics that researchers use to promote the transferability of their results within the context of qualitative research. What is meant by the term "generalizability" in the context of qualitative research is the degree to which the findings of a study may be extrapolated to a larger population than the sample that was utilized in the study.

Data gathering procedures that are standardized, statistical inference, and random sampling are the means by which this is accomplished. The purpose of this endeavor is to make certain that the results are not isolated to a particular group but rather reveal wider truths or patterns that may be utilized in a variety of different settings. Qualitative research, on the other hand, is more concerned with the depth of its findings than the breadth of its findings.

The notion of transferability is one that is focused on the reader, and it is the reader who is responsible for determining whether or not the insights gained from the research are applicable to their own situation based on the information that is supplied in that particular context. A case study that investigates the experiences of first-generation college students attending a university in Nigeria, for instance, would not be statistically generalizable to each and every student throughout Africa. Nevertheless, the research may show themes—such as the battle for belonging, financial difficulty, or institutional support—that connect with comparable student groups abroad, so making the results transportable. This might be accomplished via the use of an extensive description. Obtaining transferability in qualitative research requires a substantial amount of description. This requires the provision of specific contextual information on the study setting, the participants, the interactions they have with one another, and the meanings they ascribe to those interactions. In this way, readers are able to complete their comprehension of the setting in which the study was carried out and evaluate the extent to which the results may be applied in other contexts. As an example, in the context of ethnographic research conducted in a healthcare environment, the researcher could not only describe the behaviours of nurses during encounters with patients, but also the organizational culture, the difficulties associated with staffing, and the local views toward health. The breadth of these explanations provides readers in various healthcare settings with the information they need to determine whether or not the insights are relevant to their own institutions [29].

It is also possible to improve transferability in qualitative research by using a method called purposeful sampling. An increase in the richness and depth of data collection, which in turn aids transferability, is achieved by the selection of participants based on particular qualities that are important to the research. In order to investigate a broad variety of viewpoints and experiences, researchers are able to carefully choose participants who have prior experience with the phenomena that is being investigated. For both academics and practitioners, it is vital to improve transferability via the use of reflexivity and transparency. The idea of transferability highlights the significance of methodological rigor, extensive documentation, and ethical responsibility for researchers with regard to their work. Ensure that readers have sufficient information to judge the relevance of their study and take into consideration how their findings could enlighten others.

Practitioners, especially those working in sectors such as education, health, and social work, may benefit greatly from the results of qualitative research since they can provide significant insights into environments that are comparable. These realizations have the potential to influence policy, practice, and further study, given that they are implemented with consideration and with a knowledge of the distinctions that stem from the context. To summarize, transferability has taken the place of the traditional emphasis on generalizability in qualitative research. This shift in emphasis reflects the belief that knowledge is contextually situated and that the value of research lies in its depth, richness, and relevance to similar contexts rather than its ability to produce universal laws. In qualitative research, approaches such as dense description, purposive sampling, reflexivity, and methodological transparency are used to provide readers with the capacity to evaluate the relevance of results in their environments (Table 1.3).

1.3 Differences Between Qualitative and Quantitative Research

Qualitative and quantitative research are two fundamental approaches in academic inquiry, each with unique philosophical underpinnings, methods, and goals. Understanding the key differences between them is essential for selecting the appropriate approach based on the research questions and objectives. Qualitative research focuses on non-numerical data, such as words, narratives, images, and symbols, while quantitative research deals with numerical data and emphasizes quantification and statistical analysis. Its primary focus is to explore meaning, motivations, and experiences, while qualitative research tests hypotheses, measures variables, and analyzes relationships among variables. Qualitative research is subjective and interpretive, accepting that the researcher's background influences findings. Open-ended questions allow participants to express themselves freely, while qualitative research uses closed-ended questions for standardized data collection. Depth vs. breadth is an important aspect of qualitative research, with qualitative studies emphasizing depth and breadth. Sample size is typically small, purposive, while qualitative studies use large, random samples for statistical generalizability. Data collection methods include interviews, observations, focus groups, content analysis, surveys, tests, experiments, and structured questionnaires.

Qualitative research designs are often flexible and adaptive, while qualitative research follows a rigid structure. Data analysis techniques vary, with qualitative data being thematic analysis, coding, or narrative analysis, and quantitative data using statistical methods. Qualitative findings are context-specific and meaningful within a particular setting, while qualitative findings aim to be generalizable to larger populations. The role of the researcher is also crucial, with qualitative researchers engaging closely with participants and grounded in positivism. Qualitative methods are commonly used in social sciences, anthropology, education, and health research, while quantitative methods are more prevalent in natural sciences, psychology, economics, and engineering [30] (Table 1.4).

Table 1.3 Key characteristics of qualitative research

Characteristic	Description
1. Natural setting	Qualitative research takes place in the participant's real-world environment. This allows researchers to study behaviour and context as they naturally occur without manipulation or control
2. Researcher as key instrument	The researcher is the primary tool for data collection and analysis. Their interaction with participants shapes the data and findings. Reflexivity (self-awareness) is crucial to minimize bias
3. Multiple sources of data	Data is collected from various sources such as interviews, focus groups, observations, documents, and artifacts. This enhances the depth and validity of the findings
4. Focus on meaning and interpretation	Emphasizes understanding how people make sense of their experiences. Researchers explore participants' subjective meanings, beliefs, and emotions
5. Inductive data analysis	Patterns and themes are identified from the ground up rather than testing pre-established hypotheses. Theories emerge from the data itself
6. Emergent design	The research process is flexible and can evolve. Questions and strategies may change in response to what is learned during data collection
7. Holistic perspective	Seeks to understand the whole phenomenon, considering emotional, social, cultural, and environmental contexts
8. Thick description	Provides rich, detailed accounts of settings, events, and participant experiences to help readers grasp the context and significance
9. Participant-centered	Participants' voices are central. The goal is to understand their perspectives rather than impose external interpretations
10. Contextualization	Data is interpreted within the specific cultural, institutional, or social context in which it occurs. Generalization is not the goal
11. Subjectivity and reflexivity	Researchers acknowledge and critically reflect on their role, biases, and influence throughout the study
12. Small, purposeful samples	Sampling focuses on depth rather than breadth. Participants are selected based on relevance to the research question, not to represent a population

1.4 Qualitative Research Application Areas

Qualitative research is a method of inquiry that focuses on understanding the meaning individuals or groups ascribe to social or human problems. Unlike qualitative research that relies on numerical data and statistical analysis, qualitative research seeks to

Table 1.4 Differences between qualitative and quantitative research

S. No.	Qualitative research	Quantitative research
1	Focuses on understanding meaning and experiences	Focuses on measuring and quantifying data
2	Data is non-numerical (words, images, observations)	Data is numerical (numbers, statistics)
3	Subjective and interpretive	Objective and statistical
4	Uses open-ended questions	Uses closed-ended questions
5	In-depth, detailed exploration	Broad, generalizable measurements
6	Sample sizes are usually small	Sample sizes are usually large
7	Data collection through interviews, observations	Data collection through surveys, experiments
8	Flexible and emergent design	Structured and fixed design
9	Analysis through coding, themes, narratives	Analysis through statistical tools and formulas
10	Results are descriptive and interpretive	Results are statistical and predictive
11	Findings are context-specific	Findings aim to be generalizable
12	Researcher is involved in the process	Researcher maintains distance from subjects
13	Emphasizes depth over breadth	Emphasizes breadth over depth
14	Commonly used in social sciences, humanities	Commonly used in natural sciences, economics
15	Examples: case studies, ethnographies, grounded theory	Examples: surveys, experiments, correlation studies

explore phenomena through the lens of human experience, behaviour, and culture. It uses open-ended questions, interviews, focus groups, observation, and textual analysis to collect data. Due to its exploratory and interpretative nature, qualitative research has found wide application across numerous academic, professional, and social fields. This section explores the major application areas of qualitative research, highlighting its importance in various domains [31–33].

1. **Social Sciences**: One of the primary fields where qualitative research thrives is the social sciences. Disciplines such as sociology, anthropology, psychology, and political science rely heavily on qualitative methods to study complex human behaviours, social norms, values, institutions, and interactions. For instance, ethnography, a qualitative research method, is widely used in anthropology to study cultures and communities through immersive observation and participation. In sociology, qualitative methods help in understanding societal issues such as poverty, crime, gender inequality, and racism. Researchers conduct interviews and focus groups to gather data on how individuals perceive and experience social phenomena, providing rich, contextual insights that qualitative methods may overlook.

2. **Education**: Qualitative research is extensively used in educational studies to explore teaching practices, learning experiences, curriculum implementation, and classroom dynamics. Teachers, administrators, and policymakers use qualitative data to evaluate student engagement, teacher effectiveness, and the impact of educational reforms. For example, researchers may observe a classroom setting to assess the interaction between teachers and students, or conduct interviews with students to understand their learning challenges and motivation. This type of research is invaluable for developing inclusive, student-centred educational policies and practices.
3. **Health and Medicine**: In healthcare and medical research, qualitative approaches are employed to explore patient experiences, treatment outcomes, and health behaviours. While qualitative data might reveal how many patients recovered from a treatment, qualitative research can explain how they felt during the treatment process, what factors influenced their compliance, or how cultural beliefs affected their healthcare choices. Fields such as nursing, public health, and mental health rely heavily on qualitative studies. For example, a qualitative study might investigate how cancer patients cope with emotional distress or how communities perceive vaccination campaigns. These insights are essential for designing patient-centred care and effective public health interventions.
4. **Business and Management**: In business research, qualitative methods are increasingly used to study organisational behaviour, employee satisfaction, leadership styles, customer preferences, and market trends. Through case studies, interviews, and focus groups, businesses can uncover the underlying motivations of employees and consumers. For instance, a company might conduct qualitative interviews to understand why employees feel disengaged or to explore customer experiences with a new product. These insights can then inform strategic decisions, marketing approaches, or human resource practices. Qualitative research also supports innovation by uncovering unmet needs and generating new ideas.
5. **Communication and Media Studies**: Qualitative research is vital in communication and media studies, where the focus is on how messages are created, transmitted, and interpreted by audiences. Content analysis, discourse analysis, and narrative analysis are common techniques used to study television programs, social media content, advertisements, and news reports. Researchers analyze how the media influences public opinion, constructs identity, or perpetuates stereotypes. For example, a qualitative study might examine how gender roles are portrayed in popular TV shows or how social media users engage in political discussions. This helps media organizations and policymakers understand the impact of media on society.
6. **Law and Criminology**: In legal and criminological research, qualitative methods are applied to understand the social contexts of crime, the experiences of victims and offenders, and the functioning of legal institutions. Interviews with inmates, observations of courtroom procedures, and case studies of criminal investigations provide deep insights into the justice system. Such research is crucial for reforming legal policies, improving rehabilitation programs, and addressing systemic biases. For instance, qualitative research can reveal how marginalized

communities experience policing or how youth offenders perceive their chances of reintegration into society.

7. **Cultural and Gender Studies**: Cultural studies and gender studies use qualitative research to explore identity, representation, and power relations. Through narrative and discourse analysis, researchers examine how cultural norms shape individual experiences and how gender roles are constructed and challenged in different societies. For example, feminist researchers may use life history interviews to understand women's experiences in patriarchal societies, while cultural studies scholars might analyze popular music lyrics to uncover themes of resistance and identity. This kind of research fosters critical thinking and social change.
8. **Development Studies**: In development and humanitarian work, qualitative research helps organizations understand the needs, priorities, and perspectives of the communities they serve. It is used in needs assessments, program evaluations, and impact studies, especially in complex settings like conflict zones or rural communities. For example, NGOs may conduct focus group discussions with villagers to learn about barriers to education for girls or interview local farmers about the effects of climate change. Such context-rich information is essential for designing culturally appropriate and effective development interventions.
9. **Information Technology and Human–Computer Interaction (HCI)**: Qualitative research is also used in IT and user experience (UX) design to understand how people interact with technology. Researchers observe users as they navigate websites or apps and conduct interviews to understand frustrations, preferences, and behaviours. This approach helps in creating user-centred designs that are intuitive, accessible, and satisfying. For example, a software company might use qualitative feedback from beta testers to improve the usability of a new application.

In summary, the qualitative research application areas are illustrated in Fig. 1.2.

1.5 Qualitative Researchers' Mindset

Qualitative research is based on the investigation of human experiences, social processes, and cultural phenomena via the collection of data that is rich in descriptive and interpretative capabilities. One of the most important aspects of this research paradigm is the qualitative researcher, who is an investigator who is not only directed by methodological expertise but also by a particular attitude. A qualitative researcher's mentality is comprised of a collection of intellectual orientations, personal dispositions, and ethical convictions that together affect the research process from the beginning stages of conception to the end stages of conclusion. Throughout this book, the essential features that form the mentality of a qualitative researcher are investigated, and an example of how this mindset facilitates meaningful engagement with complicated and context-bound human realities is provided [19, 32, 33].

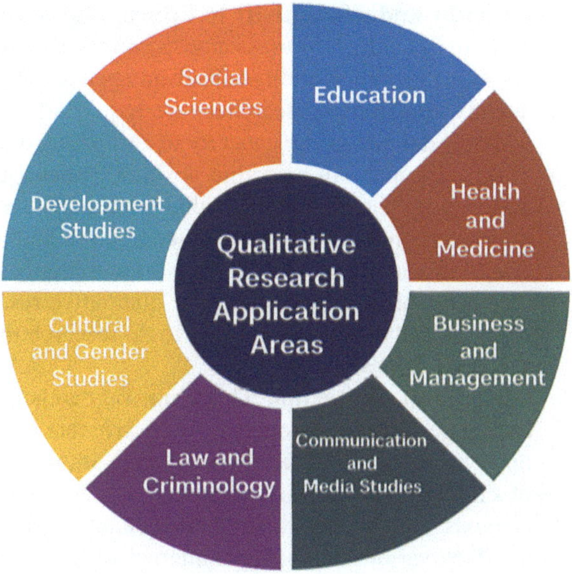

Fig. 1.2 The qualitative research application areas [34]

1. **Adopting a Reflexive and Subjective Attitude**: An acknowledgement and acceptance of subjectivity is one of the most distinguishing characteristics of the mentality of a qualitative researcher. When compared to the positivist tradition of qualitative research, which places a premium on objectivity and detachment, qualitative researchers acknowledge that they are active participants in the study process. Their perceptions, values, and experiences will inevitably affect the enquiries that they conduct. In close connection with this is the practice of reflexivity, which may be defined as the ongoing process of self-examination and critical reflection on the researcher's role, assumptions, and influence on the study. Instead of attempting to eradicate prejudice, a qualitative researcher will focus on gaining an understanding of it and discovering ways to account for it. In this mentality, some of the techniques that are often used include methodological transparency, positionality declarations, and reflexive journaling.
2. **Curiosity and an Openness to Exploration and Discovery**: Curiosity is the driving force behind qualitative research, which is characterised by a profound want to investigate the meanings that lay beneath people's behaviours, beliefs, and relationships with others. They do not go into the field with predetermined hypotheses to test, but rather with open-ended questions that provide participants the opportunity to express themselves and share their experiences. A mentality of openness is required for this approach, which means being attentive to unexpected discoveries, fresh viewpoints, and information that differs from what was previously believed. It requires the capacity to listen attentively and with empathy, as well as the suspension of judgement. The process of doing research is seen to be emergent rather than linear, and adaptability is essential to accommodate the intricacies of human behaviour in the actual world.

3. **Empathy and a Sensitivity to Ethical Issues**: Participation in the lived experiences of individuals is an essential component of qualitative research, which often involves the participation of disadvantaged or marginalised target groups. Consequently, qualitative researchers are required to have a high sense of empathy, which is defined as the capacity to comprehend and identify with the emotions of other people. An empathic position encourages ethical sensitivity, which is an essential component of qualitative research methodology. The provision of informed consent, the maintenance of confidentiality, and the respectful depiction of the narratives of participants are not only technical requirements; rather, they are ethical imperatives. The qualitative researcher is conscious of power relations and makes it a priority to do research that is courteous, inclusive, and socially responsible.
4. **Global and Contextual Thinking**: Researchers who use qualitative methods look at occurrences through a perspective that is both holistic and contextual. As opposed to isolating factors, they take into account the complete complexity of circumstances, which includes personal, cultural, societal, and historical contexts. Having this frame of mind enables individuals to comprehend the ways in which various components interact with one another and impact one another within a particular environment. As an instance, a qualitative researcher would not only document observable behaviours in a study that investigates interactions between teachers and students, but they would also dive into the institutional culture, personal beliefs, and socio-political background that affect such relationships. This all-encompassing approach allows insights that are deeper and subtler than those obtained by reductionist techniques.
5. **An Unwavering Dedication to Meaning and Interpretation**: One of the most important aspects of the qualitative mentality is the dedication to discovering significance. Researchers who use qualitative methods are interested in how people make meaning of their lives, the connections in their lives, and the world around them. Through the interpretation of books, discussions, and actions, they can comprehend more profound symbolic and emotional levels that go beyond superficial assessments. When doing qualitative research, interpretation is not mechanical but rather hermeneutic. Hermeneutic interpretation is a dynamic and iterative process of understanding that includes the interaction between the researcher's viewpoint and the experience of the participant. The researcher must strike a balance between interpretation and integrity to the participants' voices. To maintain authenticity, the researcher often uses direct quotes and narrative formats.
6. **Methodological Adaptability and Creative Thinking**: A significant amount of originality is required in the process of data gathering, analysis, and presentation while doing qualitative research. In contrast to qualitative approaches that are standardised, qualitative methods are flexible and may often be tailored to match the research issue and the circumstances around it. The use of a wide variety of methods, including but not limited to interviews, focus groups, participant observation, ethnography, case studies, and narrative analysis, is supported by this creative attitude. A qualitative researcher has to be able to create instruments with

expertise, conduct in-depth investigations, and interpret data in ways that are both inventive and believable. To explain results in a manner that is both captivating and evocative, it is common practice to use storytelling, visual approaches, and metaphors.

7. **Patience and Tolerance for Ambiguity Are Essential**: Inherently time-consuming, unpredictable, and sometimes untidy, qualitative research is a method of doing research. It requires a significant amount of time spent in the field, the collection of data from a variety of sources, as well as intricate coding and thematic analysis. The researchers need to have patience to completely immerse themselves in the data and wait for themes to emerge naturally over time. Furthermore, a qualitative worldview accepts ambiguity and acknowledges that various interpretations are not only feasible but also desirable. When doing qualitative research, there is no one "truth" that can be identified; rather, there are several realities and a variety of experiences that both need to be respected. That researchers can deliver their results with honesty, subtlety, and integrity is made possible by their tolerance for ambiguity.
8. **An Orientation That Emphasises Collaboration and Participation**: A method to research that is collaborative or participatory is used by a significant number of qualitative researchers. Participants are not just seen as subjects by them; rather, they are seen as co-constructors of knowledge. Within the context of the research process, this attitude places a high importance on discussion, mutual learning, and shared authority. The terms "Participatory Action Research" (PAR) and "Community-Based Participatory Research" (CBPR) are examples of research methodologies that include researchers and communities working together to identify issues, gather data, and devise solutions. Through the use of this methodology, empowerment, equality, and relevance in the research outputs are fostered.
9. **An Analytical and Revolutionary Perspective**: Some qualitative researchers embrace a critical or transformational perspective to question power systems and promote social justice. From a variety of theoretical frameworks, including critical theory, feminism, postcolonial studies, and critical race theory, this approach takes inspiration. Researchers who use this frame of mind are not happy with just describing the world; rather, they want to bring about change in it. They make use of their research to elevate the voices of those who are marginalised, to challenge the prevalent narratives, and to fight for policy change. The endeavour of qualitative research is given a political and ethical dimension by virtue of this viewpoint.

In summary, when it comes to qualitative research, the attitude of the researcher is just as much about who they are as it is about the work that they do. Empathy, curiosity, ethical sensitivity, and an openness to complexity are some of the values that are included in this concept. Intellectual orientations such as reflexivity, contextual awareness, and interpretative reasoning are required for it to be successful. In addition to this, it encourages creative and flexible methodological approaches, as well as a dedication to comprehending the human experience in all of its complexities and variations. The mentality of a qualitative researcher is in a position that is distinctively positioned to investigate, comprehend, and contribute to the improvement of the human condition in a world that is characterised by variety, inequality, and fast change. In the pursuit of knowledge, qualitative researchers offer a vital depth of insight and compassion to the table, regardless of whether they are working in the fields of education, health, social sciences, or development.

1.6 Qualitative Formulation of Research Questions and Hypotheses

Qualitative research is an exploratory and interpretive approach that seeks to understand human experiences, social interactions, and cultural phenomena through rich, descriptive data. At the core of any qualitative study lies a well-formulated research question. Unlike qualitative studies that often begin with testable hypotheses, qualitative studies emphasize open-ended inquiry rooted in context, meaning, and the complexity of lived experiences. This essay examines the distinctive nature of research questions and the nuanced role of hypotheses in qualitative research, emphasizing their formulation, purpose, and methodological implications [35, 36].

Nature of Qualitative Research Questions

Qualitative research questions are designed to explore the "how" and "why" of phenomena rather than measure frequency or predict outcomes. These questions aim to uncover deeper insights into the meanings, perceptions, beliefs, and social processes that shape human behaviour. They are inherently open-ended, flexible, and iterative, reflecting the dynamic nature of qualitative inquiry. For instance, a qualitative study might ask, "*Does teacher feedback improve student performance?*", while a qualitative version of the question could be, "*How do students perceive and respond to different types of teacher feedback?*" The latter seeks to understand the experience and meaning behind the interaction, rather than test a direct relationship.

Qualitative research questions often focus on:

1. Understanding experiences (e.g., *What is it like to live with a chronic illness?*)
2. Exploring meanings (e.g., *What does success mean to first-generation college students?*)
3. Investigating processes (e.g., *How do communities cope with disaster recovery?*)
4. Examining cultural or social practices (e.g., *How is gender expressed in youth subcultures?*)

Key Characteristics of Qualitative Research Questions

1. Open-Ended: Qualitative questions allow participants to respond freely, encouraging detailed and nuanced data.
2. Exploratory: The questions are designed to explore areas where little is known or where complexity needs unpacking.

3. Contextual: They are grounded in the specific context or setting being studied, acknowledging that meaning is often context-dependent.
4. Emergent: Questions may evolve over time as the researcher gains deeper understanding through fieldwork and engagement with participants.

Formulating Good Qualitative Questions

The development of strong qualitative questions involves several stages, including reviewing existing literature, identifying gaps in understanding, and grounding questions in theoretical or conceptual frameworks. Good qualitative questions are:

1. Clear and focused, yet open enough to invite discovery.
2. Relevant to the participants and meaningful within their social world.
3. Aligned with qualitative methods such as interviews, observations, focus groups, or document analysis.

A commonly used format for formulating qualitative questions is:

1. What are the experiences of …?
2. How do individuals make sense of …?
3. In what ways do people perceive …?

Example:

1. Topic: Mental health in adolescent refugees.
2. Qualitative Question: *How do adolescent refugees describe their experiences of accessing mental health services in urban resettlement areas?*

This question invites narrative and depth, allowing researchers to uncover patterns, themes, and meanings that may not emerge in closed-ended surveys.

The Role of Hypotheses in Qualitative Research

As basic components, hypotheses play an important role in conventional research paradigms, notably in the field of qualitative research. They are claims that are intended to be tested by statistical processes and are in the form of predictions. Nevertheless, qualitative research, which is founded on interpretivist and constructivist paradigms, takes a different philosophical perspective from qualitative research. Even though hypotheses are not always in the forefront of qualitative investigations, they are nevertheless capable of playing a part, although in a more adaptable and exploratory capacity. The purpose of this book is to investigate the complex function that hypotheses play in qualitative research, focusing on their relevance, limits, and how they contribute to a more profound knowledge in human-centred investigations [35, 37, 38].

1. **Philosophical Underpinnings**: Understanding Qualitative Inquiry: Qualitative Research is primarily based on the concept that reality is socially created, subjective, and dependent on the environment in which it is seen. When doing qualitative research, the objective is not to discover universal truths but rather to get an understanding of complicated events from the point of view of the participants.

Because of this emphasis, a flexible research design that develops during the study is required. As a result, inflexible assumptions are somewhat misaligned with the nature of qualitative inquiry.

2. **The Traditional View: Hypotheses in Qualitative Versus Quantitative Research**: For qualitative research, a hypothesis is a prediction that can be tested and is derived from theory. In doing so, it provides direction for the construction of variables, measuring tools, and procedures for data processing. According to this positivist perspective, objectivity and the capacity to quantify connections are assumed to be present. On the other hand, qualitative research often starts with broad research questions rather than hypotheses that are carefully specified. The researchers can avoid making hasty conclusions because to the open-ended nature of the study, which enables them to stay receptive to new data and ideas. However, this does not always imply that hypotheses are completely disregarded in qualitative research. In its place, its function is modified so that it is compatible with the exploratory and inductive processes that are characteristic of qualitative approaches.
3. **Hypotheses as Guiding Concepts, Not Predictions**: The hypotheses that are developed in qualitative research are often conceptual rather than predictive. Rather than being in the form of rigid claims that need to be tested, they can be in the form of propositions or assumptions that serve to direct the investigation. In the context of a study that investigates the lived experiences of nurses working in palliative care, for instance, a researcher can hold the hypothesis that emotional resilience is impacted by the support systems that are provided by institutions. Instead of putting this hypothesis to the test with numerical data, the researcher investigates it via the use of interviews, observations, and theme analysis. These guiding notions provide direction without limiting the researcher's ability to be receptive to the narratives provided by the participants. Additionally, they serve as sensitising ideas, which are phrases or conceptions that assist researchers in keeping open to the unexpected while being aware of certain trends that are present in the data.
4. **Hypotheses in Grounded Theory**: Among the several subfields of qualitative research, grounded theory is one in which hypotheses play a significant role. Inductive theory development is a strategy that academics use to build hypotheses based on facts. In the first stages of the process, there are no assumptions involved, but as the coding and categorisation processes progress, theoretical notions start to develop. Following this, these newly arising hypotheses are subjected to more investigation, refinement, and incorporation into a coherent theory. In grounded theory, hypotheses are developed from the ground up, developing from patterns discovered in the data rather than being imposed at the beginning of the investigation. Using hypotheses in a manner that is both flexible and methodical is highlighted by this iterative hypothesis creation, which is in line with the emergent design of qualitative research approaches.
5. **Role in Literature Review and Conceptual Frameworks**: To have a better knowledge of the study topic, qualitative researchers often use previously published material in their work. In the context of this discussion, theories may

originate from gaps or contradictions that have been observed in earlier investigations. It is possible that these basic assumptions may assist in the formation of the conceptual framework and will serve as a lens through which to investigate the experiences of the participants. For instance, a researcher doing a study on leadership styles in nonprofit organisations would hypothesise, based on the existing body of literature, that transformational leadership is more widespread and successful than transactional leadership. This assumption may be used to direct interview questions and the emphasis of coding, but it is still open to reinterpretation when additional data becomes available.

6. **The Emergent Nature of Qualitative Hypotheses**: In contrast to qualitative research, which often seeks to validate or invalidate hypotheses, qualitative research views hypotheses as being in a state of precariousness and undergoing development. They are flexible and may be altered when the researcher engages in conversation with the participants and collects further data. When it comes to sustaining the integrity of qualitative research, which emphasizes the viewpoints of participants and the intricacies of the actual world, flexibility is very necessary. Researchers can maintain their reflexivity and awareness of their own biases and preconceptions with the assistance of emergent hypotheses. Researchers can critically evaluate whether their original assumptions are correct or if they need to be adjusted as they reflect on their continuously developing understandings.
7. **Avoiding Hypotheses in Purely Exploratory Studies**: When conducting some qualitative studies, particularly those of an exploratory type, researchers may choose to refrain from generating hypotheses at the beginning of the study process. Because of this, they can join the study environment with a minimum amount of prejudice and have an open mind throughout the process. Studies such as ethnographies and phenomenological research, for instance, often place more of an emphasis on in-depth absorption and comprehension than they do on verifying hypotheses. In situations like these, hypotheses may develop naturally during or after the fieldwork, as a component of the process of analysis and interpretation. Following that, these post-hoc hypotheses may be used to either frame conversations centred on theoretical contributions or to influence future research investigations.
8. **Hypotheses and Methodological Rigor**: When used appropriately, hypotheses in qualitative research may still add to the rigour of the methodology, although they are flexible. As a result, the topic of the study is made more clear, the data gathering and analysis processes are guided, and the research process is made more visible. They assist readers in comprehending the researcher's line of thinking and how interpretations were generated when they are presented clearly and concisely. Furthermore, hypotheses have the potential to fulfil the role of a bridge between qualitative and qualitative research, hence providing support for mixed-methods designs. In the case of qualitative research, for instance, the results may result in the generation of hypotheses, which may then be tested in a qualitative phase, therefore enriching the research cycle.

Hence, when doing qualitative research, hypotheses have a complex function that is based on the circumstances. However, they may serve as conceptual tools that drive investigation, influence data gathering, and enhance interpretation. Although they are not fundamental in the same way as they are in qualitative studies, they are central nonetheless. Maintaining a fluid, emergent, and reflexive approach is essential for qualitative hypotheses, regardless of whether they originate from literature reviews, develop via grounded theory, or serve as a source of inspiration for conceptual frameworks. Instead of being put to the test in a statistical sense, their function is to provide assistance for the in-depth, context-rich investigation that is characteristic of qualitative research. Hypotheses, when developed with careful consideration, have the potential to improve the clarity, coherence, and contribution of qualitative research without sacrificing the interpretative integrity of the findings [39] (Table 1.5).

Integration with Theoretical Frameworks

When it comes to qualitative research, theoretical frameworks are quite important in terms of being able to shape the direction and depth of the investigation. As they serve as lenses through which phenomena are interpreted, these frameworks guide the formulation of broad, exploratory research questions. They also ensure that the research remains focused and grounded in established bodies of knowledge, while at the same time allowing for the openness and flexibility that qualitative inquiry requires. When it comes to qualitative research, the objective of theoretical frameworks is to provide a methodical and organised way to comprehend and interpret the social environment. These factors contribute to the researcher's viewpoint, assisting in the concentration of the investigation while preserving awareness of the complexities and contexts involved. In qualitative research, theoretical frameworks are not strict moulds into which data must fit; rather, they serve as conceptual maps that assist in navigating the research process, from framing questions to interpreting results. This is because theoretical frameworks aid in the interpretation of findings.

The researchers realise that all information is created from certain vantage points by establishing a study inside a theoretical framework at the beginning of the research process. By using this method, critical reflection is encouraged, and it ensures that researchers continue to be conscious of their assumptions and how these assumptions may impact the data gathering and analysis processes. To summarise, theoretical

Table 1.5 Differences between qualitative and quantitative hypotheses

Aspect	Qualitative hypotheses	Qualitative "Hypotheses" or ideas
Purpose	Predict relationships between variables	Explore meanings, processes, and experiences
Structure	Formal and testable	Informal, flexible, and evolving
Based on	Deductive reasoning	Inductive reasoning
Use of variables	Clearly defined and measurable	Concepts are fluid and contextual
Outcome	Confirmation or rejection of hypotheses	Theory generation or deep understanding

frameworks in qualitative research can provide coherence, depth, and scholarly rigour without imposing the rigidity that is typical of hypothesis-driven investigations. When it comes to qualitative research, theoretical frameworks largely have an impact on the framing of research questions, rather than on the formation of hypotheses that may be tested. The purpose of these questions is to elicit rich, comprehensive replies that reflect the meanings that individuals attribute to their experiences. They are open-ended and exploratory. Phenomenology, feminist theory, critical theory, constructivism, and symbolic interactionism are some examples of theoretical frameworks that are utilised in qualitative research. Researchers are encouraged to ask a variety of questions and pay attention to a variety of elements of the human experience, which enriches the research process. Each framework provides new perspectives and priorities, which encourage researchers to ask various kinds of questions.

One further essential characteristic of qualitative research is the constant interaction that takes place between theoretical frameworks and actual evidence. The researchers do not just push a theory onto the data; rather, they allow the theory to shape their interpretations while maintaining an openness to discoveries that may expand or question the frameworks that are already in place. As an instance, a study that is guided by feminist theory may initially concentrate on the challenges that are encountered by female CEOs. However, via the use of interviews, the researcher might uncover unanticipated themes, which might result in feminist theory being refined or new understandings being contributed to ongoing discussions. Reflexivity is an essential practice in qualitative research, particularly when working within a theoretical framework. Researchers are required to engage in ongoing self-reflection on how their theoretical assumptions influence their interactions with participants, their interpretation of data, and the whole process of doing research. Researchers may guarantee that their theoretical convictions enrich rather than hinder their knowledge of the facts by participating in the practice of reflexivity [40].

One of the difficulties associated with incorporating theoretical frameworks into qualitative research is the tendency to develop an excessive attachment to a certain framework, which may result in selective attention being paid to data that conforms to theoretical expectations and the disregarding of information that is in conflict with the framework. Furthermore, in order to guarantee that the theoretical framework continues to be suitable all during the course of the investigation, it is necessary to strike a balance between theoretical direction and empirical openness. Since this is the case, theoretical frameworks in qualitative research serve as essential scaffolding, offering consistency, depth, and direction without limiting the openness that is required for rich, context-sensitive inquiry. Researchers that use qualitative methods are able to generate nuanced and insightful studies that make a significant contribution to the comprehension of the complexity of human existence. This is accomplished via the meticulous and reflective integration of theoretical views [41].

Importance of Flexibility in Question Formulation

The dynamic and interpretative aspect of qualitative research is one of its defining characteristics. This nature enables this kind of research to be flexible in terms of the design and development of research questions. It is important to note that qualitative inquiry thrives on flexibility, in contrast to qualitative research, which typically adheres to predetermined questions that are aligned with hypotheses. Adaptability is not a weakness; rather, it is a strength that increases the depth and authenticity of results by helping researchers to stay alert to the lived experiences and complexity that arise throughout the process of doing research. The questions that are asked in qualitative research are often broad, open-ended, and exploratory. They are centred on gaining an understanding of the "how" and "why" of human experiences rather than attempting to quantify facts. For instance, the question "How do refugees experience resettlement in a new country?" might potentially serve as the starting point for a research study. As the process of data gathering continues, unexpected patterns may emerge, which will inspire researchers to modify their questions in order to investigate these new insights in more depth.

Flexibility in question formulation is crucial for several reasons:

1. A Receptiveness to the Voices of Participants: Qualitative research places a high priority on the views of participants, enabling their experiences and observations to have an impact on the enquiries that are conducted. The richness of the data that is obtained may be negatively impacted by a strict adherence to prescribed questions. Being flexible allows researchers to modify their questions according to what the participants consider to be relevant, which ensures that the method is centred on the participants.
2. Alignment with Emerging Themes: In qualitative research, the time spent collecting data and analysing it often occurs concurrently. The researchers may uncover new areas of interest or realise that particular questions are no longer relevant when themes emerge from the interviews or focus groups that they have conducted earlier. The modification of research questions in response to these themes makes it possible to conduct a more nuanced investigation into the subject matter.
3. Contextual Sensitivity: Qualitative research often takes place in intricate social situations, where the cultural and interpersonal dynamics of the participants are of great importance. Adjusting questions in response to changes in the environment, such as discomfort experienced by participants or ethical concerns, ensures that the study continues to be conducted in a manner that is both courteous and morally sound.
4. Support for Grounded Theory and Inductive Approaches: Researchers are required to construct hypotheses based on the data that they have gathered to practise several qualitative approaches, such as grounded theory. Because this is an inductive process, it is necessary to take a progressive approach to the formulation of questions. This means that the first broad enquiries should become increasingly detailed as the data gathering process continues.

An illustration of this flexibility is provided by examples of flexible question development. Research on instructors' experiences with digital learning tools may alter emphasis based on early interviews, indicating infrastructure issues. In a similar vein, research that investigates the perspectives of senior patients towards home-based healthcare may extend to incorporate trust and cultural attitudes depending on the results of the first investigation. It is not the case that a lack of rigour is implied by the fact that qualitative research is characterised by its flexibility. The maintenance of methodological openness and the documentation of the development of research topics are equally important for establishing credibility. To ensure that the modifications are based on data rather than arbitrary shifts, researchers are required to explain the reasons behind the adjustments.

For instance, an initial question like "*How do teachers support students with learning disabilities?*" might evolve into "*How do teachers' beliefs about inclusion influence their support strategies for students with learning disabilities?*" as deeper themes emerge.

When it comes to the framing of flexible questions, ethical concerns are essential as well. Researchers can modify their enquiries to prevent participants from experiencing any harm or discomfort, thereby ensuring a humane and participant-centred approach. To allow for responsiveness to participants, sensitivity to context, and openness to new insights, qualitative research needs to have a flexible approach to the design of questions. This adaptive technique captures the complexity of real-world experiences, boosting the depth, authenticity, and ethical integrity of qualitative investigation. To do good qualitative research, acknowledging the need for flexibility is vital.

The formulation of research questions in qualitative research is a thoughtful, iterative, and creative process. Rather than rigid hypotheses, qualitative studies are guided by open-ended inquiries that seek to understand meanings, behaviours, and contexts. These questions reflect the epistemological foundations of qualitative research, emphasizing depth over breadth, context over generalization, and human experience over statistical inference. While hypotheses as traditionally defined may not play a central role in qualitative inquiry, the researcher often works with emerging ideas, interpretive frameworks, and guiding assumptions that evolve through deep engagement with data. The flexibility, richness, and insight offered by well-crafted qualitative questions are what make this approach indispensable for exploring the complexity of the human condition.

1.7 Selecting a Research Design

Selecting an appropriate research design is a crucial step in the research process. The research design serves as the blueprint for conducting a study, outlining how data will be collected, analyzed, and interpreted to answer the research questions or test the hypotheses. It ensures that the evidence obtained enables the researcher to address the problem logically, unambiguously, and as efficiently as possible. The choice of research design must align with the research objectives, the nature of the study, available resources, and ethical considerations. In both qualitative and quantitative traditions, selecting the right design determines the strength, credibility, and impact of the research findings [42, 43].

Understanding Research Designs

A research design is not merely about selecting specific methods or techniques; it represents the strategic blueprint that holds an entire study together. It provides a structured framework that links the research questions to the processes of data collection, analysis, and ultimately to the conclusions drawn from the study. A carefully chosen research design ensures that the study remains coherent, systematic, and focused, enhancing the reliability and validity of the findings. Research designs can be broadly categorised into two primary types: Qualitative Research Designs and quantitative Research Designs. Each type serves different goals, addresses different kinds of research questions, and follows unique approaches to inquiry (Fig. 1.3).

Thus, research designs can broadly be categorized into two main types:

1. **Qualitative Research Designs**

Qualitative research designs are typically used when the aim is to test hypotheses, measure variables, and determine relationships among those variables. They often seek to make generalizations from a sample to a larger population. The emphasis is on objectivity, replicability, and statistical analysis. Common types of qualitative designs include:

a. **Experimental Designs**: These involve the manipulation of one or more independent variables to observe their effect on a dependent variable. Participants are randomly assigned to groups to ensure that differences are due to the treatment and not other factors.

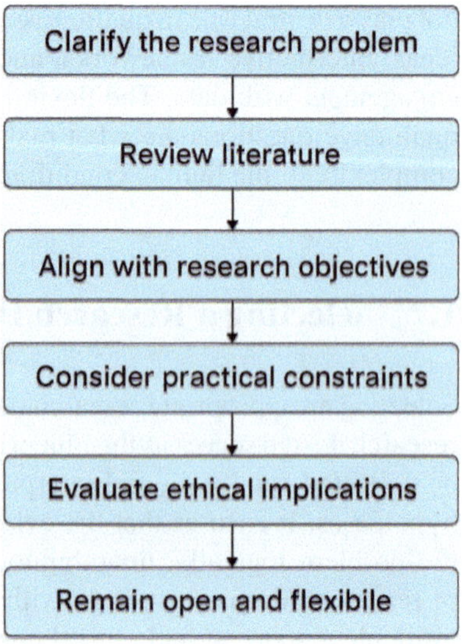

Fig. 1.3 Understanding research designs

b. **Quasi-Experimental Designs**: Similar to experimental designs but without random assignment. These are often used when it is impractical or unethical to randomly assign participants to conditions.
c. **Correlational Designs**: These assess the relationships between variables without manipulating them. While they can identify associations, they do not establish causality.

Qualitative designs are best suited for research questions that require precision, comparison, and measurement, such as "What is the effect of a new teaching method on student performance?" or "Is there a relationship between social media usage and anxiety among teenagers?"

2. **Qualitative Research Designs**

In contrast, qualitative research designs are used to explore complex phenomena that are not easily reduced to numbers or measurements. They focus on understanding meanings, experiences, perspectives, and social processes in their natural context. The goal is to gain deep insights rather than to generalize findings. Common types of qualitative designs include:

a. **Ethnography**: This design involves immersive observation and participation to understand cultural groups or social practices.
b. **Phenomenology**: Focuses on exploring and describing the lived experiences of individuals around a particular phenomenon.
c. **Grounded Theory**: Aims to generate or discover a theory that emerges from systematically gathered and analyzed data.
d. **Narrative Inquiry**: Centres around collecting and interpreting the stories people tell about their lives and experiences.
e. **Case Study Research**: Involves an in-depth investigation of a single case (an individual, group, event, or organization) within its real-world context.

Qualitative designs are chosen when researchers seek to answer questions like, "How do teachers experience the integration of digital tools in their classrooms?" or "What are the coping strategies of refugees adjusting to a new country?"

Choosing the Right Research Design

The selection of a research design is fundamentally tied to the nature of the research questions. Qualitative designs are appropriate when the focus is on testing a theory, measuring variables, and making predictions. Qualitative designs are suitable when the aim is to understand meanings, develop theories, or capture the richness of lived experiences. A good research design considers not only the type of questions but also practical factors such as the availability of participants, resources, time constraints, and ethical considerations. In some cases, researchers may even employ mixed-methods designs, combining both qualitative and qualitative approaches to take advantage of the strengths of both traditions. Ultimately, understanding research designs enables researchers to align their goals, methodology, and outcomes effectively, ensuring that their work contributes meaningfully to knowledge in their field [44].

Factors Influencing the Selection of Research Design

Several key factors influence the selection of an appropriate research design:

1. **Nature of the Research Problem**: The research problem is the foundation for selecting an appropriate research design, shaping the direction, structure, and methodology of a study. Researchers must understand the type of question they are trying to answer before data collection begins, as this decision will guide every subsequent step of the research process. Qualitative research designs are often the best choice for exploring "how" or "why" a phenomenon occurs, as they delve into meanings, interpretations, motivations, and processes. For example, a study exploring "How do cancer survivors experience changes in their identity?" requires a design that captures personal narratives, emotions, and subjective experiences, which qualitative metrics cannot adequately measure. On the other hand, qualitative designs are preferred for measuring "how much," "how many," or to establish cause-and-effect relationships, aiming for results that can be generalized across populations. Examples of qualitative designs include experimental, quasi-experimental, and correlational designs. The distinction often lies in the kind of data and the depth versus breadth of understanding sought. Qualitative designs aim for depth, exploring a smaller number of cases in great detail, while qualitative designs aim for breadth, surveying larger populations to identify patterns, trends, and statistically significant results. Mixed-methods approaches may also be required, combining both qualitative and qualitative elements. Careful consideration of the research problem's nature ensures methodological alignment, increasing the validity, reliability, and credibility of the findings. Ignoring this step or mismatching the design to the problem can lead to weak conclusions, misinterpretation of results, or ethical issues [45].
2. **Research Objectives**: Research objectives are crucial in determining the most appropriate research design, as they articulate the purpose of a study and guide decisions during the research process. Exploratory research is generally suitable for exploring new insights, understanding perspectives, or generating new theories, often in fields with limited existing knowledge or complex phenomena. Qualitative research designs, such as interviews, focus groups, or ethnographic observations, focus on understanding feelings, processes, and relationships that cannot be adequately captured through numerical data. On the other hand, qualitative research designs are typically preferred when the objective is to predict, control, or test existing theories. These objectives are precise, structured, and often seek to establish cause-and-effect relationships. For example, an objective like "determining the effect of a leadership training program on employee productivity" requires statistical measurement and hypothesis testing. Research objectives also influence design choices within these broad categories. For qualitative research, grounded theory methodology may be used to develop a new theory grounded in data, while phenomenology is appropriate for understanding lived experiences. In qualitative research, correlational designs are used to find relationships between variables without manipulating them, while experimental designs with random assignment are best for determining causality. Clarity in

objectives is essential, as vague or conflicting objectives can lead to methodological confusion and weak results. Mixed-methods designs, which combine the depth of qualitative inquiry with the generalizability of quantitative analysis, are often adopted in contemporary research. Aligning the design with the objectives ensures the study remains coherent, focused, and impactful.

3. **Type of Data Needed**: The type of data needed is a crucial factor in selecting a research design. Qualitative research designs are typically used when numerical data is required, such as surveys or tests, can be counted, measured, and statistically analyzed. These designs focus on questions like "how many," "how much," or "to what extent." On the other hand, qualitative research designs are more suitable for data that captures meaning, experiences, emotions, or social phenomena. Qualitative research deals with open-ended information, collected through methods like interviews, focus groups, participant observations, video recordings, or document analysis. For example, a study exploring students' experiences of online learning would focus on narratives, emotions, and personal reflections. The distinction between numeric and non-numeric data is not just about format; it reflects the nature of the knowledge the research aims to produce. Qualitative data tends to answer "what" and "how much" questions, emphasizing breadth, generalizability, and objectivity, while qualitative data answers "how" and "why" questions, focusing on depth, context, and subjective understanding. The type of data needed affects not only the initial design choice but also the way research instruments are constructed, how participants are engaged, and how findings are reported. In qualitative studies, data collection tools must be designed with scales, categories, and structured response options, while in qualitative studies, tools like interview guides must allow for open-ended responses, storytelling, and exploration of participants' perspectives. Technological advances have further expanded the types of qualitative data researchers can collect, including photographs, social media posts, podcasts, videos, and virtual reality experiences. Therefore, choosing the right data type is essential for research design selection, as choosing the wrong type can undermine the validity and relevance of the study [46].
4. **Resource Availability**: Resource availability is a critical factor in choosing the right research design. It involves assessing practical constraints such as time, budget, access to participants, and researcher expertise before deciding on a methodological approach. Longitudinal studies, experimental designs, and qualitative research designs require significant resources, such as long-term funding, sustained participant engagement, and specialized equipment. These resources may not always be available, making them impractical for some projects. Qualitative research designs, while flexible, also demand resources, such as time investment for in-depth interviews, ethnographic fieldwork, and case studies. Participants' availability can also impact design decisions, with hard-to-reach or vulnerable populations often making large-scale qualitative surveys impractical. Budget constraints can also influence research decisions, with researchers opting for cross-sectional studies, online surveys, or secondary data sources to reduce costs. Ultimately, resource availability requires a careful balancing act, with a

well-resourced but poorly designed study being ineffective, while a modestly funded but strategically planned study can yield valuable insights. Therefore, resource considerations must align with research objectives to ensure project feasibility and success.

5. **Ethical Considerations**: Ethical considerations play a crucial role in research design, as researchers have an obligation to protect the rights, dignity, and welfare of their participants. For studies involving vulnerable populations, qualitative designs are often preferred, as they allow researchers to adapt to participants' needs and comfort levels. The sensitivity of the research topic also influences design choice, as topics like domestic violence, mental health, or grief require careful ethical handling. Foundational ethical principles include informed consent, confidentiality, and the right to withdraw. Covert ethnography poses ethical dilemmas and requires strong justifications and ethical review board approvals. Qualitative studies involving randomized control trials (RCTs) must ensure participants are not exposed to harm by being denied beneficial treatment, which can limit the feasibility of certain experimental designs, especially in fields like healthcare and education. Ethical frameworks also encourage researchers to consider the potential impact of their findings, such as stigmatizing a group or reinforcing negative stereotypes. By thoughtfully selecting a design that prioritizes participants' wellbeing, researchers uphold their moral and professional responsibilities [47].
6. **Philosophical Assumptions**: Philosophical assumptions significantly influence research design decisions. Epistemology and ontology are crucial in understanding the nature and scope of knowledge and reality. Researchers with positivism, which believes in objective and measurable reality, often use qualitative research designs like experiments, surveys, and statistical analysis. Conversely, constructivist researchers believe in subjective reality, constructed through human interaction and interpretation. They prefer qualitative designs like ethnography, phenomenology, and narrative inquiry to understand individuals' lived experiences and perspectives. The interpretivist stance emphasizes the co-construction of knowledge between researchers and participants, acknowledging that researchers bring their own backgrounds and perspectives into the study. For instance, a constructivist study might explore how students define success differently based on cultural, social, and personal factors. Critical theorists may use qualitative designs to uncover power dynamics, inequalities, and hidden structures in society. Research underpinned by feminist, postcolonial, or Marxist perspectives often employs qualitative methods to highlight marginalized voices. Ultimately, philosophical assumptions are not just academic concerns but fundamentally shape every decision in the research process, from framing questions to choosing methods and interpreting results. Recognizing these beliefs ensures methodological coherence and enriches the depth and relevance of the research [48].

Therefore, selecting the right research design is fundamental to the success of any study. It requires careful consideration of the research problem, objectives, resources, philosophical assumptions, and ethical standards. A well-chosen design strengthens the study's validity, relevance, and contribution to knowledge. Researchers must approach this task thoughtfully, balancing theoretical understanding with practical realities. Whether the study is qualitative, quantitative, or mixed-methods, a sound research design provides the necessary foundation for credible and impactful research outcomes.

1.8 Developing a Research Proposal Using a Qualitative Approach

The research proposal is the scholar's first act of creation. It is where the seeds of inquiry are planted, where the broad curiosity of the mind is channelled into a purposeful investigation. In the qualitative tradition, such a proposal is more than a technical plan; it is a narrative—a carefully woven story about a problem worth exploring, about people whose voices must be heard, and about contexts that shape their experiences.

At the heart of a qualitative proposal lies the conviction that knowledge is not merely a matter of counting and measuring. Instead, it emerges from engaging with the richness of human perception, behaviour, and interaction. The qualitative researcher seeks to see the world as participants see it, to interpret meanings in their natural settings, and to illuminate dimensions of a problem that numbers alone cannot reveal.

The opening of a qualitative proposal, therefore, must draw the reader into the world of the research problem. It paints the landscape in which the study will take place: the social, organisational, or theoretical backdrop against which questions arise. This background is not mere ornamentation; it is the stage upon which the drama of inquiry will unfold. Here, the researcher explains why the problem matters—not only to the academic community but also to practitioners, policymakers, or the public at large.

Following this comes the articulation of purpose. In qualitative research, the aim is rarely to predict or control; rather, it is to explore, to understand, to interpret. Objectives are framed as invitations to discovery: to explore the lived experiences of a particular group, to understand the processes behind a social phenomenon, or to interpret the meanings people assign to their actions and environments. The language here is open, avoiding the fixed outcomes of qualitative hypotheses, for the journey of qualitative research is one of following the thread wherever it leads.

The literature review that accompanies a qualitative proposal is equally purposeful. It is not simply a compendium of what others have written; it is a conversation with existing scholarship. Through it, the researcher identifies gaps—those spaces where the lived realities of people have been underexplored or misrepresented—and positions the proposed study as a bridge between what is known and what remains to be discovered.

Methodology in qualitative research is the architecture of meaning-making. Whether the design is phenomenological, ethnographic, grounded theory, or narrative inquiry, it reflects a deliberate choice about how best to encounter the phenomenon of interest. Data collection methods such as interviews, focus groups, observations, and document analysis are not chosen for convenience but for their capacity to capture the textures of human experience. Sampling, too, is purposeful: participants are selected not for their statistical representativeness but for their ability to illuminate the research question in rich and varied ways.

Analysis in a qualitative proposal is a matter of interpreting patterns and themes from the data. It requires both creativity and rigour—creativity to see connections that are not immediately obvious, and rigour to ensure that interpretations are grounded in the evidence. The researcher describes how coding will be carried out, how themes will be constructed, and how the integrity of the process will be preserved.

Ethics are at the core of this kind of inquiry. Because qualitative research often delves into personal experiences, the proposal must convey a deep respect for participants' dignity, privacy, and autonomy. Informed consent, confidentiality, and sensitivity to emotional or social risks are not afterthoughts; they are embedded in the fabric of the research design.

Finally, a qualitative research proposal acknowledges the importance of trustworthiness. It outlines how credibility, transferability, dependability, and confirmability will be upheld—whether through member checking, triangulation, reflective journaling, or maintaining an audit trail. These commitments signal to the reader that the researcher values not only insight but also integrity.

In the end, a qualitative research proposal is a promise: a promise to listen, learn, and faithfully represent the voices and experiences of those who participate. It is an invitation to join in the search for understanding, to walk alongside the researcher as they navigate the winding paths of human experience, and to discover, in the particular, something of the universal.

Structuring the Proposal

A well-structured research proposal is not simply a formal requirement; it is a bridge between an idea and its realisation. In the world of scholarship, structure is what transforms a collection of thoughts into a coherent plan, making it possible for others to see the logic, feasibility, and significance of the intended study. For qualitative research in particular, this structure must balance academic rigour with the flexibility needed to capture the complexities of human experience [49, 50].

1. The opening of the proposal begins with a **title**—not just a label, but a distilled essence of the research. A good title signals both the focus and the scope of the inquiry, inviting readers into its conceptual world.
2. Immediately following this, the **abstract** serves as a concise portrait of the study, offering a glimpse of the problem, objectives, methods, and anticipated contributions. Though brief, it sets the tone and creates the first impression on assessors, committees, or potential funders.

3. The **introduction** follows, acting as the gateway into the subject matter. It situates the research within its broader context, explains the motivation behind it, and sets out the research problem with clarity. In qualitative research, this section often draws upon narratives, illustrative examples, or preliminary observations that highlight the human dimensions of the problem. It is here that the reader is persuaded of the study's relevance and necessity.
4. Next comes the **literature review**, which forms the intellectual foundation of the proposal. Far from being a mechanical listing of sources, this section weaves together key theories, empirical studies, and ongoing debates, revealing where the proposed research fits within the scholarly landscape. In qualitative proposals, the literature review also helps refine the conceptual framework, identifying which lenses will be most useful in interpreting the findings.
5. The **research questions or objectives** stand at the centre of the proposal's architecture. They must be open enough to allow for discovery, yet precise enough to guide the inquiry. Qualitative research thrives on questions that begin with "how" or "why," signalling a focus on processes, meanings, and interpretations rather than numerical outcomes.
6. The **methodology section** is where the plan takes concrete shape. Here, the researcher explains the choice of qualitative design—be it grounded theory, ethnography, phenomenology, or narrative analysis—and shows how it aligns with the research questions. This section describes the data collection methods, sampling strategy, and analysis procedures in sufficient detail to assure the reader of their appropriateness and rigour. Importantly, it also addresses practicalities: the setting of the research, the role of the researcher, and the anticipated challenges of the fieldwork.
7. **Ethical considerations** follow naturally, for they are integral to the design rather than a separate obligation. This section outlines how participants will be protected through informed consent, confidentiality safeguards, and sensitivity to potential harm. Qualitative research often enters deeply personal or sensitive areas, so ethical reflexivity—awareness of how the researcher's presence and actions affect participants—is key.
8. A well-structured proposal also includes a **timeline**. This transforms the research from an abstract idea into a tangible plan of action. By breaking the project into stages—literature review, data collection, analysis, and writing—the timeline demonstrates both foresight and feasibility.
9. Finally, the proposal concludes with **expected outcomes** and, where relevant, a **budget**. The outcomes section does not predict results in the qualitative sense; instead, it anticipates contributions to knowledge, policy, or practice. The budget, if required, should be realistic, itemising resources needed for travel, transcription, software, or dissemination.

When well-structured, a proposal is more than a checklist of components; it is a persuasive narrative. It guides the reader step-by-step from the recognition of a problem, through the justification for studying it, to a carefully considered plan for doing so. For the qualitative researcher, this structure does not constrain creativity—it provides the framework within which insight, empathy, and scholarly discipline can flourish. Developing a research proposal using a qualitative approach is an intellectually demanding but rewarding endeavor. It requires balancing structure with flexibility, rigor with openness, and planning with responsiveness to participants' realities. A strong qualitative proposal lays the groundwork for meaningful, insightful, and ethically responsible research that gives voice to participants' experiences and enriches understanding of complex human phenomena. By thoughtfully constructing each element of the proposal, researchers position themselves to conduct studies that are both methodologically sound and deeply impactful.

1.9 Ethics in Qualitative Research

Qualitative research is a critical field that involves close interactions with individuals, exploration of sensitive topics, and interpretation of personal experiences. Maintaining high ethical standards is crucial for the credibility of the research and the dignity of participants. Ethical principles guide researchers in designing studies, interacting with participants, collecting and analyzing data, and reporting findings. They ensure that participants are respected, protected, and treated fairly throughout the research process.

Informed consent is a foundational ethical principle in qualitative research, as participants must voluntarily agree to participate, fully understanding the nature, purpose, risks, and benefits of the study. This is an ongoing process, and participants should feel empowered to withdraw at any time without penalty. Researchers must present information clearly, using language that participants can understand, avoiding technical jargon or coercive pressure. In some cases, dealing with vulnerable populations, such as children, refugees, or people with disabilities, obtaining consent may involve additional layers of protection, such as parental consent or assent procedures. Confidentiality and anonymity are other critical ethical concerns in qualitative research. Researchers must take steps to anonymize data, such as removing identifying details from transcripts or using pseudonyms. Maintaining confidentiality can be challenging, as detailed descriptions of contexts and experiences might inadvertently reveal participants' identities. Researchers must carefully consider how much detail to include when reporting findings and be transparent with participants about the limits of confidentiality, particularly when legal obligations (such as reporting abuse) might require disclosure.

Respect for participants is essential in qualitative research, as it often explores sensitive aspects of participants' lives, such as trauma, grief, discrimination, or illness. Researchers must show empathy, sensitivity, and respect throughout the research process, creating a safe, non-judgmental environment where participants feel heard and valued. Respect also involves acknowledging participants as active collaborators rather than passive subjects, and sometimes even involves them in shaping the study or interpreting findings (a practice known as participatory research). Power dynamics and researcher reflexivity are essential ethical practices in qualitative research, as they help minimize power imbalances between researchers and participants. Researchers must constantly reflect on their own biases, assumptions, and positions of privilege, enhancing the ethical and interpretive quality of their work. Cultural sensitivity and inclusivity are also crucial in qualitative research, as they should strive to understand participants' cultural contexts and avoid imposing their cultural assumptions. Researchers must adapt communication styles, data collection methods, and analysis approaches to honor participants' traditions and values, while also ensuring that marginalized voices are represented and that findings do not perpetuate stereotypes or biases. Ethical approval and institutional oversight are necessary for qualitative studies, ensuring that the study design protects participants' rights and well-being. By embedding ethical principles in every stage of their work, researchers not only protect participants but also enhance the credibility, richness, and impact of their findings.

References

1. Dodds, S., Hess, A.C.: Adapting research methodology during COVID-19: lessons for transformative service research. J. Serv. Manag. **32**(2), 203–217 (2021). https://doi.org/10.1108/JOSM-05-2020-0153
2. Obi, K.C., Iwuanyanwu, O.: Innovation in the assessment of primary school mathematics: an approach to revamping Nigerian education. J. Teach. Perspect. **18**(1), 51–66 2023 [Online]. Available: http://repositorio.unan.edu.ni/2986/1/5624.pdf%0Ahttp://fiskal.kemenkeu.go.id/ejournal%0A10.1016/j.cirp.2016.06.001%0A10.1016/j.powtec.2016.12.055%0A10.1016/j.ijfatigue.2019.02.006%0A10.1
3. Al Shawabkeh, R.K.: Evaluating the satisfaction rate for affordable housing in non-gated residential area (NGR): the case of Al-Sharq housing project in Zarqa-Jordan. Int. J. Hous. Mark. Anal. **14**(1), 192–217 (2021). https://doi.org/10.1108/IJHMA-10-2019-0105
4. Assi, J.P., Antoniolli, P.D., Wolfshorndl, D.A., de C. Junior, J.B.: Proposal for information technology strategic planning approach. Eur. J. Bus. Soc. Sci. **5**(02), 90–105 (2016)
5. Musungwini, S., Mugoniwa, B., Furusa, S.S., Rebanowako, T.G.: An analysis of the use of cloud computing among university lecturers: a case study in Zimbabwe. Int. J. Educ. Dev. Using Inf. Commun. Technol. **12**(1), 53–70 (2016) [Online]. Available: http://files.eric.ed.gov/fulltext/EJ1099577.pdf
6. Preston, S., Anderson, A., Robertson, D.J., Shephard, M.P., Huhe, N.: Detecting fake news on Facebook: the role of emotional intelligence. PLoS ONE **16**(3), 1–13 (2021). https://doi.org/10.1371/journal.pone.0246757
7. Cheah, C.W., Low, B., Lee, C.K.C.: Sustainable housing development: the legitimacy-seeking perspective. J. Bus. Ind. Mark. **36**(6), 1027–1041 (2020). https://doi.org/10.1108/JBIM-07-2020-0318
8. Ajagba, C.O., Gberevbie, D.E., Agbu, O.: Rebranding the electoral process in Nigeria's Fourth Republic (1999–2019): constraints and prospects of the independent national electoral commission. Acad. J. Interdiscip. Stud. **9**(1), 56–69 (2020). https://doi.org/10.36941/ajis-2020-0005
9. Sweeney, A., Clarke, N., Higgs, M.: Shared leadership in commercial organizations: a systematic review of definitions, theoretical frameworks and organizational outcomes. Int. J. Manag. Rev. **21**(1), 115–136 (2019). https://doi.org/10.1111/ijmr.12181
10. Prasad, P.: Crafting Qualitative Research : Beyond Positivist Traditions (2017)

11. Talwar, S., Dhir, A., Singh, D., Virk, G.S., Salo, J.: Sharing of fake news on social media: application of the honeycomb framework and the third-person effect hypothesis. J. Retail. Consum. Serv. **57**, 102197 (2020). https://doi.org/10.1016/j.jretconser.2020.102197
12. Chandra, R., Jatmika, S.: Unilever Surabaya Corporate Social Responsibility (CSR) policy in maintaining environmental sustainability in Surabaya in 2014–2020. In: Proceedings of the International Conference on Public Organization (ICONPO 2021), vol. 209, pp. 444–452 (2022) [Online]. Available: https://www.atlantis-press.com/proceedings/iconpo-21/125970904
13. Singh, V.: Digitalization, BIM ecosystem, and the future of built environment: how widely are we exploring the different possibilities? Eng. Constr. Archit. Manag. (2019). https://doi.org/10.1108/ECAM-01-2018-0004
14. Arifin, J.A., Juned, M.: Nigeria's compliance with the Yaoundé code of conduct in the cases of piracy and armed piracy of the Nigerian State in the Gulf of Guinea in 2016–2021. J. Soc. Polit. Sci. **6**(1), 110–120 (2023). https://doi.org/10.31014/aior.1991.06.01.398
15. Liu, W.S.K.: Global Leadership Dynamics: Refining Executive Selection in Multinational Corporations, No. 0123456789. Springer, US (2024). https://doi.org/10.1007/s13132-024-01794-3
16. Intarapanich, C.: Teaching methods, approaches and strategies found in EFL classrooms: a case study in Lao PDR. Procedia Soc. Behav. Sci. **88**, 306–311 (2013). https://doi.org/10.1016/j.sbspro.2013.08.510
17. Wu, C.W., Monfort, A.: Role of artificial intelligence in marketing strategies and performance. Psychol. Mark. **40**(3), 484–496 (2023). https://doi.org/10.1002/mar.21737
18. Dobiesz, V.A.: Maintaining health professional education during war: a scoping review. Med. Educ. **56**(8), 793–804 (2022). https://doi.org/10.1111/medu.14808
19. Assadpour, H., Ghalehnoee, M., Bahramian, A.: Developing the model of research method in urban landscape studies emphasizing the Saunders' research onion. Motaleate Shahri **12**(45), 3–18 (2023). https://doi.org/10.34785/J011.2022.010
20. Jin, S.: Identifying barriers to sustainable apple production: a stakeholder perspective. J. Environ. Manage. **302**(PB), 114082 (2022). https://doi.org/10.1016/j.jenvman.2021.114082
21. Mukumbang, F.C.: Retroductive theorizing: a contribution of critical realism to mixed methods research. J. Mix. Methods Res. **17**(1), 93–114 (2021). https://doi.org/10.1177/15586898211049847
22. Gentry, W.A., Griggs, T.L., Deal, J.J., Mondore, S.P., Cox, B.D.: A comparison of generational differences in endorsement of leadership practices with actual leadership skill level. Consult. Psychol. J. **63**(1), 39–49 (2011). https://doi.org/10.1037/a0023015
23. Ntuli, L.T., Mahlangu, V.P.: Legal execution of curriculum in primary schools: school management teams' perspectives. Res. Educ. Policy Manag. **5**(2), 177–206 (2023). https://doi.org/10.46303/repam.2023.16
24. Nasir, M.J., Ciroma, A.: Impact of media and information literacy in Nigerian colleges of education curriculum: a case of detecting fake news. Am. J. Commun. **5**(1), 28–37 (2023). https://doi.org/10.47672/ajc.1385
25. Garrido, R., Paloma, V., Benítez, I., Skovdal, M., Verelst, A., Derluyn, I.: Impact of COVID-19 pandemic on the psychological well-being of migrants and refugees settled in Spain. Ethn. Heal. **28**(2), 257–280 (2023). https://doi.org/10.1080/13557858.2022.2035692
26. Kim, S., Park, C.H., Chin, S.: Assessment of BIM Acceptance Degree of Korean AEC Participants, pp. 1–15 (2015). https://doi.org/10.1007/s12205-015-0647-y
27. Civitillo, R., Ricci, P., Simonetti, B.: Management and performance of non-profit institutions: finding new development trajectories—evidence from Italy. Qual. Quant. **53**(5), 2275–2290 (2019). https://doi.org/10.1007/s11135-018-0719-4
28. Saprikis, V., Avlogiaris, G., Katarachia, A.: A comparative study of users versus non-users' behavioural intention towards M-Banking apps' adoption. Information **13**(1) (2022). https://doi.org/10.3390/info13010030
29. Aydin, H., Ustaoğlu, F., Tepe, Y., Soylu, E.N.: Assessment of water quality of streams in northeast Turkey by water quality index and multiple statistical methods. Environ. Forensics **22**(1–2), 270–287 (2021). https://doi.org/10.1080/15275922.2020.1836074

30. Alcaraz, A.O.: Politicians and racism on Twitter: citizen reactions and new forms of social participation. Int. Multidiscip. J. Soc. Sci. **11**(1), 82–103 (2022). https://doi.org/10.17583/rimcis.9909
31. Muraina, Ajímátanraẹjẹ, A.J.: Gender relations in indigenous Yoruba culture: questioning feminism action and advocacy. Afric arXiv Preprint, pp. 1–19 [Online]. Available: https://osf.io/preprints/africarxiv/x6vma/ (2022)
32. Nyamasege, O.J., Evans Biraori, O.: Effect of supplier relationship management on the effectiveness of supply chain management in the Kenya public sector. Int. J. Manag. Value Supply Chain **6**(1), 25–32 (2015). https://doi.org/10.5121/ijmvsc.2015.6103
33. Amegayibor, G.K.: Leadership styles and employees' performance: a case of family-owned manufacturing company, Cape Coast. Int. J. Financ. Account. Manag. **3**(2), 149–164 (2021). https://doi.org/10.35912/ijfam.v3i2.624
34. Diana Situmorang, E., Kusuma Wardhani, N.: The influence of transformational leadership style, workload and job satisfaction on turnover intention. Dinasti Int. J. Manag. Sci. **3**(3), 463–476 (2022). https://doi.org/10.31933/dijms.v3i3.1092
35. Delgado-Santos, P., Stragapede, G., Tolosana, R., Guest, R., Deravi, F., Vera-Rodriguez, R.: A survey of privacy vulnerabilities of mobile device sensors. ACM Comput. Surv. 1–18 (2022). https://doi.org/10.1145/3510579
36. Lemi, K., Alemu, D.: Effects of leadership styles on Ethiopian banking industry performance. Eur. J. Bus. Manag. **12**(16), 11–20 (2020). https://doi.org/10.7176/ejbm/12-16-02
37. Bouacida, N., Mohapatra, P.: Vulnerabilities in federated learning. IEEE Access **9**, 63229–63249 (2021). https://doi.org/10.1109/ACCESS.2021.3075203
38. Barrett, P.: Intersections between housing affordability and meanings of home: a review. Kotuitui 1–18 (2022). https://doi.org/10.1080/1177083X.2022.2090969
39. Anderson, S.: Public, private, neither, both? Publicness theory and the analysis of healthcare organisations. Soc Sci Med **74**(3), 313–322 (2012). https://doi.org/10.1016/j.socscimed.2011.07.021
40. Lyon, C.: Climate change research and action must look beyond 2100. Glob. Chang. Biol. **28**(2), 349–361 (2022). https://doi.org/10.1111/gcb.15871
41. Kloda, L.A., Boruff, J.T., Cavalcante, A.S.: A comparison of patient, intervention, comparison, outcome (PICO) to a new, alternative clinical question framework for search skills, search results, and self-efficacy: a randomized controlled trial. J. Med. Libr. Assoc. **108**(2), 185–194 (2020). https://doi.org/10.5195/jmla.2020.739
42. Olatoye, T.A.: Analysis of modal choice of residents in Lagos State. J. Hum. Ecol. **65**(1–3) (2019) https://doi.org/10.31901/24566608.2019/65.1-3.3157
43. Gangi, F., Mustilli, M., Varrone, N.: The impact of corporate social responsibility (CSR) knowledge on corporate financial performance: evidence from the European banking industry. J. Knowl. Manag. **23**(1), 110–134 (2019). https://doi.org/10.1108/JKM-04-2018-0267
44. Israel, O.U., Ikem, Oh.C., Nduka, O.R.: Expenditure in human resource and financial performance of quoted manufacturing companies in Nigeria. East Afr. Sch. Econ. Bus. Manag. **5**(2), 48–80 (2022). https://doi.org/10.36349/easjebm.2022.v05i02.003
45. Kaynak, H.: The relationship between total quality management practices and their effects on firm performance. J. Oper. Manag. **21**(4), 405–435 (2019). https://doi.org/10.1016/S0272-6963(03)00004-4
46. Hilty, D.M.: A scoping review of social media in child, adolescents and young adults: research findings in depression, anxiety and other clinical challenges. BJPsych Open **9**(5), 1–12 (2023). https://doi.org/10.1192/bjo.2023.523
47. Obi, L., Arif, M., Goulding, J.: Low-cost housing: a cost management model for process integration and evaluation. Archit. Eng. Des. Manag. **16**(6), 411–425 (2020). https://doi.org/10.1080/17452007.2020.1738993
48. Dries, N.: The psychology of talent management: a review and research agenda. Hum. Resour. Manag. Rev. **23**(4), 272–285 (2013). https://doi.org/10.1016/j.hrmr.2013.05.001

49. Hwang, K.-E., Kim, I.: Post-COVID-19 modular building review on problem seeking framework: function, form, economy, and time. J. Comput. Des. Eng. 1369–1387 (2022). https://doi.org/10.1093/jcde/qwac057
50. Sezer, S.: Introduction: what is software-defined networking? Future carrier networks are we ready for SDN? Implementation challenges for software-defined networks background: WHY SDN? Futur. Carr. Netw. **51**(7), 36–43 (2013). https://doi.org/10.1109/MCOM.2013.6553676

Chapter 2
Qualitative Sampling Techniques

2.1 Introduction

In qualitative research, the act of sampling is not simply a technical step—it is a deliberate choice about whose voices will be heard, which settings will be explored, and what experiences will be brought into the light. Unlike qualitative approaches, which often seek large, statistically representative samples, qualitative sampling is purpose-driven. It aims to gather the richest, most relevant, and most diverse insights that can illuminate the phenomenon under study. At its core, qualitative sampling is guided by the principle of information richness. The researcher selects participants, sites, or cases that are most likely to yield meaningful and nuanced data. This choice is never arbitrary; it is anchored in the research questions and the context in which they are situated. The aim is not to generalise to an entire population, but to develop a deep, textured understanding of specific perspectives and experiences. One of the most widely used approaches is purposive sampling. Here, participants are selected because they meet specific criteria relevant to the research focus. For instance, in a study on cybersecurity practices in heterogeneous environments, purposive sampling might target network administrators, security analysts, and IT managers who have direct experience with intrusion detection systems. This method ensures that every participant contributes meaningfully to the inquiry [1].

Closely related is criterion sampling, which involves setting predetermined conditions that participants must meet. This is particularly useful when the study requires first-hand accounts of a particular situation, such as professionals who have managed cyber incidents involving both IoT and cloud-based systems. Snowball sampling is another valuable technique, especially when studying specialised or hard-to-reach populations. In this approach, initial participants refer others who meet the study's criteria, creating a network of recruitment. This is often employed in research where trust is essential, as referrals come from known and respected peers. Maximum variation sampling seeks to capture a wide range of perspectives by deliberately including participants who differ in key characteristics—such as organisational size, technical

F. A. Alaba et al., *Fundamentals of Qualitative Research*, Studies in Systems, Decision and Control 663, https://doi.org/10.1007/978-3-032-18427-6_2

background, or geographical location. By embracing diversity, this technique helps reveal patterns that cut across varied contexts, as well as unique insights tied to specific conditions. In certain cases, theoretical sampling is used, particularly in grounded theory research. Here, sampling decisions evolve as the study progresses, driven by emerging concepts and themes. The researcher may decide to seek out new participants to test, refine, or expand developing ideas. Finally, there is convenience sampling, which selects participants based on accessibility and willingness to participate. While less rigorous, it can be a pragmatic choice in exploratory phases or when constraints such as time and resources are significant. However, researchers using this method must be transparent about its limitations. Whatever the technique, the qualitative researcher remains attentive to the concept of data saturation—the point at which additional interviews or observations no longer yield new themes. This marker ensures that the sample is sufficient in size and scope to answer the research questions without unnecessary redundancy [2].

Qualitative sampling is as much an art as a science. It demands reflexivity, cultural sensitivity, and an awareness of how the researcher's own position might shape the process. Above all, it is a commitment to selecting those participants and cases that can offer the deepest, most authentic insight into the world the research seeks to understand (Fig. 2.1).

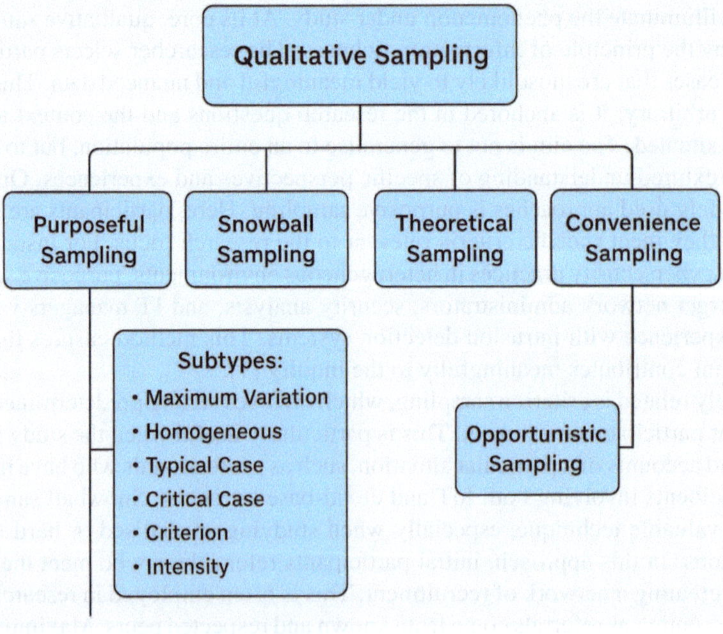

Fig. 2.1 Qualitative sampling techniques [3]

Purposeful Sampling

Purposeful sampling lies at the heart of qualitative research, embodying the deliberate, thoughtful selection of participants, cases, or settings that will best illuminate the phenomenon under investigation. Unlike random sampling in qualitative studies, which seeks statistical representativeness, purposeful sampling is concerned with depth, relevance, and richness of information. It is about asking: Who can tell me the most about what I need to know? The essence of purposeful sampling is intentionality. The researcher identifies individuals or groups based on specific characteristics, experiences, or knowledge that are directly relevant to the research questions. This deliberate focus ensures that every participant adds value to the inquiry, contributing perspectives that advance understanding rather than dilute it. For example, a study examining the effectiveness of intrusion detection systems in heterogeneous computing environments would not seek the views of all IT users; instead, it would target cybersecurity specialists, network engineers, and system administrators who have first-hand experience managing complex, multi-platform infrastructures. Purposeful sampling is not a single technique but an umbrella concept encompassing several strategies. Criterion sampling, for instance, involves setting defined eligibility requirements to ensure participants meet the study's needs. Maximum variation sampling deliberately seeks a diversity of experiences to uncover both shared patterns and unique insights. Expert sampling focuses on individuals with specialised knowledge or skills, while homogeneous sampling narrows the scope to a group with similar characteristics, useful for in-depth exploration of a shared experience (Fig. 2.2).

The strength of purposeful sampling lies in its capacity to yield information-rich cases—instances that offer detailed, nuanced accounts capable of generating deeper insights than broad surveys might achieve. This richness, however, comes with responsibility. The researcher must justify their choices clearly, explaining why each participant or site is relevant, and how the combination of cases will provide a balanced, credible picture of the phenomenon. Purposeful sampling also aligns closely with the concept of data saturation, the point at which additional interviews or observations no longer produce new themes or insights. In practice, this means that sample size in qualitative research is determined not by statistical formulas but by the sufficiency of data to fully address the research questions. Ultimately, purposeful sampling reflects a fundamental truth of qualitative inquiry: that understanding complex social or organisational realities requires depth rather than breadth. By choosing participants strategically, the researcher ensures that the voices captured are those most capable of shedding light on the intricate, contextual, and often hidden dimensions of human experience [4].

Snowball Sampling

Snowball sampling is a distinctive qualitative sampling technique built on the power of human networks. It begins with a small group of initial participants—often referred to as "seeds"—who meet the study's criteria. These individuals are then asked to refer others they know who also fit the research profile. In this way, the sample grows organically, much like a snowball rolling downhill, gathering both size and

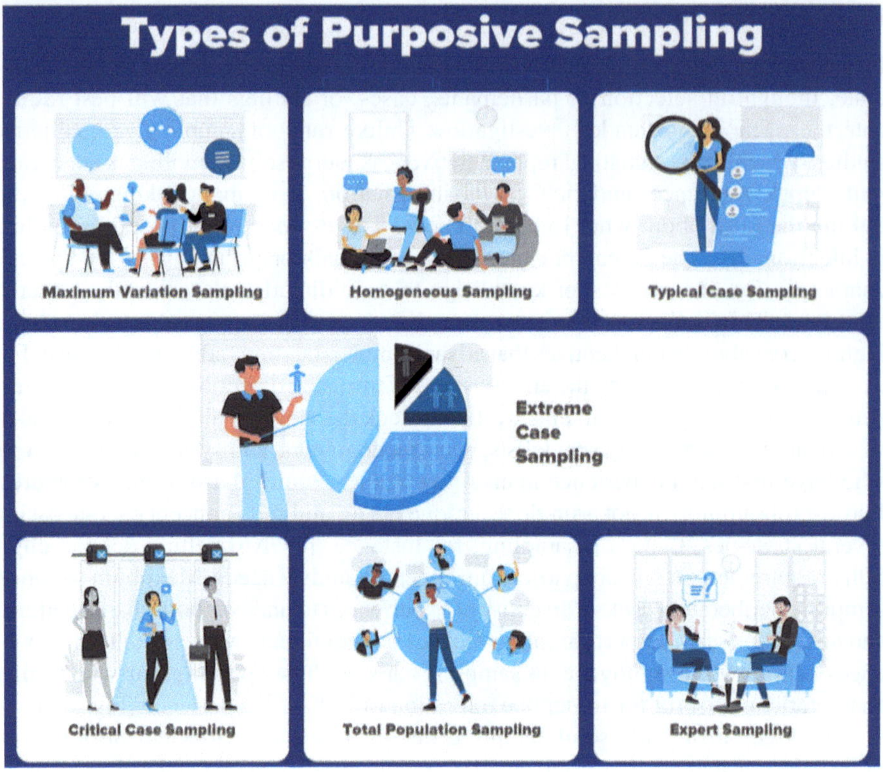

Fig. 2.2 Purposeful sampling [5]

momentum as the study progresses. This method is particularly valuable when the population of interest is hard to reach, specialised, or hidden. In such cases, traditional recruitment strategies may prove ineffective because potential participants are scattered, protective of their privacy, or difficult to identify through public channels. For example, in a study exploring the experiences of cybersecurity analysts dealing with zero-day attacks in heterogeneous computing environments, direct recruitment could be challenging due to the sensitive nature of the work and confidentiality agreements. Snowball sampling allows access to these participants through personal trust networks, where a recommendation from a known contact can open doors. One of the key strengths of snowball sampling lies in its capacity to tap into trust-based relationships. When referrals come from known and respected peers, potential participants may feel more comfortable sharing their experiences. This can lead to richer, more candid data than might be obtained through cold outreach. Additionally, the technique can be efficient, as each participant acts as both a source of data and a recruitment point for the next. However, snowball sampling is not without limitations. Because the sample grows through personal networks, it may over-represent certain subgroups while under-representing others, leading to potential bias. The researcher

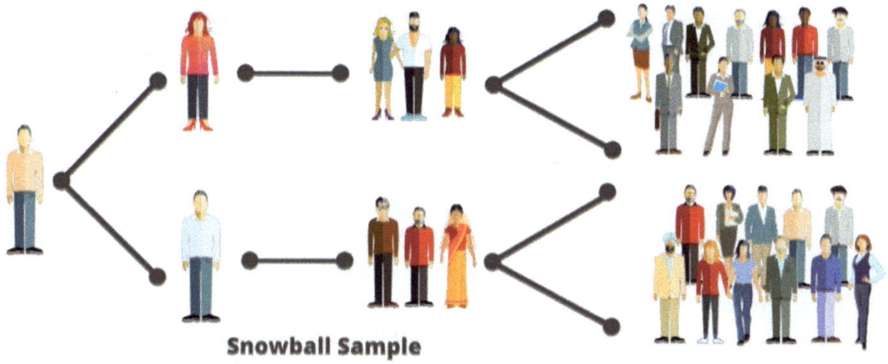

Fig. 2.3 Snowball sampling [6]

must remain alert to this possibility, actively seeking diverse entry points into the network to ensure a wider range of perspectives. Ethics also play a crucial role in snowball sampling. Referrals should be made with the potential participant's consent, and the researcher must protect the confidentiality of both the referrer and the referred. Trust, once given, must be handled with care to preserve both the integrity of the data and the relationships that make the method possible. Ultimately, snowball sampling is a tool of both practicality and sensitivity. It acknowledges that some knowledge lives within close-knit circles, and that to access it, one must move respectfully from one trusted connection to the next. When used with awareness and methodological care, it can open pathways into communities and professional spheres that would otherwise remain closed to the researcher's gaze [5] (Fig. 2.3).

Theoretical Sampling

Theoretical sampling is a hallmark of grounded theory research, representing a dynamic and iterative approach to participant selection. Unlike many sampling strategies that are planned entirely before data collection begins, theoretical sampling evolves as the study progresses. It is driven not by pre-set numbers or rigid criteria, but by the developing needs of the emerging theory. In essence, it is the process of deciding what data to collect next, and where to find it, based on what the analysis so far is revealing. The journey often begins with a small set of participants or cases chosen for their relevance to the research problem. As data is gathered and coded, patterns begin to emerge—initial concepts, categories, or relationships that hint at the contours of the phenomenon under study. The researcher then deliberately seeks out new participants, situations, or documents that can deepen, refine, or challenge these early insights. For example, in a study on the adoption of IDS in heterogeneous computing environments, the researcher might initially interview network administrators from large enterprise organisations. Early analysis could reveal that perceptions of IDS usability vary widely depending on the diversity of device types managed. This insight might prompt the researcher to purposefully seek out participants from small and medium-sized enterprises, or from cloud service providers, to compare and contrast experiences.

Theoretical sampling is therefore a responsive and adaptive process. It is guided by questions like: Where can I find more information to elaborate on this concept? Who can confirm or disconfirm this emerging pattern? What cases might offer exceptions that test the robustness of my interpretation? This approach ensures that the theory being developed is grounded in evidence that has been deliberately sought to cover its complexity and variation. One of its strengths is that it prioritises conceptual completeness over representativeness. The goal is not to create a statistically generalisable sample, but to construct a theory that is rich, nuanced, and applicable across a range of relevant contexts. This requires ongoing reflection, flexibility, and an openness to follow the data where it leads—even into unexpected domains. Ethical considerations remain important. Because theoretical sampling can lead to shifting recruitment plans mid-study, researchers must ensure that changes in participant groups do not compromise informed consent or the protection of sensitive information. Transparency in documenting these decisions also enhances the study's credibility. In the end, theoretical sampling is as much a way of thinking as it is a method. It treats data collection as a conversation between the researcher's emerging ideas and the world of lived experience. Each new participant or case is not just another data point, but a strategic step toward building a theory that reflects the complexity, diversity, and depth of the reality being studied [7] (Fig. 2.4).

Convenience Sampling

Convenience sampling is perhaps the simplest and most readily available sampling strategy in qualitative research. As its name suggests, it involves selecting participants who are easily accessible and willing to take part in the study. This method is often employed when time, resources, or access to the target population are limited, allowing the researcher to begin data collection quickly without the logistical

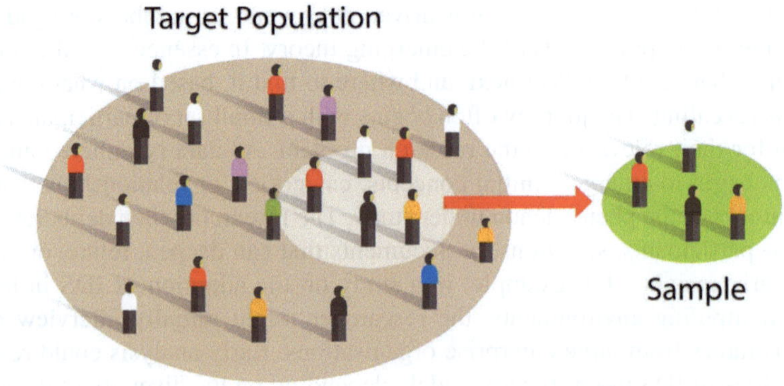

Fig. 2.4 Theoretical sampling [8]

complexities of more selective approaches. In many cases, convenience sampling serves as a practical starting point, particularly during exploratory stages of a project. For example, a researcher examining the communication challenges in healthcare settings might initially interview colleagues within their own organisation, simply because they are readily available and familiar with the study's context. Similarly, in the field of cybersecurity, an investigator exploring intrusion detection systems in heterogeneous environments could start with contacts in their professional network—such as IT staff or security practitioners already known to them—before expanding the sample to other sources. The primary strength of convenience sampling is its efficiency. It enables the rapid gathering of data, which can be particularly valuable when the research is under tight deadlines or when initial findings are needed to refine the scope of the inquiry. It is also cost-effective, as it minimises travel, recruitment expenses, and the administrative burden associated with more targeted sampling methods. However, the ease of convenience sampling comes with clear limitations. Because participants are chosen for their accessibility rather than their representativeness or information-rich qualities, there is a greater risk of bias. The views obtained may reflect the experiences of a narrow group and fail to capture the diversity of perspectives within the broader population. In qualitative research, where depth and variation are prized, this can limit the richness and transferability of findings [9] (Fig. 2.5).

Ethical considerations remain central. Even when participants are conveniently accessible—such as colleagues, classmates, or friends—the researcher must ensure

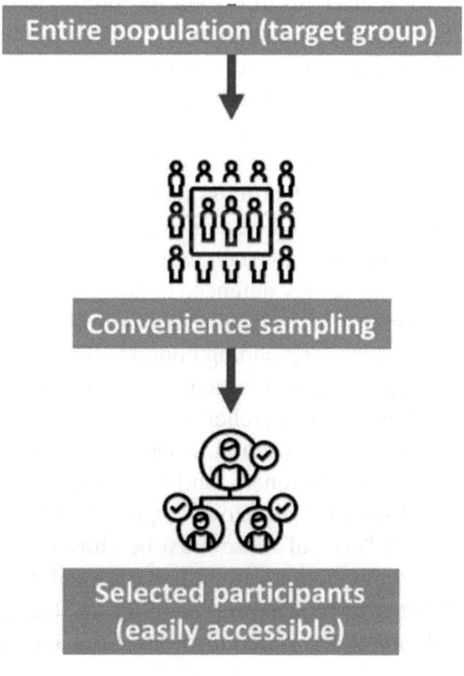

Fig. 2.5 Convenience sampling [10]

that informed consent is obtained, that participation is voluntary, and that power dynamics do not pressure individuals into taking part. Confidentiality is equally important, especially when working within close professional or social circles. In essence, convenience sampling is a tool best used with awareness and caution. While it offers undeniable practicality, it is most effective when paired with other strategies—such as purposive or snowball sampling—that can help balance ease of access with diversity and depth of insight. By acknowledging its limitations and applying it strategically, the researcher can still draw valuable, if contextually bound, understandings from the participants at hand.

Convenience sampling involves selecting participants who are easily accessible to the researcher. While this method can be practical and cost-effective, especially in exploratory studies, it risks producing data that is less representative or comprehensive. Researchers using convenience sampling must be cautious not to overgeneralize their findings. This technique is best used when resources are limited or when the study's purpose is to generate preliminary insights rather than draw broad conclusions.

Critical Case Sampling

Critical case sampling is a purposeful sampling strategy built on the principle that studying a small number of strategically chosen cases can yield insights of exceptional value. The premise is simple yet powerful: if findings are true for these "critical" cases, they are likely to apply to many other situations. Conversely, if the phenomenon does not hold in these crucial cases, it is unlikely to hold elsewhere. This approach is particularly useful when resources are limited and the researcher must maximise the potential impact of every interview, observation, or document analysed. Instead of seeking a wide variety of participants, the researcher identifies those cases that are most likely to produce significant, transferable insights. These may be cases that are extreme in some way, exemplary in their performance, or situated in contexts where the stakes are unusually high. For example, in a study exploring the resilience of intrusion detection systems (IDS) in heterogeneous computing environments, a critical case might be an organisation that operates in an exceptionally complex, high-risk setting—such as a national cybersecurity agency managing multi-domain networks across defence, finance, and public services. If the IDS framework can function effectively in this environment, it is reasonable to assume it might also work in less demanding contexts (Fig. 2.6).

The strength of critical case sampling lies in its efficiency and persuasive power. It allows the researcher to focus on cases that serve as "litmus tests" for a theory, model, or intervention. Such cases can generate compelling evidence for policymakers, practitioners, and scholars, making the findings highly relevant despite a small sample size. However, the strategy requires careful judgement in case selection. A "critical" case must be chosen for a defensible reason, grounded in theory or prior evidence, rather than personal preference or convenience. The logic linking the case to broader applicability must be transparent, so that readers can assess the validity of any generalisations. Ethical considerations are also heightened in critical

Fig. 2.6 Critical case sampling [12]

case sampling because the individuals or organisations involved often occupy prominent or sensitive positions. Maintaining confidentiality, securing informed consent, and handling potentially high-profile findings with care are essential to preserving trust and integrity. In qualitative research, critical case sampling is best seen as a way to focus the investigative lens with precision. It is not about breadth, but about strategic depth—choosing the case that, by its very nature, will reveal the most about the question at hand. When executed thoughtfully, this method can transform a small-scale study into one with broad significance and lasting impact [11].

Critical case sampling focuses on selecting participants or cases that are particularly important, unusual, or revealing. The idea is that if something is true for this critical case, it is likely to be true for others as well. For example, a researcher studying leadership resilience during crises might choose a CEO who successfully navigated a company through a major disaster. Insights gained from such critical cases can provide powerful lessons, though they may not be widely generalizable.

Criterion Sampling

Criterion sampling is a purposeful sampling strategy in which participants, cases, or settings are selected because they meet a clearly defined set of conditions relevant to the research question. The logic is straightforward: by applying specific criteria, the researcher ensures that all selected participants share an essential characteristic or experience that makes them directly relevant to the study. This approach is especially valuable in qualitative research, where the goal is to gather rich, detailed data from individuals who can speak meaningfully to the phenomenon under investigation. The process begins by establishing the inclusion criteria. These criteria may be based on demographic factors, professional roles, lived experiences, or exposure to particular events or contexts. For instance, in a study on IDS in heterogeneous computing environments, the criterion could be that participants must have at least three years' experience managing security protocols across multiple device types and network architectures. This ensures that every participant can offer informed, experience-based insights into the challenges and opportunities of such systems. Criterion sampling's greatest strength lies in its focus and relevance. By narrowing the participant pool to those who meet predefined conditions, the researcher minimises the collection of irrelevant or superficial data. This approach also facilitates comparability, as all participants can be examined against the same contextual backdrop. However, the method is not without limitations. If the criteria are too restrictive, the resulting sample may be too small or too homogeneous, limiting the diversity of perspectives. Conversely, overly broad criteria may dilute the depth of insight. Striking the right balance requires careful consideration, often informed by a preliminary review of the literature or exploratory fieldwork [13] (Fig. 2.7).

Ethical considerations are essential in criterion sampling, particularly when the criteria involve sensitive or identifiable characteristics. In such cases, confidentiality and respectful recruitment become paramount to protect participants from unintended exposure or stigma. Ultimately, criterion sampling is about precision. It ensures that

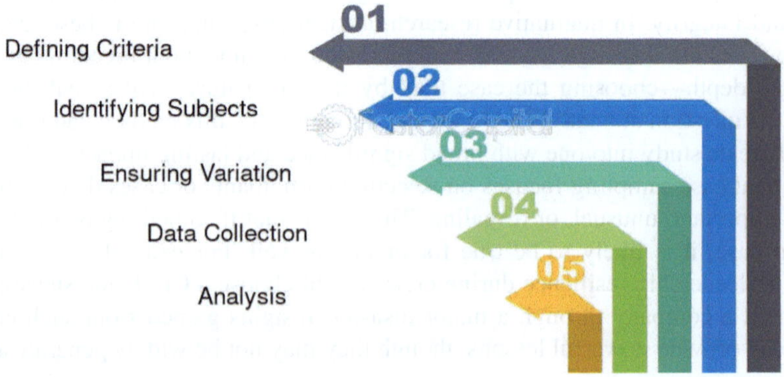

Fig. 2.7 Criterion sampling [14]

every voice included in the study is both relevant and capable of contributing meaningfully to the research objectives. When applied thoughtfully, it can produce a focused, coherent dataset that supports deep analysis and credible findings. In criterion sampling, participants are selected because they meet a specific predetermined criterion or set of criteria. This method is often used in evaluation research or case studies where it is important that participants have certain experiences or characteristics. For instance, in a study exploring the experiences of first-generation university graduates, participants would be selected based on the criterion of being the first in their family to complete higher education. Criterion sampling ensures relevance and consistency across cases.

Intensity Sampling

Intensity sampling is a purposeful sampling strategy that focuses on selecting information-rich cases that strongly represent the phenomenon of interest, but without being extreme or atypical. These cases are chosen because they exemplify the issue under study in a clear and pronounced way, allowing the researcher to explore it in depth while avoiding the distortions that may arise from highly unusual situations. The key idea behind intensity sampling is to seek strength without extremity. Whereas extreme case sampling deliberately targets the most unusual or dramatic examples, intensity sampling works with cases that are both illustrative and representative, offering a solid basis for deep analysis. This method is particularly useful when the researcher wants to understand a phenomenon in its "normal but strong" form—situations where it is clearly visible and well-developed, but not so exceptional that the findings risk being dismissed as anomalies. For example, in a study examining the effectiveness of IDS in heterogeneous computing environments, intensity sampling might focus on organisations where IDS deployment has been well-integrated into daily operations and where security incidents occur regularly enough to provide meaningful data. These organisations are not necessarily dealing with the most catastrophic breaches, but their experiences are intense enough to reveal important patterns, challenges, and best practices.

One advantage of intensity sampling is that it allows for rich, nuanced understanding while maintaining a degree of generalisability within qualitative boundaries. Because the cases are neither too rare nor too ordinary, the insights gained often resonate across a broader range of contexts. However, identifying "intense" cases requires preliminary knowledge of the field. The researcher may need to conduct informal conversations, review records, or analyse secondary data to determine which potential participants or sites meet the desired intensity threshold. Ethical considerations in intensity sampling are similar to those in other purposeful approaches: informed consent, respect for confidentiality, and sensitivity to the context. Since the cases are not extreme, participants may feel more comfortable sharing their experiences openly, which can further enrich the data. In essence, intensity sampling provides a balanced lens through which to explore a phenomenon—strong enough to offer vivid insights, yet grounded enough to avoid overemphasis on the unusual. It is a strategy that respects both depth and relevance, making it a valuable tool in the qualitative researcher's methodological repertoire. Intensity sampling focuses on selecting participants who strongly exhibit the phenomenon of interest but are not extreme cases. Researchers choose individuals whose experiences are rich and intense but within a normal range. This technique allows researchers to explore the phenomenon deeply without being skewed by the extremities that could come with highly unusual cases. For example, in a study on teacher burnout, researchers might select teachers who are experiencing significant stress but are still actively working, rather than those who have already left the profession [15].

Opportunistic Sampling

Opportunistic sampling, also known as emergent sampling, is a flexible qualitative research strategy in which the researcher takes advantage of unexpected opportunities for data collection as they arise in the field. Unlike more structured approaches, opportunistic sampling recognises that real-world research rarely unfolds exactly as planned. New leads, participants, or contexts may present themselves unexpectedly, and when they do, the researcher adapts to incorporate them into the study. This method is especially valuable in complex or dynamic environments where conditions change rapidly, and access to information can be unpredictable. Rather than adhering rigidly to a predetermined recruitment plan, the researcher remains open to situations that can enrich the dataset. For example, in a study on IDS in heterogeneous computing environments, an unplanned opportunity might occur when a cybersecurity conference or emergency response simulation suddenly becomes accessible. Speaking with practitioners in such settings—though not initially planned—could yield unique insights into the operational realities of IDS use under pressure. The strength of opportunistic sampling lies in its adaptability. It allows the researcher to capture rich, timely data that might otherwise be missed if they were confined to pre-arranged channels. It also enables exploration of emerging themes that become apparent only after initial data collection begins. However, this flexibility requires strong ethical awareness. Opportunities must be assessed quickly, but the researcher must still secure informed consent, respect confidentiality, and ensure that the inclusion of new participants or contexts does not compromise the study's focus. Documentation is also critical—explaining why and how new cases were added builds transparency and strengthens credibility. In practice, opportunistic sampling works best as a complementary strategy, used alongside more structured approaches such as purposive or criterion sampling. This combination preserves the rigour of deliberate selection while allowing the agility to capture unexpected, yet valuable, perspectives [16] (Fig. 2.8).

Ultimately, opportunistic sampling reflects the responsive nature of qualitative inquiry. By staying open to chance encounters and unplanned openings, the researcher can gather data that not only deepens the study but also captures the fluid, often unpredictable, realities of the world under investigation.

Opportunistic sampling takes advantage of unforeseen opportunities during data collection. As the research progresses, the researcher might encounter unexpected situations, events, or participants that are highly relevant to the study and decide to include them. While opportunistic sampling can enrich the study with valuable

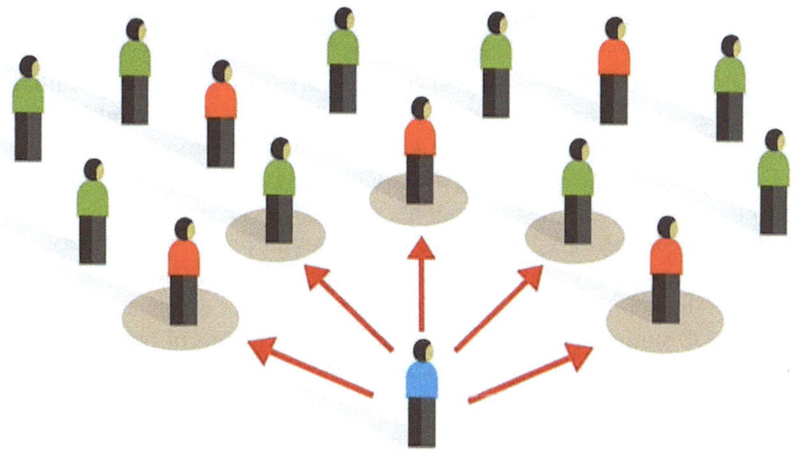

Fig. 2.8 Opportunistic sampling [17]

insights, it must be handled carefully to avoid deviating too far from the study's original focus or objectives.

2.2 Probability Sampling Techniques

Probability sampling is a cornerstone of qualitative research, built on the principle that every member of a population has a known, non-zero chance of being selected. By using random selection methods, probability sampling minimises selection bias and supports statistical generalisation from the sample to the wider population. Unlike non-probability sampling, which focuses on depth and context, probability sampling emphasises representativeness and precision, making it essential for studies that seek to produce findings with broad applicability. At its core, probability sampling requires a clearly defined sampling frame—the complete list of all units in the population from which the sample will be drawn. From this foundation, several techniques can be employed [18, 19]:

1. **Simple Random Sampling (SRS)**: In simple random sampling, each individual in the population has an equal chance of being selected. This method is straightforward and unbiased, often implemented using random number generators or lottery systems. For example, in cybersecurity research, SRS could be used to select networked devices from a registry for testing an intrusion detection algorithm's performance.
2. **Systematic Sampling**: Systematic sampling selects every kth element from a list after a random starting point. It is efficient and easy to implement but assumes that the list has no underlying periodic pattern that could bias results. In an IDS

deployment study, this could involve selecting every tenth network log file from a chronological list for analysis.

3. **Stratified Sampling**: Stratified sampling divides the population into subgroups, or strata, based on shared characteristics, and then draws random samples from each stratum. This ensures that key subgroups are proportionately represented. In heterogeneous computing environments, strata might include network type (cloud, IoT, on-premises), enabling targeted yet representative sampling.
4. **Cluster Sampling**: Cluster sampling divides the population into groups (clusters) and then randomly selects entire clusters for inclusion. This is often more practical when the population is geographically dispersed or organisationally segmented. For IDS studies, clusters could be entire departments or branch offices, with all systems within the selected clusters included in testing.
5. **Multi-stage Sampling**: Multi-stage sampling combines two or more probability methods in stages, allowing for flexibility and cost-efficiency in large-scale studies. For instance, a researcher might first use cluster sampling to select organisations, then stratified sampling to choose specific IT teams within those organisations.

The strength of probability sampling lies in its statistical validity. It allows researchers to estimate sampling error, apply inferential statistics, and make reliable generalisations. However, it requires a complete and accurate sampling frame, which may be difficult to obtain in certain contexts, especially in fields like cybersecurity where confidentiality and access restrictions are common. In practice, probability sampling is particularly suited to the qualitative phases of mixed-methods research, where the goal is to test hypotheses, validate models, or measure the prevalence of certain behaviours or characteristics. In studies involving intrusion detection systems, probability sampling could be used to generate a representative dataset of network traffic across diverse environments, forming a robust basis for algorithm evaluation. Ultimately, probability sampling embodies the qualitative researcher's commitment to rigour, fairness, and representativeness. By giving every unit in the population a fair chance of inclusion, it creates findings that are both empirically sound and widely applicable—qualities that are essential in research areas where precision is as important as discovery.

2.2.1 Simple Random Sampling in Qualitative Research

Simple Random Sampling (SRS) is one of the most fundamental and widely recognised techniques in probability sampling. It is grounded in the principle that every unit in the population has an equal and independent chance of being selected for inclusion in the study. This method is often regarded as the "gold standard" for ensuring fairness and minimising selection bias, providing a solid foundation for statistical inference and generalisation.

Defining Simple Random Sampling

In its purest form, SRS requires two essential elements:

1. **A clearly defined sampling frame**—A complete list of all members or units in the population of interest.
2. **A method of random selection**—Such as a lottery draw, random number generator, or computerised algorithm, ensuring that the selection process is free from human interference or subjective judgement.

When these two conditions are met, the resulting sample is considered representative of the population within a known margin of error, enabling researchers to apply inferential statistics with confidence.

The Logic Behind SRS

The core logic of SRS is grounded in probability theory. Because each unit has the same likelihood of being chosen, the sample's composition will, on average, mirror the characteristics of the population as a whole. While random chance may produce some variability—especially in smaller samples—statistical laws such as the Law of Large Numbers and the Central Limit Theorem ensure that, as sample size increases, the distribution of characteristics in the sample approximates that of the population. This theoretical grounding makes SRS particularly appealing in qualitative research, where unbiased parameter estimation is a central goal. It also underpins the calculation of sampling error, allowing researchers to express the reliability of their findings in probabilistic terms.

Implementation Steps

Implementing SRS involves a sequence of structured steps:

1. **Define the Population**: Clearly specify the group to which the research findings will be generalised. For example, in a cybersecurity study on IDS, the population might be *all enterprise-level networks operating in a specific country.*
2. **Construct the Sampling Frame**: Create an exhaustive list of all units in the population. In the IDS example, this might involve compiling a registry of organisations that operate heterogeneous computing environments, ensuring the list is current and accurate.
3. **Assign Identification Numbers**: Each unit in the sampling frame is assigned a unique numerical identifier.
4. **Select the Sample Randomly**: Using a randomisation method—such as a random number generator, lottery system, or statistical software—the required number of units is selected. The independence of selection is critical to maintaining the integrity of the method.
5. **Contact and Recruit Participants**: After selection, participants are contacted, provided with study details, and invited to take part.

Illustrative Example: IDS Evaluation in Heterogeneous Environments

Consider a research project aiming to evaluate the performance of a newly proposed conceptual IDS framework across different types of organisational networks. Suppose the researcher identifies 2000 eligible organisations in the national registry. Each organisation is assigned a number from 1 to 2000. Using a random number generator, the researcher selects 200 organisations for participation. Because each organisation had the same chance of selection, the resulting dataset is statistically representative, allowing performance results to be generalised to the broader population.

Advantages of SRS

1. **Elimination of Selection Bias**: Since the selection is random, personal preferences or systematic patterns in recruitment are avoided.
2. **Simplicity and Transparency**: The method is straightforward to implement and easy to explain to stakeholders, which supports transparency in research design.
3. **Foundation for Statistical Inference**: The randomness of selection allows for the calculation of sampling error and confidence intervals, facilitating robust statistical testing.
4. **High Representativeness**: When implemented correctly, SRS tends to produce samples that closely match the population in terms of key characteristics, particularly in large samples.

Limitations of SRS

1. **Dependence on a Complete Sampling Frame**: SRS requires a full list of the population, which may be difficult or impossible to obtain in certain contexts—especially in cybersecurity, where organisations may be reluctant to disclose operational details.
2. **Cost and Practicality**: When the population is geographically dispersed or logistically difficult to reach, implementing SRS can be expensive and time-consuming.
3. **Potential for Sampling Error in Small Samples**: While SRS is unbiased, random variation can still produce samples that differ from the population, particularly if the sample size is small.
4. **Non-response Bias**: Even with perfect random selection, if certain groups within the sample are less likely to respond, bias can still occur.

Ethical Considerations in SRS

Ethics play an important role in SRS, particularly when dealing with sensitive topics such as cybersecurity threats or patient health records. Key considerations include:

1. **Informed Consent**—Selected participants must be fully briefed and voluntarily agree to participate.
2. **Confidentiality**—Personal or organisational data must be stored securely and used only for the stated research purposes.
3. **Minimising Harm**—Especially important when research involves discussing security vulnerabilities or sensitive operational details.

Strategies to Improve SRS in Practice

1. **Ensuring an Accurate Sampling Frame**: Regularly updating and verifying the list of population members is essential to prevent coverage errors.
2. **Enhancing Response Rates**: Using reminders, incentives, and clear communication can reduce non-response bias.
3. **Combining with Stratification**: In cases where important subgroups exist, SRS can be applied within strata to maintain proportional representation.
4. **Random Number Integrity**: Using reputable software or certified randomisation processes ensures the selection process is genuinely random.

Relevance to Cybersecurity and IDS Research

In cybersecurity research, particularly when testing the effectiveness of IDS frameworks, SRS can play a crucial role in generating representative datasets. For example:

1. **Network Traffic Analysis**—Randomly selecting data packets or logs from a large dataset ensures that the evaluation is not skewed toward specific periods or event types.
2. **User Behaviour Monitoring**—Random selection of user accounts for behavioural analysis can help assess whether the IDS detects anomalies across diverse activity profiles.
3. **Incident Response Evaluation**—Selecting a random subset of security incidents for detailed post-mortem analysis can yield unbiased insights into system performance.

By ensuring that each organisation, log file, or event has an equal chance of being chosen, researchers can more confidently generalise their findings to the broader cybersecurity environment. SRS remains one of the most powerful and widely respected techniques in research methodology. Its strength lies in its fairness, transparency, and ability to produce statistically valid results when implemented correctly. While it demands a complete sampling frame and may face logistical challenges, its capacity to eliminate selection bias and underpin reliable statistical inference makes it indispensable in many research contexts—including advanced domains such as IDS evaluation in heterogeneous computing environments. In practice, SRS works best when paired with careful planning, robust randomisation tools, and proactive measures to maintain participation rates. When these conditions are met, it provides a clear pathway from unbiased data collection to credible, generalisable findings—qualities that are especially critical in fields where accuracy and trust are paramount [20].

Advantages and Disadvantages of Simple Random Sampling

See Table 2.1.

Applications of Simple Random Sampling

SRS is a versatile probability sampling technique applicable across a broad range of research domains. Its defining principle—that every unit in the population has an

Table 2.1 Advantages and disadvantages of simple random sampling

Advantages	Disadvantages
1. Eliminates selection bias due to equal probability for all units	1. Requires a complete and accurate sampling frame, which may be hard to obtain
2. Produces a representative sample, especially with large sample sizes	2. Can be costly and time-consuming for large or geographically dispersed populations
3. Simple to understand and easy to explain to stakeholders	3. May lead to sampling error if the sample size is too small
4. Facilitates accurate estimation of sampling error and confidence intervals	4. Non-response from selected participants can introduce bias
5. Random selection supports strong statistical inference	5. Less efficient than stratified sampling if subgroups are of interest
6. Avoids systematic patterns that could distort findings	6. Random selection may still result in underrepresentation of rare subgroups
7. Works well as a baseline method for other probability sampling techniques	7. Data collection logistics may be complex in large-scale studies
8. Transparent selection process enhances research credibility	8. May include units that are difficult or impossible to contact
9. Can be implemented with readily available randomisation tools	9. Selection process must be strictly randomised to avoid bias, requiring care in execution
10. Suitable for both small and large populations	10. Does not automatically account for differences between population subgroups
11. Flexible—can be applied to people, documents, events, or objects	11. Inefficient when certain strata are small but important
12. Lays the foundation for the generalisability of findings	12. Risk of over-representing or under-representing some characteristics purely by chance

equal and independent chance of selection—makes it a valuable tool wherever representativeness, fairness, and statistical validity are priorities. Because it is unbiased in design, SRS is especially suitable for studies aiming to generalise findings to an entire population [21, 22].

1. **Academic and Social Science Research**: In education, psychology, and sociology, SRS is frequently used to obtain representative samples of students, households, or communities. For example, in an educational survey exploring the impact of digital learning tools, researchers might randomly select students from a district's enrolment database to ensure that all demographic groups are fairly represented. This approach helps prevent overrepresentation of any particular school or socio-economic category.
2. **Public Health and Epidemiology**: Public health researchers rely on SRS to estimate disease prevalence, assess vaccination coverage, and evaluate health interventions. For instance, in an epidemiological study on the spread of diabetes within a city, a list of all residents could be used to randomly select individuals for

health screenings. This ensures that results reflect the overall population rather than specific clusters, reducing bias in prevalence estimates.

3. **Government and Policy Research**: Government agencies often use SRS for census sampling, national surveys, and policy evaluation. For example, in labour market studies, SRS can be used to select households for unemployment data collection. Because every household has an equal chance of being chosen, the resulting statistics are credible for policy formulation.
4. **Business and Market Research**: In business contexts, SRS helps organisations make evidence-based decisions. Market researchers can randomly select customers from a database to evaluate product satisfaction or brand perception. This ensures that feedback reflects the broader customer base rather than a self-selected group, leading to more accurate strategic planning.
5. **Cybersecurity and Technology Research**: In information security, particularly in evaluating IDS in heterogeneous environments, SRS can be applied to select datasets, system logs, or participating organisations. For example, when testing a new IDS algorithm, researchers might randomly choose network logs from different time periods and devices to ensure that the evaluation is unbiased. Similarly, organisations could be randomly selected from a registry to participate in field trials, ensuring that results are generalisable to varied operational environments.
6. **Quality Control and Manufacturing**: Manufacturers apply SRS in quality assurance processes to test a subset of products from a production batch. By giving every unit an equal chance of selection, they can reliably estimate defect rates without inspecting every item, saving time and resources while maintaining accuracy.
7. **Environmental Studies**: Environmental researchers use SRS to sample water bodies, soil sites, or air quality monitoring locations. For instance, in a pollution study, researchers might randomly select sampling points across a river system to gain an unbiased picture of contamination levels.

The broad applicability of SRS stems from its fairness, transparency, and statistical rigour. Whether in public health, market research, environmental science, or cybersecurity, the method supports the production of credible, generalisable findings. By ensuring every unit has an equal opportunity to be included, SRS enables researchers to draw conclusions that truly reflect the population under study.

2.2.2 Systematic Sampling in Qualitative Research

Systematic sampling is a probability sampling method in which elements from the population are selected at regular intervals after a randomly chosen starting point. This approach combines the simplicity of implementation with the principles of randomness, making it a practical alternative to simple random sampling (SRS), especially when working with large populations or when the sampling frame is already ordered.

Definition and Process

The systematic sampling process begins by identifying the sampling frame—a complete list of all population units. The researcher then determines the sampling interval (k) by dividing the total population size (N) by the desired sample size (n):

$$k = N/n$$

Once k is calculated, a random starting point between 1 and k is chosen. From there, every kth element on the list is selected until the sample is complete. For example, if the population consists of 1000 entries and the desired sample size is 100, k would be 10, meaning every tenth unit is selected after the initial random start.

Illustrative Example: IDS in Heterogeneous Environments

In a study evaluating the effectiveness of IDS in heterogeneous computing environments, suppose the researcher has access to a dataset of 50,000 network log entries. If they wish to select a sample of 5000 logs, the interval k would be 10. After selecting a random starting point—say, the 7th log entry—they would then choose every 10th log thereafter (e.g., 7, 17, 27, and so on). This method ensures that the selection covers the dataset evenly, which is particularly useful for identifying patterns across time.

Advantages of Systematic Sampling

1. **Ease of Implementation**—It is simpler and faster to execute than SRS because only the first selection is randomised; the rest follow a fixed pattern.
2. **Even Coverage**—Ensures that the sample is distributed evenly across the population list, reducing clustering of selected units.
3. **Cost-Effectiveness**—Requires less time and fewer resources than generating a fully random sample for large populations.
4. **Suitable for Ordered Data**—Works well when the sampling frame is already logically ordered, such as time-stamped logs or production sequences.
5. **Transparency**—Easy to explain to stakeholders and replicate in subsequent studies.

Disadvantages of Systematic Sampling

1. **Risk of Periodicity Bias**—If there is a hidden periodic pattern in the population list that coincides with the sampling interval, the sample may be biased.
2. **Dependence on List Ordering**—The quality of the sample depends heavily on how the sampling frame is arranged.
3. **Less Flexibility**—Once k is fixed, it may not allow for targeted oversampling of rare subgroups.
4. **Incomplete Sampling Frame Issues**—Like SRS, it requires a complete and accurate list of the population.

Ethical and Practical Considerations

Systematic sampling requires the same ethical safeguards as other probability methods. In cybersecurity research, especially when dealing with sensitive IDS data, informed consent, confidentiality, and secure handling of datasets are paramount. Care must also be taken to ensure that the sampling interval does not inadvertently exclude specific subgroups or time periods that are critical to the research objectives.

Relevance to Cybersecurity Research

In IDS research, systematic sampling is particularly advantageous for large datasets where uniform coverage is essential. For example:

1. **Network Log Analysis**—Sampling logs at fixed intervals ensures that events are captured across different operational periods, avoiding bias toward peak activity hours.
2. **Incident Trend Monitoring**—Selecting security alerts at regular intervals from a chronological database provides balanced insight into detection rates over time.
3. **Dataset Reduction**—When dealing with massive datasets, systematic sampling offers a practical way to create manageable subsets without compromising representativeness.

Systematic sampling offers a structured yet efficient approach to probability sampling, bridging the gap between the rigour of random selection and the practical needs of large-scale research. While it must be applied cautiously to avoid periodicity bias, its even coverage, simplicity, and cost-effectiveness make it highly valuable in contexts such as IDS performance evaluation, where large, ordered datasets are common. With careful planning and awareness of potential pitfalls, systematic sampling can produce reliable, representative results that support sound statistical inference and practical decision-making (Fig. 2.9).

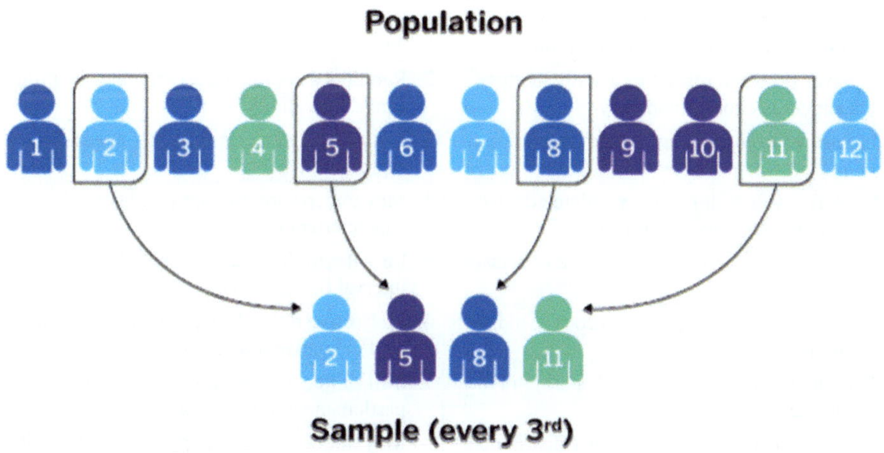

Fig. 2.9 Overview of systematic sampling [23]

Some of the Benefits and Drawbacks of Using Systematic

Because it provides a number of useful advantages, systematic sampling is a well-liked option in the field of research, especially in situations where simplicity and effectiveness are of greater significance [24] (Table 2.2).

Application of Systematic Sampling

Systematic sampling is a widely used probability sampling technique that offers both simplicity and efficiency, making it suitable for research across various disciplines. Its defining feature—the selection of every *k*th unit after a random start—ensures even coverage of the sampling frame. This approach is especially valuable when dealing with large or ordered datasets, where a fully random selection process might be logistically cumbersome. The applications of systematic sampling span the social sciences, public health, business, engineering, environmental studies, and cybersecurity [25–27].

1. **Academic and Social Science Research**: In the social sciences, systematic sampling is often applied in large-scale surveys, educational research, and behavioural studies. For example, a researcher studying student engagement in online learning could use systematic sampling to select every 20th student from an enrolment list after a random start. This method ensures that participants are evenly distributed across different institutions or course types. Because many educational databases are pre-ordered (e.g., alphabetically or by enrolment date), systematic sampling can save time without compromising representativeness—provided there is no periodic pattern that aligns with the interval. In sociology and political science, systematic sampling can also be used to

Table 2.2 Benefits and drawbacks of systematic sampling

Benefits	Drawbacks
1. Easy to understand and implement compared to simple random sampling	1. Risk of periodicity bias if the population list has repeating patterns
2. Saves time and resources as only the first element is chosen randomly	2. Requires a complete and accurate sampling frame for validity
3. Ensures even coverage of the population list	3. List ordering can significantly affect representativeness
4. Works well with naturally ordered datasets, such as time logs or sequences	4. May underrepresent rare or clustered characteristics
5. Suitable for large-scale studies and datasets	5. Less flexibility to adjust once the sampling interval (k) is fixed
6. Transparent and replicable process, aiding credibility	6. Ineffective if the dataset contains hidden cyclical patterns matching k
7. Reduces the likelihood of clustering in the sample	7. Still subject to non-response bias if participants fail to respond
8. Cost-effective for large, evenly distributed populations	8. May not be ideal for populations with high internal variability

select households for opinion polls or demographic surveys. For instance, selecting every 5th household from a census list in a particular region can yield a representative sample for studying voting patterns or public opinion on policy issues.

2. **Public Health and Epidemiology**: In public health, systematic sampling is frequently used for surveillance studies, prevalence estimation, and programme evaluations. Suppose a national health agency wishes to assess the vaccination coverage rate among school-aged children. Starting from a random point in a registry, they might select every 50th child for inclusion in the survey. This method reduces the time required to draw the sample compared to simple random sampling and ensures coverage across the entire registry. Epidemiologists also use systematic sampling when conducting screenings in hospitals or community clinics. For instance, in a study on hypertension prevalence, researchers could screen every 10th patient arriving at a clinic during a specified period. This approach ensures that the sample is drawn evenly over time, reducing bias that might occur if sampling were limited to certain hours or days.
3. **Government and Policy Research**: Government agencies often rely on systematic sampling in population censuses, administrative audits, and programme monitoring. When conducting an employment survey, for example, a government labour department could select every 15th business from a registry of employers. This guarantees that large and small businesses across all sectors are proportionally included. Systematic sampling is also useful for quality assurance in public service delivery. An oversight body might review every 25th case file in a child welfare system to assess compliance with safeguarding protocols. The regular interval helps ensure that records from different geographical areas and case types are represented.
4. **Business and Market Research**: In business contexts, systematic sampling is a practical choice for customer satisfaction surveys, product testing, and market trend analysis. For example, a retail company with a database of 200,000 customers could select every 100th customer to invite for a feedback survey. This ensures that the feedback is spread across the entire customer base rather than concentrated in one demographic group. In product quality research, systematic sampling allows for efficient monitoring of items on a production line. For instance, every 50th product might be inspected for defects. This method provides a representative overview of production quality while avoiding the need to inspect every single unit, which would be resource-intensive.
5. **Cybersecurity and Technology Research**: In cybersecurity research—particularly in evaluating IDS in heterogeneous environments—systematic sampling can be a strategic tool for dataset selection. Consider a large network log containing millions of entries; analysing every entry would be impractical. Instead, researchers could select every 500th log entry after a random start to create a manageable dataset for algorithm testing. This approach ensures that the sample covers different time periods, network events, and traffic patterns, reducing the risk of bias toward specific operational windows (e.g., peak hours or low-activity periods). Systematic sampling can also be applied in user behaviour

monitoring by selecting every 10th user session for anomaly detection testing, ensuring balanced coverage of activity across the entire user base. Furthermore, in simulation-based cybersecurity studies, researchers can use systematic sampling to choose test cases from a larger set of simulated attack scenarios, ensuring even representation across different attack vectors.

6. **Quality Control and Manufacturing**: In industrial and manufacturing settings, systematic sampling is a cornerstone of quality assurance processes. It is common practice to inspect products at fixed intervals—such as every 20th item—on a production line. This ensures that quality checks are distributed evenly throughout production and can quickly identify defects arising from specific stages or machinery. For example, in the production of electronic components, every 100th circuit board might be tested for functionality. This method not only ensures coverage across batches but also helps identify recurring quality issues linked to particular time frames or production shifts.
7. **Environmental Monitoring and Research**: Environmental scientists use systematic sampling for field surveys, pollution monitoring, and biodiversity assessments. For instance, when sampling soil quality in a large agricultural area, a researcher might take samples every 500 m after an initial random start point. This ensures that the study covers the entire area evenly rather than concentrating on certain sections. Similarly, in water quality monitoring, samples might be taken from every 10th location along a river. This interval-based approach helps track spatial variations in pollution levels while maintaining a manageable sample size.
8. **Archival and Document Analysis**: In historical and archival research, systematic sampling can be used to review large collections of documents or records. For example, in a study of news coverage on cybersecurity threats, researchers might review every 15th article published over ten years. This method ensures that findings are based on a broad and evenly spaced dataset, preventing overrepresentation of any specific period.
9. **Healthcare Service Evaluation**: Healthcare providers often use systematic sampling to audit patient records, review treatment outcomes, or monitor service delivery quality. For example, a hospital might review every 30th patient record to assess compliance with infection control protocols. In doing so, they ensure that different departments, patient conditions, and treatment types are proportionately represented.
10. **Large-Scale Data Reduction**: In big data analytics, systematic sampling offers a practical way to reduce massive datasets to manageable sizes while preserving representativeness. For example, in machine learning model training for IDS, researchers might use systematic sampling to create balanced subsets of normal and abnormal traffic data for faster processing.

The applications of systematic sampling are broad, cutting across academic, industrial, government, environmental, and technological contexts. Its structured approach ensures even coverage of the population, while its simplicity makes it an attractive choice for large-scale studies. In cybersecurity research—particularly in IDS evaluation—it allows for unbiased dataset selection, efficient data reduction, and balanced representation of diverse conditions. While systematic sampling must be applied carefully to avoid periodicity bias, its efficiency, transparency, and adaptability make it a valuable tool in both traditional research and emerging fields such as big data analytics and cyber threat detection. When the sampling frame is well-prepared and potential biases are addressed, systematic sampling delivers results that are both statistically valid and operationally practical.

2.2.3 Stratified Sampling in Qualitative Research

Stratified sampling is a probability sampling technique that divides a population into distinct subgroups, or *strata*, based on shared characteristics before selecting participants from each group. While stratified sampling is most often associated with qualitative research, it can be adapted for qualitative studies when the aim is to ensure that specific perspectives, experiences, or contexts are deliberately represented in the dataset. This approach is particularly valuable in studies where diversity across certain dimensions is essential to understanding the phenomenon in depth [28].

Purpose in Qualitative Contexts

In qualitative research, the objective is not to achieve statistical generalisation but rather to capture rich, varied insights. Stratification can help achieve representational balance across key subgroups, preventing overrepresentation of easily accessible participants and underrepresentation of harder-to-reach voices. For example, in a qualitative study exploring the lived experiences of cybersecurity professionals working with IDS in heterogeneous environments, researchers might stratify participants by organisational role (e.g., analysts, managers, engineers) or type of computing infrastructure (e.g., cloud-based, IoT networks, on-premises).

Process

The stratified approach in qualitative research generally involves:

1. **Identifying Relevant Strata**—Determining the most critical categories that reflect diversity in the phenomenon under study.
2. **Defining Inclusion Criteria**—Establishing clear definitions for each stratum to avoid overlap and ambiguity.
3. **Selecting Participants from Each Stratum**—Using purposive, criterion, or snowball sampling within each group to identify participants who can provide relevant, detailed perspectives.

For instance, in a healthcare-focused qualitative study on infection prevention, strata might be based on healthcare role (nurses, support workers, managers) or type of care setting (acute hospital wards, community clinics, residential care homes).

Advantages in Qualitative Studies

- **Ensures Comprehensive Coverage**—Guarantees that all key subgroups are represented in the data.
- **Facilitates Comparative Analysis**—Allows researchers to examine differences and similarities across strata.
- **Reduces Sampling Bias**—Prevents dominance of voices from more accessible or vocal groups.
- **Strengthens Transferability**—By including diverse contexts, findings may be more applicable across similar settings.

Challenges and Limitations

- **Time and Resource Intensive**—Requires upfront work to define strata and recruit from each.
- **Risk of Over-Segmentation**—Too many strata may dilute the depth of data from each subgroup.
- **Access Barriers**—Certain strata may be harder to reach, leading to delays or imbalances.
- **Potential for Overemphasis on Structure**—Excessive focus on stratification may overshadow the emergent, flexible nature of qualitative research.

Example in Practice

Consider a qualitative case study examining the barriers to implementing AI-assisted IDS across different organisational environments. The researcher could stratify the sample by industry sector (e.g., finance, healthcare, manufacturing, government) and then select a small number of participants from each sector. This ensures that perspectives reflect the unique regulatory, technical, and operational challenges in each setting, enabling more nuanced thematic comparisons. In another context, stratification could be applied in a study on patient experiences of post-operative care in orthopaedic wards. Strata might include age groups, mobility levels, and post-surgery recovery timeframes, ensuring that the sample reflects varied recovery trajectories. Although stratified sampling is primarily a qualitative method, it has important applications in qualitative research when diversity of perspective is critical. By ensuring that all relevant subgroups are represented, researchers can produce richer, more balanced accounts of the phenomenon under study. When implemented thoughtfully and combined with qualitative selection methods such as purposive or snowball sampling, stratification can enhance the depth and credibility of findings without compromising the flexibility that is central to qualitative inquiry [29].

Applications of Stratified Sampling

Stratified sampling has a wide range of applications across research domains, particularly when it is important to ensure that key subgroups within a population are adequately represented. While it is primarily associated with qualitative research, its principles can be adapted to qualitative studies that require balance across categories or perspectives. In both contexts, the method ensures that the diversity present in the

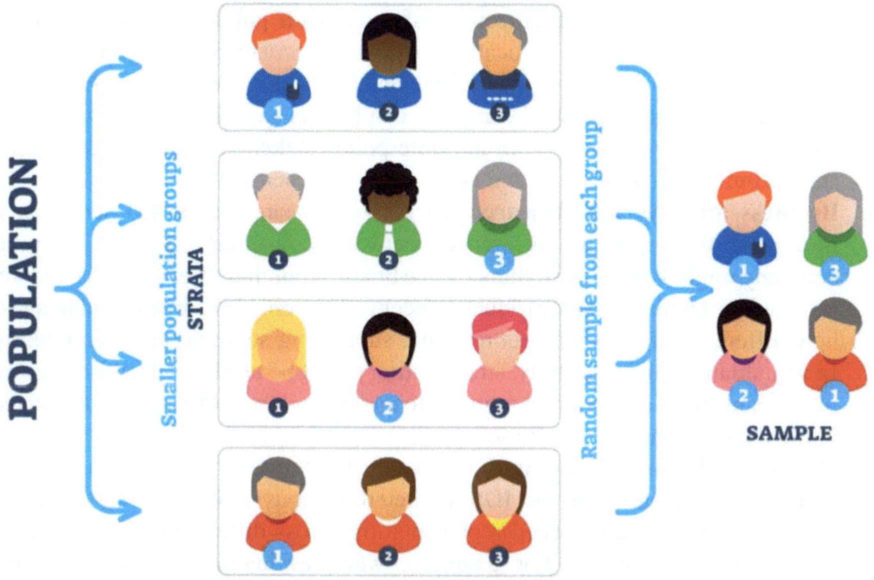

Fig. 2.10 Stratified sampling [28]

population is mirrored in the sample, enabling more meaningful comparisons and reducing the likelihood of bias [30, 31] (Fig. 2.10).

1. **Academic and Educational Research**: In educational studies, stratified sampling is often used to ensure representation across demographic or institutional subgroups. For example, in a study investigating student engagement with e-learning platforms, researchers might stratify by year level, subject area, or institution type (e.g., public vs. private schools). This guarantees that the findings reflect the experiences of students across different learning environments, rather than being dominated by one subgroup. In qualitative educational research, stratified purposive sampling might be applied to select participants from each defined category to ensure that varied voices are included in in-depth interviews or focus groups.
2. **Public Health and Epidemiology**: Stratified sampling is crucial in health research for obtaining representative data across demographic and risk-related strata. For example, in a study assessing community perceptions of vaccination programmes, participants may be stratified by age group, gender, or geographical location. This approach ensures that insights reflect diverse population segments, making public health interventions more targeted and effective. In qualitative studies, stratification may be applied to ensure that individuals from high-risk groups, healthcare workers, and policy decision-makers are all represented in a thematic analysis of vaccination hesitancy.
3. **Government and Policy Research**: Government agencies frequently use stratified sampling in population censuses, national surveys, and policy evaluations.

For instance, in assessing citizen satisfaction with public services, stratification by region, urban/rural location, and socio-economic status ensures that results account for regional disparities and structural inequalities. In qualitative policy analysis, stratified sampling can help ensure that interviews or case studies cover perspectives from multiple policy stakeholders, such as frontline staff, administrative officers, and community representatives.

4. **Business and Market Research**: In market research, stratified sampling is invaluable for capturing differences between consumer segments. A retail company might stratify customers by age bracket, purchase frequency, or income level when testing a new product concept. This enables targeted analysis of how different market segments respond, supporting strategic decision-making. In a qualitative business study, stratification might be used to select focus group participants from each market segment to understand the motivations and barriers influencing purchasing behaviour.
5. **Cybersecurity and Technology Research**: In technology-focused research—especially in areas such as IDS in heterogeneous environments—stratified sampling can ensure balanced representation across different types of systems or operational contexts. For example, when evaluating IDS performance, researchers might stratify by network type (e.g., corporate LAN, cloud infrastructure, IoT networks) and then select data samples or case studies from each stratum. This approach helps avoid overrepresentation of any single environment and allows for meaningful cross-comparison of system performance in varying contexts. In a qualitative setting, interviews could be stratified by role—security analysts, IT managers, and compliance officers—to capture a holistic view of challenges in IDS implementation.
6. **Quality Control and Manufacturing**: In industrial contexts, stratified sampling is often used in quality control to ensure that products from all production lines, shifts, or batches are tested. For example, a manufacturing plant producing multiple models of a product could stratify by model type before selecting units for inspection. This ensures that defect rates are assessed fairly across all product categories. While most quality control applications are qualitative, qualitative research might stratify inspection teams by experience level to explore how training impacts quality assurance practices.
7. **Environmental and Agricultural Studies**: Environmental research frequently uses stratified sampling to account for geographical or ecological variation. For instance, a study on soil health across a farming region might stratify by soil type or land use before selecting sampling sites. This guarantees that the results reflect conditions across different environmental contexts. In qualitative research, stratification might be used to ensure that farmers from different land management systems—such as organic, intensive, and mixed farming—are included in interviews about sustainable practices.
8. **Social Inclusion and Equity Research**: In studies that address inequality or social exclusion, stratified sampling ensures that marginalised or minority groups are adequately represented. For example, a project on employment barriers could stratify participants by disability status, ethnicity, or educational background.

This approach prevents findings from being skewed towards majority experiences and supports more inclusive policy recommendations. In qualitative projects, this technique can be combined with purposive selection within each stratum to amplify underrepresented voices while still maintaining structured coverage.

The versatility of stratified sampling lies in its ability to produce samples that truly reflect the diversity of the population under study. Whether applied to large-scale qualitative surveys or adapted for qualitative case selection, the method enhances representativeness, facilitates subgroup comparisons, and strengthens the relevance of findings. In cybersecurity, for example, it ensures that an IDS evaluation considers variations in system architecture, organisational roles, and operational environments. In public health, it ensures that interventions address the needs of all demographic groups. In education and business, it supports nuanced understanding of diverse learner or consumer segments. When used carefully—balancing the need for representation with the depth of analysis—stratified sampling becomes a powerful tool for both qualitative and qualitative research. It bridges the gap between the need for systematic coverage and the richness of context-driven inquiry, making it especially valuable in studies where diversity is central to the research question [32].

2.2.4 Cluster Sampling in Qualitative Research

Cluster sampling is a probability sampling technique in which the population is divided into naturally occurring groups, or *clusters*, and a selection of these clusters is chosen for study. Unlike stratified sampling, where all strata are represented, cluster sampling involves studying only the selected clusters in their entirety. This approach is particularly valuable when the population is geographically dispersed or when creating a complete list of all individuals is impractical. While traditionally associated with qualitative survey research, cluster sampling can be adapted to qualitative studies, especially those involving case study analysis, ethnography, or multi-site interviews. Its qualitative application focuses on studying entire clusters in depth to generate rich, context-specific insights rather than aiming for statistical generalisation [31, 33, 34].

Purpose in Qualitative Contexts

In qualitative research, cluster sampling is often used for logistical efficiency and to capture contextual variation across different settings. For example, in a study of healthcare workers' experiences with infection prevention protocols, researchers might treat each hospital ward or care home as a cluster. By selecting a few representative wards, they can explore the organisational culture, work practices, and communication styles within each setting.

Process

The typical qualitative cluster sampling process involves:

1. **Identifying Clusters**—Clusters are naturally occurring units such as schools, clinics, neighbourhoods, or organisations.
2. **Selecting Clusters**—A subset of clusters is chosen using random or purposive criteria, depending on the study design.
3. **Studying All or Most Members Within Clusters**—In qualitative research, all relevant participants in the chosen clusters may be invited for interviews, focus groups, or observation.

For instance, in an ethnographic study of cybersecurity operations, clusters could be defined as individual companies or security operation centres (SOCs). The researcher might choose four SOCs with different organisational structures to capture varied operational approaches.

Advantages in Qualitative Studies

1. **Cost and Time Efficiency**—Reduces the need to travel to widely dispersed sites by concentrating on selected clusters.
2. **Context-Rich Data**—Allows researchers to immerse themselves in the social, cultural, or organisational environment of each cluster.
3. **Feasible for Large Populations**—Particularly useful when a complete list of all individual members is unavailable.
4. **Flexibility in Selection**—Can be combined with purposive or theoretical selection of clusters for targeted inquiry.

Challenges and Limitations

1. **Risk of Homogeneity**—Selected clusters may not fully capture the diversity of the wider population.
2. **Cluster-Level Bias**—Characteristics unique to chosen clusters may influence findings disproportionately.
3. **Reduced Generalisability**—Especially in small qualitative samples, findings may not be transferable to all clusters.
4. **Potential Access Barriers**—Gaining permission to study entire clusters can be challenging.

Example in Practice

Consider a qualitative study exploring barriers to implementing AI-assisted IDS in different sectors. Clusters could be defined as organisations grouped by industry—such as finance, healthcare, manufacturing, and government. The researcher might select one organisation from each industry and conduct in-depth interviews with its cybersecurity team. This approach enables comparison between sectors while still allowing for deep exploration of each organisational context. In healthcare research, cluster sampling could be applied to study post-operative care practices in orthopaedic wards across different hospitals. Each ward acts as a cluster, and all staff members—nurses, support workers, and physiotherapists—within the selected wards participate, providing a comprehensive view of that setting's practices. Cluster sampling offers a practical and contextually rich approach for qualitative research when the population is large, dispersed, or organised into natural groups. By focusing on entire clusters, researchers gain in-depth insights into the dynamics, interactions, and environmental factors influencing the phenomenon under study. While this method can introduce bias if clusters are not carefully chosen, it remains an efficient and effective way to explore diversity across settings when resources or logistics limit broader coverage. In qualitative inquiry, the strength of cluster sampling lies not in statistical representation but in its ability to illuminate how context shapes experience—a perspective that is invaluable in fields ranging from education and healthcare to technology and organisational studies [35].

Applications of Cluster Sampling

Cluster sampling is a versatile probability sampling technique with broad applications across research fields, particularly when populations are large, widely dispersed, or organised into natural groups. It is especially valuable when the creation of a complete list of all individual members is impractical or too resource-intensive. While cluster sampling is most often associated with large-scale qualitative surveys, it can also be adapted effectively for qualitative studies, where it facilitates in-depth exploration of specific sites or communities (Fig. 2.11).

Its key strength lies in its ability to concentrate research activities within selected clusters, reducing logistical challenges while still capturing meaningful variation between groups.

1. **Educational Research**: In educational studies, schools, colleges, or classrooms often serve as clusters. For example, in a national survey of teaching practices, researchers might randomly select a group of schools (clusters) and then collect data from all teachers within each selected school. In qualitative educational research, cluster sampling might be used to select a subset of schools from different regions or socio-economic contexts. A researcher studying the integration of digital learning tools could choose four schools—two urban, one

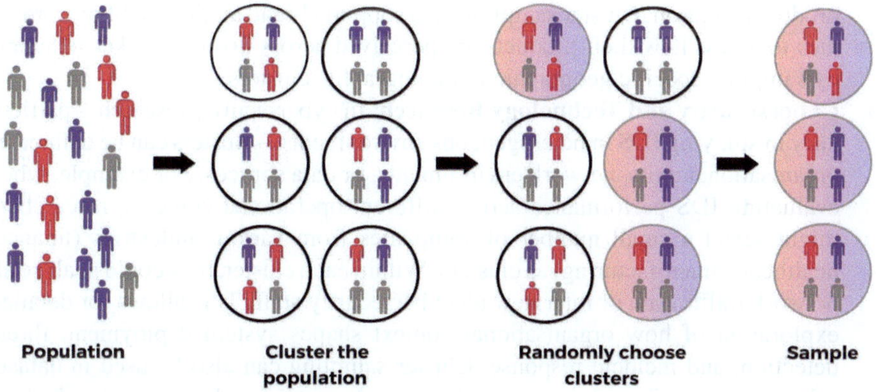

Fig. 2.11 Overview of cluster sampling [36]

suburban, and one rural—and conduct interviews with teachers, administrators, and students in each. This approach ensures that the findings reflect a range of environmental and cultural contexts while keeping the study manageable.

2. **Public Health and Epidemiology**: Public health research frequently uses cluster sampling for health surveys, disease surveillance, and programme evaluations. Villages, households, or clinics can serve as clusters for both qualitative and qualitative purposes. For example, in a vaccination coverage survey, researchers might select specific villages and then collect data from all eligible households within those villages. In a qualitative study on community perceptions of vaccination, cluster sampling could be used to select clinics from different districts, allowing in-depth interviews with healthcare workers and patients in each cluster. In epidemiological outbreak investigations, this method can speed up data collection when quick responses are needed to understand local transmission patterns and implement control measures.
3. **Government and Policy Research**: Government agencies use cluster sampling in censuses, national household surveys, and programme monitoring. Administrative districts, municipalities, or census blocks often serve as clusters. For policy evaluation, qualitative researchers may select clusters such as specific service delivery centres or community hubs. For instance, in evaluating a job-training programme, researchers could select several employment centres from different regions and conduct focus groups with staff and participants at each. This approach captures regional differences while allowing for rich, site-specific analysis.
4. **Business and Market Research**: In market research, cluster sampling helps companies study customer behaviour across different locations. For example, retail outlets, sales territories, or store branches can serve as clusters. A business evaluating customer satisfaction with a new product could select certain store branches and collect data from all customers in those stores during a specified period. In qualitative studies, focus groups could be held in selected branches to capture customer experiences and identify location-specific factors affecting product reception. Cluster sampling also supports brand audits, enabling companies to assess how their products are perceived across diverse market segments by sampling specific geographic or demographic clusters.
5. **Cybersecurity and Technology Research**: In cybersecurity research—particularly in studying IDS in heterogeneous environments—clusters can be defined as organisational units, network environments, or data sources. For example, when evaluating IDS performance across different operational contexts, researchers might select a small number of companies from various industries (finance, healthcare, manufacturing) as clusters. Within each cluster, they could analyse all network traffic data or interview all cybersecurity staff. This allows for detailed exploration of how organisational context shapes system deployment, threat detection, and incident response. Cluster sampling can also be used in dataset selection by treating servers, network segments, or geographic regions as clusters, enabling efficient data reduction without losing contextual diversity.

6. **Environmental and Agricultural Studies**: Environmental researchers often use cluster sampling to study large or geographically dispersed areas. Natural regions, watersheds, or farms can serve as clusters. For example, in a study on water quality, specific river catchments could be selected as clusters, with water samples collected from multiple points within each. In qualitative studies on farming practices, clusters might be defined as agricultural cooperatives or villages, allowing researchers to explore shared practices and challenges among members of the same community. This approach ensures that diverse ecological conditions or agricultural systems are represented without requiring exhaustive coverage of the entire population.
7. **Quality Control and Manufacturing**: In manufacturing, cluster sampling is used to monitor production quality across multiple production lines, plants, or shifts. A researcher or quality inspector might select certain production lines (clusters) and inspect all products from those lines during a specified time period. In qualitative contexts, cluster sampling could involve selecting certain factories and conducting in-depth interviews with production staff to explore process improvements, safety practices, or skill development needs.
8. **Social and Cultural Research**: Anthropologists and sociologists often work with clusters defined by communities, neighbourhoods, or cultural groups. For example, a qualitative study on cultural adaptation among immigrant communities might select a few towns with significant immigrant populations and explore the experiences of all households within these clusters. This method enables deep engagement with the social environment and facilitates observation of group-level cultural dynamics that might not be visible in dispersed individual interviews.

Cluster sampling is a powerful and adaptable tool for both large-scale surveys and in-depth qualitative studies. It reduces logistical burdens, makes research more cost-effective, and enables researchers to immerse themselves in the unique contexts of the selected clusters. In qualitative research, the strength of cluster sampling lies in its ability to deliver context-rich, holistic data from naturally occurring groups, making it well suited to studies in education, public health, government, business, cybersecurity, environmental science, and cultural studies. When applied thoughtfully—with careful attention to cluster selection and diversity—it offers a balance between practicality and depth, ensuring that findings reflect the realities of specific settings while providing insights that can inform broader discussions and applications [37].

2.3 Non-probability Sampling Techniques

Sampling is a fundamental process in research that allows investigators to study a subset of a population and draw conclusions that may inform understanding of the whole. In many situations, especially in qualitative, exploratory, or case-specific studies, researchers rely on non-probability sampling techniques—methods where participants are selected based on criteria other than random chance. Unlike probability sampling, where every member of the population has a known and equal chance of selection, non-probability sampling uses deliberate, convenient, or referral-based approaches to identify participants. These techniques are often employed when the target population is difficult to access, when time and resources are limited, or when the research focus requires specialised knowledge or experiences. While non-probability sampling can provide rich, context-specific insights, it also carries limitations in terms of generalisability and potential researcher bias. Understanding these techniques is essential for designing effective studies in contexts where statistical representativeness is less critical than depth, accessibility, or practicality [38].

2.3.1 Convenience Sampling in Qualitative Analysis

Convenience sampling, also known as accidental or grab sampling, is a non-probability sampling method where researchers select participants based on their accessibility rather than random selection. This approach is characterized by ease of access, speed of data collection, low cost, and a non-random nature that increases the risk of selection bias. For instance, a lecturer might survey students from their own class due to their immediate availability, rather than randomly selecting from the entire student body. The process of convenience sampling is straightforward. It begins with identifying the target population relevant to the research question. Researchers then select participants who are easily reachable and collect data from those willing to respond. The analysis of results must consider potential biases and limitations inherent in the sample. For example, a researcher investigating customer satisfaction at a retail store may only survey shoppers present at the time of the study, rather than a random selection from the store's entire customer base. The advantages of convenience sampling include time efficiency, cost savings, practicality, and its utility in pilot testing. The method allows for quick data collection without the need for a complex sampling frame, making it ideal for studies with tight deadlines. It also reduces logistical costs associated with random selection and is particularly useful when the research population is hard to define or access. Researchers often use convenience samples to test research instruments before conducting larger studies. However, convenience sampling has notable disadvantages and limitations. The primary concern is selection bias, as the sample may over represent certain groups while underrepresenting others, leading to distorted results. Additionally, findings from convenience samples often lack generalizability, meaning they cannot be confidently applied to the broader population. Volunteer bias is another issue, as those who agree to participate may differ significantly from those who decline, further influencing the data. Moreover, convenience sampling may overlook important subgroups, limiting the depth of analysis. Practical examples of convenience sampling can be found across various fields. In market research, a small business owner might solicit feedback on a new product from customers visiting the store during a specific week. In education, a university lecturer may survey students in their own classes rather than sampling from the entire university. In healthcare, a nurse might conduct a patient satisfaction survey by approaching patients waiting in an outpatient clinic rather than all patients treated in the hospital.

Convenience sampling is particularly useful in exploratory research, pilot studies, educational settings, and marketing. It allows researchers to quickly test concepts and gather preliminary insights, making it a valuable tool in the early stages of research. However, ethical considerations must be taken into account, including obtaining informed consent, maintaining confidentiality, and being transparent about the sample's limitations. Researchers should also avoid exploiting vulnerable populations, such as students or patients, who may feel pressured to participate. To improve the validity of convenience sampling, researchers can diversify collection sites, increase sample sizes, and combine convenience sampling with other methods to ensure demographic balance. Acknowledging the limitations of the findings is crucial, as it allows for cautious interpretation of results. Thus, convenience sampling is a widely used non-probability sampling method valued for its practicality, speed, and cost-effectiveness. While it is beneficial for exploratory research and pilot testing, its reliance on easily accessible participants introduces significant bias and limits the generalizability of findings. Researchers should apply convenience sampling strategically, supplementing it with other techniques when possible and clearly communicating its limitations in their reports. When used transparently and with an awareness of its constraints, convenience sampling can be a valuable tool in a researcher's methodological toolkit, particularly for generating initial insights and refining research instruments within practical constraints [39] (Fig. 2.12).

Practical examples of convenience sampling can be considered to better understand its function in practice.

1. **University Cafeteria Survey**: A marketing researcher wants quick feedback on a new energy drink brand. Instead of creating a randomised student list, they stand in the university cafeteria during lunch and ask the first 50 students they see to fill in a short questionnaire. The students are chosen purely because they are easy to reach at that moment.
2. **Shopping Mall Interviews**: A clothing retailer wishes to gather opinions about their latest collection. They set up a small booth in a busy shopping mall and approach shoppers who pass by. The sample includes only those willing to stop and talk, making it fast but potentially biased toward certain demographics.
3. **Social Media Polls**: A blogger posts a poll on Instagram asking followers about their skincare routines. Only those who see the post and voluntarily respond are included. While this method gives quick results, it is limited to people who already follow the blogger.
4. **Health Check at a Local Gym**: A fitness researcher collects BMI and blood pressure data from members of a single gym to explore exercise habits. The participants are simply the gym-goers present at the time of the study, making the process convenient but not representative of the broader population.

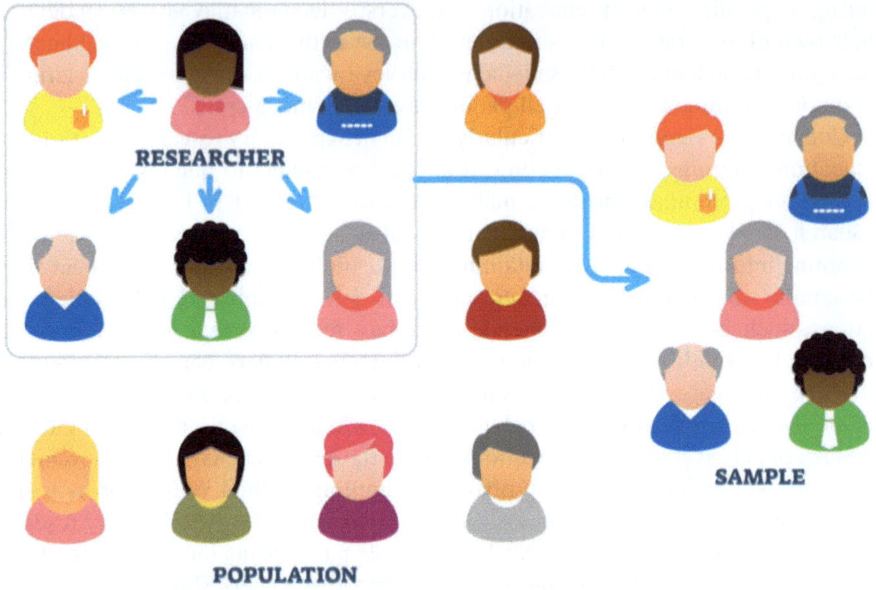

Fig. 2.12 Overview of convenience sampling in qualitative analysis [3]

2.3.2 Purposive Sampling in Qualitative Analysis

Purposive sampling, also known as judgmental or selective sampling, is a non-probability sampling technique where participants are intentionally selected based on specific characteristics relevant to the research question. This method contrasts with probability sampling, which provides each member of the population an equal chance of selection. Purposive sampling is particularly prevalent in qualitative research, case studies, and exploratory investigations, where the emphasis is on obtaining an in-depth understanding rather than broad generalizations. The core definition of purposive sampling involves the deliberate selection of participants who possess particular qualities or experiences that align with the study's objectives. Key characteristics of this sampling method include deliberate selection based on inclusion criteria, a focus on information-rich cases, non-random selection, and flexibility to adapt to various research contexts. There are several types of purposive sampling, each serving different research aims. Homogeneous sampling targets participants with similar traits to explore specific attributes in depth. Maximum variation sampling includes a diverse range of participants to capture a wide array of perspectives. Typical case sampling selects average cases to represent what is considered 'normal,' while extreme case sampling focuses on unusual or outlier cases for detailed analysis. Critical case sampling identifies cases that are crucial to the study's findings, and expert sampling involves selecting individuals with specialised knowledge in the research area.

The advantages of purposive sampling include its relevance, as it ensures data is collected from those most capable of providing valuable insights. It is efficient, directing resources toward participants likely to contribute meaningful information. This method is ideal for qualitative studies seeking depth and context-specific understanding and offers flexibility that allows adjustments as the study evolves. However, purposive sampling also has limitations. Researcher bias can influence participant selection, potentially skewing results. The findings may lack generalizability, as they may not apply to the broader population. Additionally, overly narrow selection criteria can exclude relevant perspectives, and the subjectivity in inclusion criteria may lead to a lack of transparency and consistency in the selection process.

Practical examples of purposive sampling can be found across various fields. In healthcare research, a study on cancer survivors may select participants who have been in remission for at least five years to gain insights into long-term recovery challenges. In education research, a study on inclusive education might focus on teachers with extensive experience working with students with special educational needs. In business research, a case study on corporate social responsibility could select companies recognized for their sustainability efforts. Purposive sampling is widely applied in qualitative research, case studies, policy research, and program evaluations. Researchers must consider ethical implications when using this method, ensuring informed consent, transparency in selection criteria, and protection of participants' confidentiality and privacy. Therefore, purposive sampling is a valuable research tool that emphasises depth and targeted insight over statistical representativeness. It is particularly effective in qualitative and exploratory studies aimed at understanding complex issues through the experiences of relevant individuals. Despite its potential for bias and limited generalizability, clear selection criteria and transparency can enhance the richness and meaningfulness of the data collected, contributing significantly to specific research contexts (Fig. 2.13).

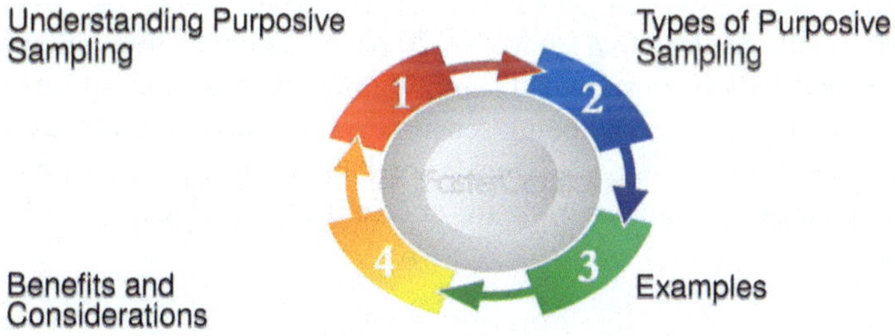

Fig. 2.13 Overview of purposive sampling in qualitative analysis [40]

2.3.3 Snowball Sampling in Qualitative Analysis

Snowball sampling is a non-probability sampling technique widely utilized in qualitative research to reach hidden or sensitive populations. This method relies on existing participants to recruit future participants from their networks, creating a chain-referral process that allows the sample size to expand as more individuals are introduced. This approach is particularly beneficial in situations where no accessible sampling frame exists. The process of snowball sampling begins with identifying a small group of eligible participants who meet the study's criteria. Initial data is collected from these participants, who are then asked to refer others with similar characteristics. This recruitment continues until the desired sample size or data saturation is achieved. In qualitative analysis, the primary aim is to gain a deep, context-rich understanding rather than to generalize findings statistically. Snowball sampling facilitates this by allowing researchers to access individuals with shared experiences, fostering trust through personal introductions, and enabling exploration of sensitive topics within participant-recommended networks. The advantages of snowball sampling in qualitative studies include access to hidden populations, such as undocumented migrants or survivors of abuse, and the ability to build trust, which can enhance participants' willingness to engage. Additionally, this method is cost and time-efficient, as it reduces the effort required to locate eligible participants independently and allows for rich data collection through in-depth interviews. However, snowball sampling also presents limitations and challenges. Sample bias can occur since networks may be homogenous, leading to a lack of diverse perspectives. There is a risk of over-representation from certain social circles, and recruitment can stall if participants are unwilling to provide referrals. Ethical concerns also arise, particularly regarding privacy, when asking for referrals, especially in sensitive contexts. An example of snowball sampling in qualitative research could involve a study on the coping strategies of refugees in an urban area, starting with three initial participants identified through a local NGO. These participants would refer others in their network, ultimately leading to a sample of 20 participants for in-depth interviews.

Ethical considerations in snowball sampling include ensuring informed consent, maintaining confidentiality, and being sensitive to the pressure that may arise when asking individuals to participate. Thus, snowball sampling is a valuable tool in qualitative research for studying small, specialized, or hidden populations [41] (Fig. 2.14).

2.3.4 Quota Sampling in Qualitative Analysis

Statistics, public opinion polls, and market research all make use of quota sampling, a non-probability sampling technique. Sorting people into smaller groups or "quotas" according to established criteria like age, gender, profession, or degree of education is what it entails. Unlike stratified random sampling, which uses a random

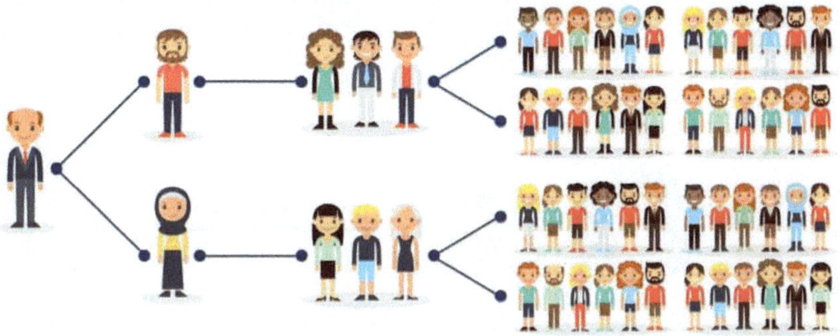

Fig. 2.14 Overview of snowball sampling in qualitative analysis

process to pick participants, quota sampling uses a non-random process to select participants within each group. Because of its ease of use, low cost, and capacity to guarantee proper representation of particular interest subgroups, this technique is commonly utilised. Problems with bias and representativeness are two areas where quota sampling falls short. The lack of randomisation in the selection process can limit the sample's capacity to accurately reflect the population's diversity, even when it guarantees proportional representation of certain groups. At each of the multiple steps that make up quota sampling, important decisions regarding the sample's structure and participant selection must be made. Defining the population of interest, or the entire group from which the sample will be drawn, is the first step in quota sampling. Age, gender, ethnicity, income, education, and geography are some of the criteria that the researcher uses to create quotas. The research goals and the elements that are considered most important to the investigation dictate the choice of characteristics. Typically, convenience sampling is used to choose participants within each quota in a non-random manner. People who are already in the area or who are shopping centre clients are examples of examples of easily accessible recruiting participants. We need to ensure that every quota is filled with the necessary number of participants [42] (Fig. 2.15).

Researchers doing qualitative studies often employ quota sampling, a non-probability sampling technique, to ensure that their sample accurately reflects the population as a whole. This method is ideal for qualitative research since it allows for the collection of multiple viewpoints while yet guaranteeing that important demographic variables such as age, gender, occupation, and level of education are represented in the sample. Quota sampling differs from probability sampling in that it depends on the researcher's discretion to fill predetermined quotas rather than random selection. Step one is to determine which traits are important, step two is to establish study objectives, and step three is to pick participants non-randomly until all subgroup quotas have been met. In qualitative research, quota sampling is essential because it guarantees a range of opinions, makes it easy to compare subgroups, and lets

Fig. 2.15 Overview quota sampling in qualitative analysis

researchers look for similarities and contrasts across different demographic or experience groups. Nevertheless, there are obstacles and constraints to consider, including the possibility of prejudice, subjectivity, and selection bias. Some subgroups may be more difficult to recruit, which could extend the data collection process, and an overemphasis on quotas could cause researchers to prioritise quantity over quality of analysis. Additionally, quota sampling must adhere to ethical standards, which include being open and honest about how quotas are selected, obtaining informed permission, and not engaging in tokenism. To sum up, quota sampling is an excellent method for collecting more comprehensive data from qualitative studies by guaranteeing a diverse sample of participants [43].

2.4 Choosing the Right Sampling Technique for Qualitative Analysis

Qualitative research is a unique approach to social science inquiry, focusing on the exploration of lived experiences, meanings, perceptions, and social contexts. The choice of sampling technique in qualitative research is crucial for the quality and credibility of the findings. Unlike qualitative sampling, which prioritizes statistical representativeness and probability-based selection, qualitative sampling often relies on non-probability approaches aimed at identifying participants or cases that can provide deep, contextually rich insights into the phenomenon under investigation. Qualitative sampling is fundamentally purposive, aiming to select cases that are information-rich and relevant to the research objectives rather than creating a sample that mirrors a larger population in a statistical sense. This emphasis on depth over breadth means that qualitative samples are typically smaller than those in qualitative studies, yet the data generated can be far more nuanced. The researcher's role in qualitative sampling is active and iterative, with sampling decisions evolving as the study progresses. Purposive sampling is the most widely used technique in qualitative research, involving the deliberate selection of participants based on their ability to provide relevant and detailed information. There are several subtypes tailored to specific research needs, such as criterion sampling, maximum variation sampling, homogeneous sampling, extreme or deviant case sampling, and critical case sampling. Purposive sampling is highly adaptable but requires careful justification and transparent reporting to address concerns about researcher bias in participant selection. Snowball sampling, also known as chain referral sampling, is often employed when studying hidden or hard-to-reach populations such as undocumented migrants, individuals engaged in illicit activities, or members of stigmatized communities. The process begins with a small number of initial participants who meet the study's criteria and refers the researcher to others in their network who might also be suitable. While this technique leverages trust within social networks to facilitate access where direct recruitment might be challenging or unsafe, researchers using this method must remain vigilant about over-representing certain perspectives and should seek to diversify their network chains where possible. In conclusion, understanding the strengths and limitations of each sampling technique allows researchers to make informed choices that maximize the validity and transferability of their work while ensuring that participants' perspectives are authentically represented.

Theoretical sampling is a method used in grounded theory methodology, where data collection and analysis occur simultaneously. It is a flexible approach that allows for the exploration of concepts, filling gaps in understanding, or testing developing theoretical categories. This method requires flexibility, methodological expertise, and ongoing engagement with the analysis to make informed sampling decisions. Convenience sampling selects participants based on their availability and willingness to participate, which can be useful in exploratory qualitative studies, pilot projects, or situations with severe constraints on time and resources. While it may produce biased or non-representative samples, convenience sampling can generate preliminary insights that inform more rigorous sampling in subsequent studies. Quota sampling involves setting specific quotas to ensure the inclusion of particular subgroups based on characteristics relevant to the research, such as age, gender, ethnicity, or occupational role. Once the quotas are filled, sampling stops, regardless of whether further diversity could be achieved. In qualitative research, quota sampling can help ensure a balanced representation of different perspectives without requiring statistical proportionality. However, the process of setting quotas must be theoretically grounded rather than arbitrary, and researchers must be mindful of the risk of oversimplifying complex social identities. The choice of sampling technique in qualitative research depends on multiple interrelated factors, including the research question, population nature, resource constraints, epistemological and methodological orientation, and ethical considerations. Researchers must ensure that participants are recruited fairly and respectfully, with full disclosure about the study's nature and their rights. Transparency in reporting how participants were selected and why is crucial for maintaining trust and credibility. Ensuring rigor in sampling is essential for qualitative researchers to demonstrate that their sampling decisions were methodologically sound and justified in relation to their research aims. This involves explaining the rationale for the technique, the criteria for participant selection, the process of recruitment, and any adaptations made during the study. Strategies such as triangulation, member checking, and maintaining an audit trail can help enhance the credibility and trustworthiness of the findings [10].

2.5 Comparing Probability and Non-probability Sampling

Sampling is a crucial process in research, determining how participants or units are selected from a population to form a study sample. The choice of sampling method directly affects the validity, reliability, and generalisability of research findings. In broad terms, sampling techniques are divided into two main categories: probability sampling and non-probability sampling. Each approach is grounded in distinct philosophical assumptions and methodological priorities, making them suitable for different types of research questions and designs. Understanding the differences between the two is essential for researchers seeking to align their sampling strategy with the aims, scope, and resources of their study.

Probability Sampling

Probability sampling refers to techniques in which every member of the target population has a known, non-zero chance of being selected. This approach relies on random selection methods, which reduce selection bias and allow statistical inferences to be made about the population from which the sample is drawn. Common forms include simple random sampling, stratified sampling, cluster sampling, and systematic sampling.

The key strength of probability sampling lies in its ability to produce representative samples, thereby enhancing the generalisability of findings. It is most commonly associated with qualitative research, where the goal is often to measure variables and test hypotheses in ways that can be extrapolated to a larger population. However, probability sampling requires a complete and accurate sampling frame—a list of all members of the target population—which can be difficult or costly to obtain. Additionally, its emphasis on representativeness sometimes comes at the expense of exploring the nuanced experiences of individuals or subgroups, which is often the focus of qualitative research.

Non-probability Sampling

Non-probability sampling, on the other hand, does not give every member of the population a known or equal chance of being selected. Selection is often based on the researcher's judgement, accessibility of participants, or other non-random criteria. Techniques under this category include purposive sampling, snowball sampling, quota sampling, and convenience sampling. This approach is widely used in qualitative research, where the objective is not statistical generalisation but rather an in-depth understanding of a phenomenon. Non-probability sampling is more flexible, often less expensive, and particularly useful for accessing hard-to-reach or specialised populations. However, because selection is not random, the resulting sample may be biased, and findings cannot be confidently generalised to the broader population. The focus is instead on transferability—the extent to which findings may be relevant in similar contexts.

Key Differences and Considerations

The choice between probability and non-probability sampling depends on the research objectives, the nature of the population, available resources, and the intended use of the findings. Probability sampling is ideal for studies aiming to make statistically valid generalisations, while non-probability sampling is better suited for exploratory research, theory development, and studies where representativeness is less important than depth and richness of data. Ethical and practical considerations also play a role. In some contexts, obtaining a complete sampling frame for probability sampling may not be feasible, and non-probability methods may be the only viable option. Conversely, where decisions must be based on population-level estimates—such as in public health policy—probability sampling provides the necessary statistical foundation. Both probability and non-probability sampling have legitimate places in the research process, each with distinct advantages and limitations. Rather than seeing them as competing approaches, researchers should consider them complementary tools, to be chosen in alignment with their research questions and constraints. The ultimate goal is to ensure that the sampling method is transparent, justified, and fit for purpose, thereby enhancing the credibility and usefulness of the research [37] (Table 2.3).

2.6 Practical Considerations for Choosing a Sampling Technique

Sampling is one of the most critical decisions in the research process, directly influencing the quality, credibility, and relevance of the findings. In both qualitative and qualitative research, the choice of sampling technique determines not only who participates in the study but also the nature of the data collected, the scope of the conclusions, and the extent to which the results can be applied beyond the study sample. While textbooks often classify sampling methods into probability and non-probability categories, the actual process of choosing a sampling technique involves far more than understanding these theoretical distinctions. It requires careful consideration of practical realities—such as research objectives, resource availability, population characteristics, ethical constraints, and the anticipated use of the findings. In practice, the researcher's choice is rarely made in an idealised vacuum.

Table 2.3 Differences between probability and non-probability sampling

Aspect	Probability sampling	Non-probability sampling
Definition	Every member of the population has a known, non-zero chance of selection	Not all members have a known or equal chance of selection
Selection method	Random selection methods	Non-random, based on judgement, convenience, or referral
Examples	Simple random, stratified, cluster, systematic sampling	Purposive, quota, snowball, convenience sampling
Purpose	To obtain a representative sample for generalisation	To gain in-depth insights or explore phenomena
Research approach	Mostly used in qualitative research	Mostly used in qualitative research
Strengths	Reduces bias, allows statistical generalisation, high external validity	Flexible, cost-effective, useful for hard-to-reach populations, allows deep exploration
Limitations	Requires complete sampling frame, can be expensive and time-consuming	May be biased, limited generalisability, lower external validity
Data quality focus	Accuracy and representativeness of population parameters	Richness, detail, and contextual understanding
Generalisation	High—Results can be generalised to the entire population	Low—Focus is on transferability rather than generalisation

Time pressures, funding limitations, access restrictions, and even the political and cultural climate of the research setting may constrain what is possible. Similarly, methodological choices must align with the philosophical orientation of the study—a grounded theory investigation of lived experience will demand a different sampling strategy from a large-scale epidemiological survey [44]. This section examines the key practical considerations that guide the selection of sampling techniques, integrating methodological theory with applied research realities to offer a nuanced understanding of this crucial decision-making process.

1. **Alignment with Research Objectives and Questions**: The starting point for any sampling decision is a clear articulation of the research objectives and questions. Sampling is not an isolated procedural choice but a direct extension of the study's purpose. In qualitative research, where the aim is often to test hypotheses or estimate population parameters, probability sampling is generally preferred because it enables statistical generalisation. For example, a national survey on vaccination rates would require a representative sample drawn using methods such as stratified or cluster sampling to ensure all subgroups are proportionately included. In qualitative research, the goal is usually to gain a deep, contextually rich understanding of a phenomenon. Here, the researcher might employ purposive or theoretical sampling, selecting participants who can provide especially relevant insights. For instance, a study on the coping strategies of refugee women may focus specifically on individuals who have resettled in the past five years, rather than attempting to cover the entire refugee population. In both

cases, the sampling technique must directly serve the research questions rather than follow a generic methodological preference.

2. **Nature and Accessibility of the Population**: Understanding the characteristics of the target population is a fundamental practical step. Some populations are well-documented, easily accessible, and geographically concentrated, making probability sampling feasible. Others are dispersed, hidden, or difficult to reach, requiring adaptive strategies. For example, researchers studying corporate executives might find it impractical to compile a comprehensive list of all eligible participants, thereby limiting the use of random sampling. Instead, purposive or snowball sampling might be employed, using professional networks to identify participants. In contrast, a study of registered university students in a single institution could easily employ simple random sampling, given the existence of a complete enrolment list. Accessibility is also shaped by cultural, political, and legal factors. In politically sensitive contexts, certain groups may be unwilling to participate unless trust is established through intermediaries, again making non-probability approaches like snowball sampling more viable.
3. **Available Resources and Time Constraints**: Practical constraints often dictate what is possible in sampling. Probability sampling, while methodologically robust, can be resource-intensive, requiring large budgets, trained personnel, and extensive time to compile sampling frames and contact participants. In time-sensitive projects—such as rapid needs assessments after a natural disaster—researchers may have no choice but to rely on convenience or quota sampling, even if this compromises representativeness. Budgetary limitations can also restrict travel and data collection options. A researcher aiming to study rural health behaviours might intend to sample multiple villages randomly but find that transportation costs make this unrealistic. In such cases, a more focused purposive sampling strategy may achieve sufficient depth without exceeding budget. Time constraints, particularly in commissioned or policy-driven research, frequently lead to pragmatic compromises that balance methodological ideals with deliverable timelines.
4. **Methodological Approach and Research Design**: The choice of sampling technique must be coherent with the broader research design and methodological orientation. Experimental designs in qualitative research, for instance, often require random assignment to ensure internal validity. Surveys intended for inferential statistics need probability-based sampling to minimise bias. Conversely, case studies, ethnographies, and phenomenological research typically rely on non-probability techniques because they prioritise depth over breadth. Grounded theory research employs theoretical sampling, where participants are chosen progressively based on emerging analytical needs rather than predetermined quotas. This iterative approach demands flexibility in recruitment and a willingness to adapt sampling criteria mid-study. By contrast, cross-sectional surveys require a fixed sample drawn at the outset, leaving little room for modification once data collection has begun.

5. **Ethical Considerations**: Ethics permeate every aspect of the sampling process. Researchers must ensure that their sampling methods respect participants' rights, minimise harm, and promote fairness. Certain sampling techniques, while methodologically sound, may raise ethical issues in specific contexts. For example, snowball sampling among marginalised groups risks breaching confidentiality if referrals unintentionally disclose someone's association with a stigmatised identity or activity. Sampling vulnerable populations—such as minors, individuals with disabilities, or survivors of trauma—demands heightened sensitivity and often additional ethical approvals. Informed consent procedures must be adapted to the literacy levels, cultural norms, and comprehension abilities of the participants. Moreover, the selection process should avoid reinforcing social inequalities by systematically excluding certain groups. For example, over-reliance on online recruitment may inadvertently exclude older adults or those without internet access.
6. **Representativeness and Generalisability**: A practical consideration, particularly in qualitative research, is the extent to which the findings need to be generalisable to the larger population. If the goal is high external validity, probability sampling is the logical choice, as it provides a statistically defensible basis for generalisation. This often entails larger sample sizes and more elaborate recruitment procedures to ensure all relevant subgroups are adequately represented. However, in qualitative research, representativeness is reframed as *transferability*. The focus shifts from statistical generalisation to the applicability of findings in similar contexts. In such cases, researchers seek diversity within the sample to capture the full range of experiences rather than proportionate representation. This might mean deliberately including extreme or deviant cases to test the boundaries of emerging themes.
7. **Data Quality and Depth of Information**: The richness of the data required also influences sampling choices. Some methods prioritise quantity over depth, while others allow for intensive engagement with fewer participants. For instance, ethnographic research may involve prolonged immersion with a small number of participants, yielding detailed qualitative data that would be impossible to obtain from a large random sample. Conversely, probability sampling in large-scale surveys allows for robust statistical analyses but may limit the opportunity for nuanced follow-up on individual responses. Researchers must therefore balance the need for statistical precision with the desire for contextual depth, often leading to mixed-methods designs that incorporate both probability and non-probability samples at different stages.
8. **Feasibility of Creating a Sampling Frame**: A sampling frame—a comprehensive list of the population from which the sample will be drawn—is a prerequisite for most probability sampling techniques. In practice, compiling such a frame can be costly, time-consuming, or simply impossible. For instance, in studies of undocumented migrant workers, there is no official list from which to randomly select participants. Similarly, populations in conflict zones or transient groups such as the homeless defy easy enumeration. In these cases, researchers turn

to adaptive non-probability methods, often combining purposive selection with referral networks to approximate diversity within the population.

9. **Flexibility and Iteration in Sampling**: Some research designs require the flexibility to adapt the sample as new insights emerge. In grounded theory or action research, sampling is iterative and evolves alongside data analysis. This allows researchers to pursue emerging themes and ensure theoretical saturation. However, such flexibility is harder to achieve in tightly scheduled, grant-funded projects that demand a fixed sampling plan from the outset. Practical flexibility also involves being prepared for non-response or attrition. Even in carefully designed probability samples, not all selected individuals will agree to participate, necessitating replacement strategies or adjustments to maintain representativeness. In qualitative studies, researchers may need to recruit additional participants to address gaps in the data or to follow up on unexpected lines of inquiry.
10. **Contextual and Cultural Factors**: Research is embedded in specific social and cultural contexts that can affect sampling decisions. In collectivist cultures, for instance, gaining access to participants may require negotiating with community leaders or elders before individual recruitment can begin. In politically unstable environments, safety concerns may limit the researcher's ability to reach certain areas or populations. Language barriers, literacy levels, and local attitudes toward research can all influence both the feasibility and appropriateness of certain sampling techniques. Cultural competence is particularly important when studying sensitive topics, where trust and rapport are essential for obtaining authentic responses. In such cases, convenience or purposive sampling through trusted intermediaries may yield better participation than more formal random selection methods, even if this limits generalisability.
11. **Use of Mixed Sampling Strategies**: In many real-world studies, researchers combine different sampling techniques to balance practical constraints with methodological goals. For example, a national health survey might use stratified random sampling to ensure proportional representation across regions, followed by purposive sampling within selected strata to conduct qualitative interviews. This mixed approach leverages the strengths of both probability and non-probability methods, enhancing both the breadth and depth of the findings. Similarly, researchers might start with convenience sampling for pilot testing, then move to a more rigorous probability sample for the main study. Such phased strategies allow for iterative refinement of recruitment methods and data collection tools, improving overall study quality.

Choosing a sampling technique is both a methodological and a pragmatic decision, shaped by the interplay of research objectives, population characteristics, available resources, ethical considerations, and contextual realities. While probability sampling offers the advantage of representativeness and statistical generalisability, it is not always feasible or even desirable, especially in studies prioritising depth, complexity, and the exploration of lived experiences. Non-probability sampling provides flexibility, accessibility, and adaptability, making it well-suited to qualitative inquiry and to research with hard-to-reach populations. Ultimately, the "right" sampling technique is fit for purpose—methodologically sound, ethically defensible, and practically achievable within the given constraints. Researchers must be transparent about their choices, acknowledging the trade-offs involved and the implications for interpreting the findings. By thoughtfully balancing theoretical ideals with practical realities, researchers can ensure that their sampling decisions enhance the credibility, relevance, and impact of their work.

2.7 Challenges in Sampling

Sampling is a cornerstone of research methodology, providing a means to draw conclusions about a larger population without having to study every individual within it. Whether in qualitative research, where sampling enables statistical generalisation, or in qualitative inquiry, where it supports in-depth exploration, the success of a study depends heavily on the appropriateness and quality of the sample. However, the process of selecting and securing an appropriate sample is rarely straightforward. Researchers face numerous challenges—methodological, logistical, ethical, and contextual—that can compromise the validity, reliability, and credibility of their findings. These challenges arise at various stages of the sampling process, from defining the target population and constructing a sampling frame to recruiting participants and maintaining engagement throughout the study. They are influenced by factors such as resource constraints, participant availability, cultural dynamics, and the sensitivity of the research topic. This essay examines the main challenges associated with sampling, illustrating how they manifest in different research contexts and discussing strategies for mitigating their impact [45].

1. **Defining the Target Population**: The first challenge in sampling is to clearly define the target population—the entire group to which the researcher intends to generalise or from which they wish to gather insights. While this might seem straightforward in theory, in practice it can be complicated by ambiguous boundaries, evolving study objectives, or incomplete population data. For instance, in a study of "small business owners," the researcher must determine whether to include only formally registered businesses or also informal traders and sole proprietors. Similarly, in public health research, defining who counts as a "smoker" may require decisions about frequency and duration of tobacco use. Poorly defined populations can lead to inappropriate sampling frames, participant selection errors, and findings that fail to answer the intended research questions.
2. **Constructing an Accurate Sampling Frame**: A sampling frame is the list or database from which the sample is drawn. The quality of this frame is a major determinant of the sample's representativeness. One of the most persistent challenges is that complete and up-to-date sampling frames are often unavailable,

especially for large or dispersed populations. In many developing countries, for example, census data may be outdated or incomplete, and official records may not capture informal or transient populations. Even in contexts with advanced data systems, privacy regulations may restrict access to certain lists, such as patient registries or customer databases. An inaccurate sampling frame can lead to *coverage error*, where some members of the target population have no chance of being included, while others may be overrepresented.

3. **Achieving Representativeness**: In probability sampling, representativeness is the goal, but achieving it in practice is difficult. Populations are often heterogeneous, with subgroups differing in ways that are relevant to the research question. Stratified sampling can help ensure proportional representation, but it requires detailed information about the population, which may not be available. Moreover, even if a sample is representative in theory, non-response can distort its composition. If certain groups—such as the elderly, non-English speakers, or low-income households—are less likely to participate, the sample may end up biased despite random selection. This *non-response bias* undermines the validity of statistical inferences and can be difficult to correct, even with weighting adjustments.
4. **Accessing Hard-to-Reach Populations**: Non-probability sampling is often used for populations that are difficult to identify, contact, or recruit. These include individuals engaged in stigmatised or illegal activities (such as sex workers or undocumented migrants), members of closed communities (such as religious sects), or those geographically dispersed (such as remote rural residents). The main challenge is gaining trust and overcoming barriers to access. Researchers may have to rely on intermediaries, gatekeepers, or snowball sampling to identify participants. While such methods can open doors, they risk producing samples that overrepresented certain subgroups—for example, those most closely connected to the initial contacts—and underrepresent others. Ethical considerations are also heightened, as revealing identities or even approaching certain individuals could expose them to harm.
5. **Balancing Sample Size with Resources**: Determining the appropriate sample size is another challenge. In qualitative research, statistical power analysis provides a guide, but achieving the calculated sample size may be impractical due to resource constraints. Large samples require more time, personnel, and funding for recruitment, data collection, and analysis. In qualitative research, the focus is on *data saturation*—the point at which additional participants no longer yield new insights. However, estimating when saturation will occur is not an exact science, and researchers risk stopping too early (resulting in incomplete findings) or too late (wasting resources on redundant data). Balancing the desire for comprehensiveness with the realities of budget and time is a constant tension in sampling.
6. **Dealing with Non-response and Attrition**: Non-response occurs when selected individuals refuse or fail to participate. In longitudinal studies, attrition—the loss of participants over time—adds another layer of complexity. Both can introduce bias if the non-respondents or dropouts differ systematically from

those who remain. For example, in a workplace satisfaction survey, employees with negative perceptions may be less inclined to participate, leading to an overly positive picture of organisational morale. In long-term health studies, participants in poorer health may be more likely to drop out, skewing the results toward healthier individuals. Mitigating these risks requires proactive follow-up, incentives, and flexible participation options, but these strategies can be costly and may still fall short of ensuring full participation.

7. **Avoiding Selection Bias**: Selection bias occurs when the method of choosing participants results in a sample that is systematically different from the population. This can arise in both probability and non-probability sampling. In convenience sampling, for example, recruiting only those who are easy to reach (such as students on a campus) may exclude important perspectives. In probability sampling, selection bias can occur if certain groups are inadvertently left out of the sampling frame or have lower probabilities of selection. This challenge is especially acute in online research, where recruitment often depends on digital access and literacy, thereby excluding those without reliable internet connectivity. Researchers must carefully evaluate their recruitment channels and adjust their methods to reduce bias wherever possible.
8. **Cultural and Contextual Barriers**: Sampling is never conducted in a social vacuum. Cultural norms, political contexts, and local attitudes toward research can all affect participant recruitment. In some communities, individuals may be reluctant to speak with outsiders, especially about sensitive topics. In others, participation may require approval from community leaders, elders, or other gatekeepers, adding complexity to the sampling process. Language barriers can also hinder recruitment, particularly when translation resources are limited. Misinterpretations of research aims, informed consent, or survey questions can result in reduced participation or inaccurate responses. In such contexts, researchers may need to adapt their sampling strategies, employ culturally competent fieldworkers, and invest time in community engagement before data collection begins.
9. **Ethics Challenges in Sampling**: Ethics and sampling are closely intertwined. Sampling decisions determine who will be exposed to the research process, and therefore to any potential risks. Ethical challenges include ensuring informed consent, protecting participant confidentiality, and avoiding exploitation of vulnerable groups. Snowball sampling, while effective for hidden populations, can compromise privacy if participants are identifiable within close-knit networks. Purposive sampling in health research may overburden certain individuals—such as patients with rare conditions—who are frequently recruited for studies. Researchers must balance the need for sufficient and relevant data with the obligation to minimise harm, respect autonomy, and distribute the burdens and benefits of research fairly.
10. **Maintaining Data Quality with Complex Sampling Designs**: Some research projects use multi-stage or mixed sampling designs to address practical and methodological needs. While these can improve feasibility and coverage, they also introduce complexity in data analysis. For example, cluster sampling

reduces travel costs but increases sampling error, requiring statistical adjustments in analysis. Stratified sampling can ensure subgroup representation, but only if the stratification variables are accurately measured and available for the entire population. Poor implementation of complex designs can result in compounded errors—such as clusters that are not truly representative of the strata they are meant to represent—leading to misleading conclusions.

11. **Adapting Sampling Plans in the Field**: Even the most carefully designed sampling plan may need to be adapted once fieldwork begins. Unexpected events—such as political unrest, natural disasters, or public health emergencies—can disrupt recruitment. Field realities may reveal that the population is more diverse or dispersed than anticipated, that access is more restricted, or that initial contact strategies are ineffective. While adaptability is a strength in research, frequent changes to sampling plans can undermine consistency and comparability. Any modifications must be well-documented, justified, and, if possible, incorporated into revised analytical frameworks to maintain transparency and credibility.

Sampling is both an art and a science, demanding not only technical knowledge of statistical and qualitative methods but also practical skill in navigating the complexities of real-world research. Challenges such as defining populations, constructing sampling frames, achieving representativeness, accessing hard-to-reach groups, managing non-response, avoiding bias, and addressing ethical concerns are common to studies across disciplines.

While these challenges can never be entirely eliminated, they can be mitigated through thoughtful planning, methodological flexibility, and ongoing reflexivity during the research process. This includes piloting recruitment strategies, engaging with communities, using mixed sampling approaches when appropriate, and maintaining transparency about limitations. By anticipating and responding to the inherent difficulties in sampling, researchers can enhance the credibility, relevance, and ethical integrity of their work—ultimately ensuring that their findings provide meaningful contributions to knowledge and practice.

References

1. Mudzielwana, N.P.: The role of parents in developing reading skills of their children in the foundation phase. J. Soc. Sci. **41**(2), 253–264 (2014)
2. Ntuli, L.T., Mahlangu, V.P.: Legal execution of curriculum in primary schools: school management teams' perspectives. Res. Educ. Policy Manag. **5**(2), 177–206 (2023). https://doi.org/10.46303/repam.2023.16
3. Gledson, B., Greenwood, D.: The adoption of 4D BIM in the UK construction industry: an innovation diffusion approach. Eng. Constr. Archit. Manag. **24**(6), 950–967 (2017). https://doi.org/10.1108/ECAM-03-2016-0066
4. Sudaryanto, S., Suroso, I., Pansiri, J., Umama, T.L., Hanim, A.: Impact of culture, brand image and price on buying decisions: evidence from East Java, Indonesia. Innov. Mark.**17**(1), 130–142 (2021). https://doi.org/10.21511/im.17(1).2021.11

5. Makasi, A., Govender, K., Munyoro, T.: The effects of corporate social responsibility (CSR) on corporate brand positioning. Mediterr. J. Soc. Sci. **5**(20), 2597–2607 (2020). https://doi.org/10.5901/mjss.2014.v5n20p2597
6. Zaidi, Z., Shukri, S.: The effects of digital marketing implementation on online consumer in Selangor during COVID-19 pandemic. Manag. Sci. Lett. **12**(1), 43–50 (2022). https://doi.org/10.5267/j.msl.2021.8.002
7. Gebeyanesh, M.: Mother tongue interference into learning English as a foreign language: analysis of Afan Oromo interference into learning EFL, orthography and pronunciation (A case of Batu Secondary School). Int. Peer-Rev. J.**26**, 95 (2016) [Online]. Available: www.iiste.org
8. Osiesi, M.P., Taiwo Sanni, K., Oluwatoyin, F.O.: Investigating the forms, causes, effects, and solutions to challenging behaviours among public primary schools in Ibadan South West, Oyo State, Nigeria. Education 3-13 Int. J. Primary Elem. Early Years Educ. **51**(4), 636–646 (2023). https://doi.org/10.1080/03004279.2021.1995459
9. Magasi, C.: The role of transformational leadership on employee performance: a perspective of employee empowerment. Eur. J. Bus. Manag. Res. **6**(6), 21–28 (2021). https://doi.org/10.24018/ejbmr.2021.6.6.1137
10. Musungwini, S., Mugoniwa, B., Furusa, S.S., Rebanowako, T.G.: An analysis of the use of cloud computing among university lecturers: a case study in Zimbabwe. Int. J. Educ. Dev. Using Inf. Commun. Technol. **12**(1), 53–70 (2016) [Online]. Available: http://files.eric.ed.gov/fulltext/EJ1099577.pdf
11. Hair, J.F.: A primer on partial least squares structural equation modeling (PLS-SEM) second edition. J. Chem. Inf. Model. **53**(9), 1689–1699 (2016). https://doi.org/10.1017/CBO9781107415324.004
12. George, A., Humphrey, O.O.: An examination of application of Vroom'S expectancy theory in the State Civil Service Commission South-South Nigeria. Br. J. Manag. Mark. Stud. **6**(11), 951–952 (2021)
13. Obi, K.C., Iwuanyanwu, O.: Innovation in the assessment of primary school mathematics: an approach to revamping Nigerian education. J. Teach. Perspect. **18**(1), 51–66 2023 [Online]. Available: http://repositorio.unan.edu.ni/2986/1/5624.pdf%0Ahttp://fiskal.kemenkeu.go.id/ejournal%0A10.1016/j.cirp.2016.06.001%0A10.1016/j.powtec.2016.12.055%0A10.1016/j.ijfatigue.2019.02.006%0A10.1
14. Chari, A., et al.: Dynamic capabilities for circular manufacturing supply chains—exploring the role of Industry 4.0 and resilience. Bus. Strateg. Environ. **31**(5), 2500–2517 (2022). https://doi.org/10.1002/bse.3040
15. Cluver, L., et al.: A parenting programme to prevent abuse of adolescents in South Africa: study protocol for a randomised controlled trial. Trials **17**(1), 1–10 (2016). https://doi.org/10.1186/s13063-016-1452-8
16. Chandra, R., Jatmika, S.: Unilever Surabaya corporate social responsibility (CSR) policy in maintaining environmental sustainability in Surabaya in 2014–2020. In: Proceedings of the International Conference on Public Organization (ICONPO 2021), vol. 209, pp. 444–452 (2022) [Online]. Available: https://www.atlantis-press.com/proceedings/iconpo-21/125970904
17. Cheah, C.W., Low, B., Lee, C.K.C.: Sustainable housing development: the legitimacy-seeking perspective. J. Bus. Ind. Mark. **36**(6), 1027–1041 (2020). https://doi.org/10.1108/JBIM-07-2020-0318
18. Artemova, A.: Engaging generation Z through social media marketing. Tour. Manag. **2**(6), 1–130 (2018) [Online]. Available: https://www.theseus.fi/bitstream/handle/10024/142658/Artemova_Alexandra.pdf?sequence=1&isAllowed=y
19. Oyedijo, A., Francois Koukpaki, A.S., Kusi-Sarpong, S., Alfarsi, F., Yang, Y.: Restraining forces and drivers of supply chain collaboration: evidence from an emerging market. Supply Chain Manag.**27**(3), 409–430 (2022). https://doi.org/10.1108/SCM-06-2020-0263
20. Wu, C.W., Monfort, A.: Role of artificial intelligence in marketing strategies and performance. Psychol. Mark. **40**(3), 484–496 (2023). https://doi.org/10.1002/mar.21737

21. Lister, K., Seale, J., Douce, C.: Mental health in distance learning: a taxonomy of barriers and enablers to student mental wellbeing. Open Learn. **38**(2), 102–116 (2023). https://doi.org/10.1080/02680513.2021.1899907
22. Arije, O., Madan, J., Hlungwani, T.: Quality of sexual and reproductive health services for adolescents and young people in public health facilities in Southwest Nigeria: a mystery client study. Glob. Health Action **15**(1) (2022). https://doi.org/10.1080/16549716.2022.2145690
23. Wahidmurni: Wordiness in academic writing: a systematic scoping review. Res. Result. Theor. Appl. Linguist. **10**(1), 2588–2593 (2024). https://doi.org/10.18413/2313
24. Maila, B., Paul, R., Mayimbo, S., Kabwita, K.: Prevalence and correlates of antenatal depression at Chelstone First Level Hospital in Lusaka, Zambia: a cross-sectional study. Afr. Health Sci. **22**(4), 334–347 (2022). https://doi.org/10.4314/ahs.v22i4.39
25. Al Rawajbeh, M., et al.: A new model for security analysis of network anomalies for IoT devices. Int. J. Data Netw. Sci. **7**(3), 1241–1248 (2023). https://doi.org/10.5267/j.ijdns.2023.5.001
26. Hampton, N., Baig, Z., Zeadally, S.: Ransomware behavioural analysis on windows platforms. J. Inf. Secur. Appl. **40**, 44–51 (2018). https://doi.org/10.1016/j.jisa.2018.02.008
27. Malinka, K., Hujnak, O., Hanacek, P., Hellebrandt, L.: E-banking security study-10 years later. IEEE Access **10**, 16681–16699 (2022). https://doi.org/10.1109/ACCESS.2022.3149475
28. Oba-Adenuga, O.A., Asikhia, O.U., Akpa, V.O.: Self-management and employee research output in selected private universities in Ogun State, Nigeria. Eur. J. Hum. Resour. Manag. Stud. **5**(4), 180–196 (2022). https://doi.org/10.46827/ejhrms.v5i4.1277
29. Nurdiansah, D., Dharmawan, I.W.E.: Spatial and temporal analysis for mangrove community healthiness in Liki Island, Papua-Indonesia. IOP Conf. Ser. Earth Environ. Sci. **944**(1) (2021). https://doi.org/10.1088/1755-1315/944/1/012017
30. Al-Rahmi, W.M., Alzahrani, A.I., Yahaya, N., Alalwan, N., Bin Kamin, Y.: Digital communication: information and communication technology (ICT) usage for education sustainability. Sustainability **12**(12), 1–18 (2020). https://doi.org/10.3390/su12125052
31. Mbon, U.F., et al.: Embracing diversity in the educational landscape: resource supply and inclusive education in secondary schools. J. Educ. Soc. Res. **13**(5), 155–169 (2023). https://doi.org/10.36941/jesr-2023-0128
32. Nasila, M., Cloete, C.: Adoption of building information modelling in the construction industry in Kenya design quality is influenced by the number of. Acta Structilia **25**(2), 1–38 (2018)
33. Olayinka, O.M.: Corporate governance and economic sustainability reporting in Nigeria. J. Account. Tax. **13**(4), 243–254 (2021). https://doi.org/10.5897/jat2021.0478
34. Musonda, I., Gambo, N.: Mediation effect of partnership on procurement strategy factors influencing sustainable smart housing development, Nigeria. Built Environ. Proj. Asset Manag. **11**(3), 454–467 (2020). https://doi.org/10.1108/BEPAM-12-2019-0139
35. Sonko, M.L., Akinlabi, H.B.: Inventory management and profitability of food and beverage manufacturing companies in Lagos State, Nigeria. IOSR J. Bus. Manag. **22**(5), 10–18 (2020). https://doi.org/10.9790/487X-2205051018
36. Ruano, P., et al.: Corporate Social Responsibility as the Organization's Commitment Against Stakeholders, p. 13. Intech (2016) [Online]. Available: https://www.intechopen.com/books/advanced-biometric-technologies/liveness-detection-in-biometrics
37. Jamali, A.R., Bhutto, A., Khaskhely, M., Sethar, W.: Impact of leadership styles on faculty performance: moderating role of organizational culture in higher education. Manag. Sci. Lett. **12**(1), 1–20 (2022). https://doi.org/10.5267/j.msl.2021.8.005
38. Soomro, Y.A., Baeshen, Y., Alfarshouty, F., Kaimkhani, S.A., Bhutto, M.Y.: The impact of Guerrilla marketing on brand image: evidence from millennial consumers in Pakistan. J. Asian Financ. Econ. Bus. **8**(4), 917–928 (2021). https://doi.org/10.13106/jafeb.2021.vol8.no4.0917
39. Kanski, L., Pizon, J.: The impact of selected components of Industry 4.0 on project management. J. Innov. Knowl. **8**(1), 100336 (2023). https://doi.org/10.1016/j.jik.2023.100336
40. Asvial, M., Admaja, A.F.S., Laagu, M.A.: Non-cooperative game leach for cluster distribution routing method on wireless sensor network (WSN). J. Commun. **18**(3), 198–206 (2023). https://doi.org/10.12720/jcm.18.3.198-206

41. Caraka, R.E., et al.: The impact of social media influencers Raffi Ahmad and Nagita Slavina on tourism visit intentions across millennials and zoomers using a hierarchical likelihood structural equation model. Sustainability **14**(1), 1–28 (2022). https://doi.org/10.3390/su14010524
42. Tjebane, M.M., Musonda, I., Okoro, C.: Organisational factors of artificial intelligence adoption in the South African construction industry. Front. Built Environ. **8** (2022). https://doi.org/10.3389/fbuil.2022.823998
43. Chovancova, M., Attor, C., Jibril, A.B.: Does brand personality mediate the link between social media usage and customer buying decisions on telecommunication's products and services? Evidence from Ghana. Innov. Mark. **18**(3), 84–98 (2022). https://doi.org/10.21511/im.18(3).2022.08
44. Naz, F., Agrawal, R., Luthra, S., Kumar, A.: Reviewing the applications of artificial intelligence in sustainable supply chains: exploring research propositions for future directions. Bus. Strateg. Environ. **31**(5), 2400–2423 (2022). https://doi.org/10.1002/bse.3034
45. Olatoye, T.A.: Analysis of modal choice of residents in Lagos State. J. Hum. Ecol. **65**(1–3) (2019). https://doi.org/10.31901/24566608.2019/65.1-3.3157

Chapter 3
Data Collection Methods in Qualitative Research

3.1 Introduction

Qualitative research is an approach that seeks to explore and understand human experiences, behaviours, and social phenomena from the perspective of those involved. Unlike qualitative research, which relies on numerical data and statistical analysis, qualitative research focuses on rich, detailed descriptions and interpretations of events, interactions, and meanings. The primary aim is to capture the depth and complexity of human life rather than to generalise findings to larger populations. The choice of data collection methods in qualitative research is therefore critical, as it influences the credibility, richness, and authenticity of the insights generated. Data collection in qualitative research is more than a technical process; it is a socially embedded activity that requires careful consideration of context, relationships, and ethics. The researcher is often the primary instrument for gathering data, engaging directly with participants and interpreting their words, behaviours, and environments. Consequently, qualitative data collection methods must be flexible enough to adapt to the unfolding nature of the research process while still maintaining methodological rigour. This section examines the key data collection methods used in qualitative research, including interviews, focus groups, observation, document analysis, case studies, and visual/creative methods. It will explore how each method functions, its strengths and limitations, and the considerations involved in selecting and applying it. The discussion will also highlight ethical concerns, technological advancements, and the integration of multiple methods through triangulation [1–3].

1. **Interviews**: Interviews are perhaps the most widely recognised and utilised method in qualitative research. They involve a purposeful conversation between the researcher and the participant, intending to explore the participant's perspectives, experiences, and interpretations in depth. Qualitative interviews are typically semi-structured or unstructured, allowing the discussion to flow naturally while still covering key topics of interest. Semi-structured interviews involve the use of an interview guide that outlines the main themes or questions to be

F. A. Alaba et al., *Fundamentals of Qualitative Research*, Studies in Systems, Decision and Control 663, https://doi.org/10.1007/978-3-032-18427-6_3

addressed, but the researcher remains open to new topics that emerge during the conversation. This balance of structure and flexibility allows for both comparability across participants and the capture of unexpected insights. Unstructured interviews, on the other hand, resemble informal conversations, with minimal preplanning, and are particularly valuable when exploring new or sensitive topics where participants' framing of the issue is important. The advantages of interviews include their ability to produce detailed, nuanced data and to clarify meanings through follow-up questions. They also allow for rapport to be built between researcher and participant, fostering trust and openness. However, interviews are time-intensive to conduct and analyse, and they rely heavily on the skills of the interviewer to probe effectively without leading responses. Additionally, the presence of the researcher may influence what participants are willing to share, raising concerns about social desirability bias.

2. **Focus Groups**: Focus groups bring together a small group of participants to discuss a specific topic, guided by a facilitator. This method is particularly valuable for exploring collective views, group dynamics, and the ways in which individuals construct meaning in interaction with others. The group setting can encourage participants to build on each other's ideas, challenge perspectives, and reveal shared cultural understandings. A typical focus group consists of six to twelve participants, though the exact size can vary depending on the research objectives. The facilitator plays a critical role in managing the discussion, ensuring that all voices are heard, and maintaining focus on the research topic. The conversation is often recorded and transcribed for later analysis. The strengths of focus groups lie in their efficiency—data from multiple participants can be collected simultaneously—and their ability to stimulate discussion that might not arise in individual interviews. They can also generate insights into social norms and collective decision-making processes. However, focus groups can be affected by dominant participants who overshadow others, and some individuals may feel reluctant to share openly in a group setting, particularly on sensitive topics. Furthermore, managing group dynamics requires considerable skill, and logistical challenges such as scheduling and securing an appropriate venue can be significant.
3. **Observation**: Observation involves systematically watching and recording behaviours, events, and interactions in their natural settings. In qualitative research, observation can be participant or non-participant. Participant observation means the researcher actively engages with the setting and the activities being studied, often over an extended period, in order to develop an insider's perspective. Non-participant observation involves the researcher maintaining a more detached role, observing without direct involvement. Observation is invaluable for capturing aspects of social life that may be difficult for participants to articulate, such as non-verbal communication, spatial arrangements, and tacit cultural practices. It is often used in ethnographic research, where the aim is to provide a holistic account of a community or organisation. The method's strengths include its ability to provide contextual richness and to triangulate self-reported data with observed behaviours. However, observation can be time-consuming and

may raise ethical concerns, particularly when participants are unaware they are being observed or when sensitive settings are involved. Observer bias—where the researcher's expectations shape what is noticed and recorded—is another challenge that must be addressed through reflexivity and careful documentation.

4. **Document and Artefact Analysis**: Documents, records, and artefacts can serve as valuable sources of qualitative data, offering insights into the historical, cultural, or organisational contexts of the research topic. These may include written texts (e.g., policy documents, letters, diaries), visual materials (e.g., photographs, artwork), or digital content (e.g., social media posts, blogs). Document analysis involves systematically reviewing and interpreting such materials to identify patterns, themes, and meanings. This method is particularly useful in case studies, historical research, and discourse analysis, where the aim is to understand how reality is constructed and represented through texts and other symbolic forms. The advantages of document analysis include its unobtrusive nature—researchers can gather data without directly engaging participants—and its potential to provide longitudinal perspectives. However, documents are often produced for purposes other than research, meaning they may be incomplete, biased, or limited in scope. Researchers must therefore critically assess the authenticity, credibility, and representativeness of each source.
5. **Case Study Method**: Case studies involve the in-depth examination of a single case or a small number of cases within their real-life contexts. A "case" might be an individual, an organisation, an event, or a community. Case study research often integrates multiple data collection methods, such as interviews, observations, and document analysis, to develop a comprehensive understanding of the case. The strength of the case study method lies in its ability to provide detailed, context-rich insights and to capture the complexity of social phenomena. It allows for the exploration of processes and causal mechanisms in a way that is difficult to achieve with other methods. However, because case studies focus on specific instances, their findings are not easily generalisable to broader populations. Additionally, the depth of data generated can be challenging to manage and analyse.
6. **Visual and Creative Methods**: Visual and creative methods, such as photo-elicitation, participatory video, mapping, and storytelling, are increasingly used in qualitative research to engage participants in new ways and to capture aspects of experience that may be difficult to express verbally. For example, in photo-elicitation, participants are shown images or asked to take photographs themselves, which then serve as prompts for discussion. These methods can be particularly valuable when working with children, marginalised communities, or cross-cultural settings where verbal communication may be limited. They also encourage participants to take an active role in the research process, aligning with participatory and emancipatory approaches. However, visual methods raise specific ethical considerations, such as consent for the use of images and the potential for unintended exposure of sensitive information.

7. **Mixed-Method and Triangulation Approaches**: Many qualitative researchers choose to combine multiple data collection methods within a single study, a strategy known as triangulation. Triangulation enhances the credibility and depth of findings by allowing different sources and types of data to corroborate or challenge each other. For example, interviews might be complemented by observation, enabling the researcher to compare what participants say with what they do. While triangulation can provide a more holistic understanding of the research topic, it also increases the complexity of data collection and analysis. Researchers must carefully plan how different methods will be integrated and ensure that the resulting data remain coherent and manageable.

Ethical Considerations in Qualitative Data Collection

Ethics is integral to qualitative data collection, as the process often involves personal interactions, sensitive topics, and the sharing of private information. Key ethical principles include informed consent, confidentiality, and respect for participants' autonomy and well-being. Informed consent requires that participants understand the nature of the research, what their involvement entails, and their right to withdraw at any time. Confidentiality involves safeguarding participants' identities and personal information, which may require anonymising transcripts and removing identifying details from visual materials. Researchers must also consider the potential impact of the research on participants, especially when discussing traumatic experiences or controversial topics. Reflexivity—ongoing critical reflection on the researcher's role, assumptions, and potential influence on the research—is essential for maintaining ethical and methodological integrity.

Advances in digital technology have expanded the range of tools available for qualitative data collection. Online interviews and focus groups, conducted via video conferencing platforms, offer opportunities to reach geographically dispersed participants. Mobile devices can be used for real-time data capture, including audio recordings, photographs, and geotagging. Social media platforms provide rich sources of naturally occurring data, although their use raises complex ethical and privacy issues. While digital methods can increase efficiency and inclusivity, they also introduce challenges related to access, digital literacy, and the quality of online interactions. Researchers must weigh the benefits against potential drawbacks and ensure that technological choices align with the study's goals and participant needs [4].

Thus, data collection in qualitative research is a dynamic and relational process, grounded in the principles of depth, context, and meaning. The methods available—ranging from interviews and focus groups to observation, document analysis, and creative approaches—offer diverse pathways for exploring human experiences and social realities. Each method has distinct strengths and limitations, and the choice of method should be guided by the research question, the nature of the phenomenon under study, the characteristics of participants, and practical and ethical considerations. In contemporary practice, qualitative researchers increasingly employ a combination of methods to enrich their understanding and to strengthen the credibility of their findings through triangulation. At the same time, technological innovations are reshaping the possibilities for data collection, offering both new opportunities and new challenges. Ultimately, the effectiveness of any qualitative data collection method depends not only on its technical execution but also on the researcher's reflexivity, ethical sensitivity, and commitment to capturing the voices and perspectives of those whose lives are being studied.

3.1.1 Surveys and Questionnaires

Surveys and questionnaires are among the most widely used data collection tools in research across social sciences, education, health, business, and other fields. They provide a systematic way of gathering information from individuals or groups, enabling researchers to collect qualitative, qualitative, or mixed-method data. While the two terms are often used interchangeably, there are subtle distinctions. A survey is generally a broader research process aimed at collecting data from a sample population to generalize findings, while a questionnaire refers specifically to the set of written questions used to gather the data. This distinction is important because a survey may involve additional methods, such as interviews or observations, whereas a questionnaire is strictly a structured instrument for recording responses. The popularity of surveys and questionnaires in research stems from their ability to capture a wide range of information quickly, efficiently, and often at relatively low cost. They are adaptable to various formats, including paper-based delivery, telephone administration, online distribution, and face-to-face sessions. Their versatility allows them to be applied in descriptive, exploratory, and explanatory research designs [5, 6].

3.1.2 Conceptual Foundations of Surveys and Questionnaires

Surveys are rooted in the principles of sampling and measurement. They rely on obtaining information from a representative group of people to infer conclusions about a larger population. This is where the quality of the sampling technique directly influences the reliability and validity of the data. A questionnaire, on the other hand, serves as the operational tool within a survey, transforming abstract research questions into measurable items. Each question or statement is designed to elicit a specific type of information, whether factual, attitudinal, behavioural, or perceptual. The conceptual underpinning of survey research is that people's responses to well-constructed questions can serve as valid indicators of their knowledge, attitudes, experiences, or characteristics. However, this assumption holds only if the questionnaire is carefully designed to minimize bias, ambiguity, and misinterpretation.

3.1.3 Designing Effective Questionnaires

The design process begins with clearly defining the research objectives. Without a precise understanding of the goals, the questionnaire can become unfocused, leading to data that is irrelevant or difficult to analyse. Once the objectives are set, the researcher must determine the type of data needed—whether qualitative (numerical and measurable) or qualitative (descriptive and explanatory). Qualitative data often require closed-ended questions with predetermined response options, while qualitative data may be better captured through open-ended questions that allow respondents to express themselves freely. The wording of questions is a critical consideration. Questions should be clear, concise, and free from leading or loaded language. Technical terms and jargon should be avoided unless the target audience is expected to understand them. In addition, questions should be neutral in tone to avoid influencing respondents' answers. For example, asking "How much do you agree that the new policy is beneficial?" presupposes that the policy is beneficial, whereas a neutral phrasing would be "What is your opinion on the new policy?" Question sequence also plays a significant role in ensuring high-quality data. A logical progression from general to specific questions helps maintain respondent engagement and reduces confusion. Sensitive or potentially intrusive questions are best placed toward the end of the questionnaire, after rapport has been established. The researcher must also decide whether to include demographic questions, such as age, gender, income, or education, which can be useful for segmenting and analysing the data. Another important aspect is the choice of measurement scales. Likert scales, semantic differential scales, ranking scales, and multiple-choice formats are commonly used in qualitative surveys. Each scale type has implications for data analysis and interpretation. For example, Likert scales provide ordinal data that can be analysed for patterns of agreement or disagreement, while semantic differential scales measure attitudes along a continuum between two opposing adjectives [7].

3.1.4 Modes of Administration

Surveys and questionnaires can be administered through various modes, each with its strengths and weaknesses. Paper-based questionnaires remain useful in contexts where digital access is limited or literacy levels vary. Online surveys, delivered via email or web-based platforms, offer the advantages of automation, rapid distribution, and easy data compilation. Telephone surveys can reach a wide geographical area but may suffer from declining response rates in the era of mobile communication. Face-to-face surveys provide opportunities for clarification and rapport building, although they can be costly and time-consuming. The choice of administration mode should consider the characteristics of the target population, budget constraints, time limitations, and the complexity of the questions. For instance, if the survey involves complex instructions or visual aids, face-to-face administration may be more effective. Conversely, if anonymity is important for eliciting honest responses on sensitive topics, online or self-administered questionnaires may be preferable [8, 9].

Advantages of Surveys and Questionnaires

1. One of the major strengths of surveys and questionnaires is their scalability.
2. They allow researchers to collect data from large populations within a relatively short period.
3. Standardization ensures that every respondent receives the same questions in the same order, enhancing the comparability of responses.
4. Surveys also provide a cost-effective way to reach geographically dispersed participants, especially when conducted online.
5. Moreover, questionnaires can be designed to capture a wide range of data types, from factual information such as employment history to subjective attitudes like job satisfaction.
6. They also facilitate anonymity, which can increase the likelihood of honest responses, particularly in studies dealing with sensitive subjects.

Limitations and Challenges

Despite their strengths, surveys and questionnaires are not without limitations.

1. One major challenge is non-response bias, which occurs when the individuals who choose not to participate differ systematically from those who do. This can lead to skewed results that do not accurately represent the population.
2. Poorly designed questions can introduce measurement error, resulting in data that is unreliable or invalid.
3. Another issue is the potential for social desirability bias, where respondents provide answers they believe are more acceptable or favourable rather than what they truly think or do.
4. Additionally, questionnaires are often limited by the respondents' memory, honesty, and comprehension skills.
5. In qualitative surveys, closed-ended questions can restrict the depth of responses, while open-ended questions may lead to difficulties in coding and analysing the data.

Ethical Considerations

Ethical principles must guide the use of surveys and questionnaires. Informed consent is essential, meaning respondents should understand the purpose of the research, what their participation involves, and how their data will be used. Confidentiality must be maintained, and personal identifiers should be separated from response data whenever possible. Researchers should also be sensitive to cultural and contextual factors that may affect how questions are perceived. Surveys and questionnaires are used in a variety of disciplines. In education, they can be employed to assess student satisfaction, teaching effectiveness, or curriculum relevance. In healthcare, they help evaluate patient experiences, health behaviours, and public health interventions. In business, they are used to measure customer satisfaction, employee engagement, and market trends. In social science research, surveys are a cornerstone method for studying attitudes, values, and social behaviours.

Hence, surveys and questionnaires remain indispensable tools for collecting data in both academic and applied research. Their utility lies in their adaptability, scalability, and ability to generate standardized data that can be analysed for patterns and trends. However, their effectiveness depends on careful planning, thoughtful design, and ethical implementation. Researchers must pay close attention to question wording, sequencing, and mode of administration to ensure valid and reliable results. While technological advances have expanded the possibilities for survey delivery, the core principles of clarity, neutrality, and respondent respect remain central to successful use. By understanding both the potential and the limitations of surveys and questionnaires, researchers can harness these tools to produce meaningful insights that inform decision-making and contribute to knowledge in their respective fields [10].

3.2 The Purpose of Surveys and Questionnaires

Surveys and questionnaires are fundamental tools in research, particularly in social sciences, education, marketing, and public health. They serve the critical function of systematically collecting information from individuals or groups to understand attitudes, opinions, behaviours, experiences, or demographic characteristics. The purpose of using these tools extends beyond merely gathering data; they are designed to facilitate structured inquiry that can lead to reliable, valid, and generalizable findings. This section explores the multiple purposes of surveys and questionnaires, considering their role in qualitative and qualitative research, their contribution to decision-making, and their practical applications across different sectors. One of the central purposes of surveys and questionnaires is to collect standardized data from a target population. Standardization ensures that each respondent receives the same set of questions in the same order, minimizing bias and allowing the researcher to compare responses across participants. This uniformity is essential for ensuring reliability and enabling statistical analysis. For example, a customer satisfaction survey administered to all clients of a service provider ensures that responses can be aggregated and interpreted meaningfully. By asking everyone the same questions, researchers can measure the prevalence of certain opinions or behaviours in the population and identify patterns that might otherwise go unnoticed.

Another key purpose is to obtain data that is representative of a population. In probability-based survey designs, researchers aim to generalize findings from a sample to a wider population. Questionnaires, when combined with appropriate sampling techniques, provide a practical means of reaching a large number of individuals efficiently. This is particularly relevant in national censuses, opinion polls, and market research, where the goal is to capture the perspectives or characteristics of diverse and geographically dispersed groups. Representation ensures that the conclusions drawn from survey data are applicable beyond the immediate group of respondents, thus enhancing the external validity of the research. Surveys and questionnaires are also crucial for measuring subjective constructs such as attitudes, perceptions, and satisfaction. Unlike observable behaviours, these constructs cannot be directly measured; they require indirect measurement through well-designed questions. Psychometric questionnaires, for example, measure traits like job satisfaction, self-esteem, or stress levels by asking respondents to indicate their agreement with a series of statements. By using established scales and question formats, researchers can quantify subjective phenomena, making it possible to compare them across groups or over time. This purpose is especially valuable in organizational development, public policy evaluation, and health psychology, where understanding how people feel or think can inform interventions [11].

Furthermore, surveys and questionnaires are often used to explore relationships between variables. Through careful question design, researchers can gather data on multiple factors and analyze how they relate to each other. For instance, a workplace engagement survey might include questions on employee motivation, training opportunities, and job satisfaction, enabling analysis of how these variables are connected. This purpose is fundamental for hypothesis testing, predictive modeling, and the development of theories in social research. By revealing correlations or causal links, surveys contribute to the advancement of knowledge in various disciplines. In addition, surveys and questionnaires support decision-making in policy and practice. Data collected through these tools provides evidence for managers, policymakers, and organizational leaders to make informed choices. For example, a public health department might use survey data to identify vaccination rates and public attitudes toward immunization, guiding targeted outreach efforts. Similarly, universities may use student satisfaction surveys to improve curriculum design, teaching methods, and campus services. In business, customer feedback questionnaires help companies tailor products and services to meet consumer needs, enhancing competitiveness and profitability.

Another important purpose is to monitor changes over time. Longitudinal surveys track the same variables or respondents across multiple points in time, enabling the study of trends, progress, or the impact of interventions. For example, an annual employee climate survey allows organizations to assess whether initiatives to improve workplace culture are having the desired effect. By comparing results over years, researchers can detect subtle shifts in attitudes, behaviours, or conditions that might otherwise remain hidden. This tracking function is especially valuable in program evaluation, policy assessment, and social change research. Surveys and questionnaires are also widely used to gather qualitative insights when they include open-ended questions. Although they are more often associated with qualitative research, they can also capture rich, descriptive data about participants' experiences and perspectives. Open-ended questions allow respondents to answer in their own words, providing depth and context that closed questions cannot achieve. This qualitative dimension is particularly useful in exploratory studies, needs assessments, and participatory research, where understanding the reasoning behind responses is as important as the responses themselves. A further purpose is to facilitate cost-effective and scalable data collection. Compared to methods like interviews or focus groups, surveys and questionnaires—especially when administered online—allow researchers to reach large samples at a fraction of the cost and time. This efficiency makes them accessible to organizations with limited resources and enables rapid collection of data for timely decision-making. The scalability of these tools is one reason they are so common in large-scale research projects, marketing studies, and institutional assessments [12].

Finally, surveys and questionnaires serve the purpose of documenting and recording information systematically. They create a written or digital record of responses that can be stored, retrieved, and reanalyzed as needed. This archival function is valuable for accountability, transparency, and future research. For example, a non-profit organization conducting an impact assessment can keep survey data on file to demonstrate outcomes to funders or to revisit results when designing new programs. In conclusion, the purpose of surveys and questionnaires is multifaceted, encompassing the collection of standardized and representative data, the measurement of subjective constructs, the exploration of variable relationships, and the provision of evidence for decision-making. They play a central role in both qualitative and qualitative research, offering a practical, scalable, and cost-effective means of gathering information. Whether the aim is to track changes over time, capture individual perspectives, or support strategic planning, surveys and questionnaires remain indispensable tools for researchers, organizations, and policymakers alike. Their continued relevance lies in their adaptability to diverse research contexts and their ability to bridge the gap between data collection and actionable insights [13].

3.2.1 Measuring Attitudes, Opinions, or Behaviours

Surveys and questionnaires are fundamental tools in research, particularly in social sciences, education, marketing, and public health. They serve the critical function of systematically collecting information from individuals or groups to understand attitudes, opinions, behaviours, experiences, or demographic characteristics. The purpose of using these tools extends beyond merely gathering data; they are designed to facilitate structured inquiry that can lead to reliable, valid, and generalizable findings. This section explores the multiple purposes of surveys and questionnaires, considering their role in qualitative and qualitative research, their contribution to decision-making, and their practical applications across different sectors.

One of the central purposes of surveys and questionnaires is to collect standardized data from a target population. Standardization ensures that each respondent receives the same set of questions in the same order, minimizing bias and allowing the researcher to compare responses across participants. This uniformity is essential for ensuring reliability and enabling statistical analysis. For example, a customer satisfaction survey administered to all clients of a service provider ensures that responses can be aggregated and interpreted meaningfully. By asking everyone the same questions, researchers can measure the prevalence of certain opinions or behaviours in the population and identify patterns that might otherwise go unnoticed. Another key purpose is to obtain data that is representative of a population. In probability-based survey designs, researchers aim to generalize findings from a sample to a wider population. Questionnaires, when combined with appropriate sampling techniques, provide a practical means of reaching a large number of individuals efficiently. This is particularly relevant in national censuses, opinion polls, and market research, where the goal is to capture the perspectives or characteristics of diverse and geographically dispersed groups. Representation ensures that the conclusions drawn from survey data are applicable beyond the immediate group of respondents, thus enhancing the external validity of the research [14].

Surveys and questionnaires are also crucial for measuring subjective constructs such as attitudes, perceptions, and satisfaction. Unlike observable behaviours, these constructs cannot be directly measured; they require indirect measurement through well-designed questions. Psychometric questionnaires, for example, measure traits like job satisfaction, self-esteem, or stress levels by asking respondents to indicate their agreement with a series of statements. By using established scales and question formats, researchers can quantify subjective phenomena, making it possible to compare them across groups or over time. This purpose is especially valuable in organizational development, public policy evaluation, and health psychology, where understanding how people feel or think can inform interventions. Furthermore, surveys and questionnaires are often used to explore relationships between variables. Through careful question design, researchers can gather data on multiple factors and analyze how they relate to each other. For instance, a workplace engagement survey might include questions on employee motivation, training opportunities, and job satisfaction, enabling analysis of how these variables are connected. This purpose is fundamental for hypothesis testing, predictive modeling, and the development of theories in social research. By revealing correlations or causal links, surveys contribute to the advancement of knowledge in various disciplines. In addition, surveys and questionnaires support decision-making in policy and practice. Data collected through these tools provides evidence for managers, policymakers, and organizational leaders to make informed choices. For example, a public health department might use survey data to identify vaccination rates and public attitudes toward immunization, guiding targeted outreach efforts. Similarly, universities may use student satisfaction surveys to improve curriculum design, teaching methods, and campus services. In business, customer feedback questionnaires help companies tailor products and services to meet consumer needs, enhancing competitiveness and profitability [15].

Another important purpose is to monitor changes over time. Longitudinal surveys track the same variables or respondents across multiple points in time, enabling the study of trends, progress, or the impact of interventions. For example, an annual employee climate survey allows organizations to assess whether initiatives to improve workplace culture are having the desired effect. By comparing results over years, researchers can detect subtle shifts in attitudes, behaviours, or conditions that might otherwise remain hidden. This tracking function is especially valuable in program evaluation, policy assessment, and social change research.

Surveys and questionnaires are also widely used to gather qualitative insights when they include open-ended questions. Although they are more often associated with qualitative research, they can also capture rich, descriptive data about participants' experiences and perspectives. Open-ended questions allow respondents to answer in their own words, providing depth and context that closed questions cannot achieve. This qualitative dimension is particularly useful in exploratory studies, needs assessments, and participatory research, where understanding the reasoning behind responses is as important as the responses themselves. A further purpose is to facilitate cost-effective and scalable data collection. Compared to methods like interviews or focus groups, surveys and questionnaires—especially when administered online—allow researchers to reach large samples at a fraction of the cost and time. This efficiency makes them accessible to organizations with limited resources and enables rapid collection of data for timely decision-making. The scalability of these tools is one reason they are so common in large-scale research projects, marketing studies, and institutional assessments. Finally, surveys and questionnaires serve the purpose of documenting and recording information systematically. They create a written or digital record of responses that can be stored, retrieved, and reanalyzed as needed. This archival function is valuable for accountability, transparency, and future research. For example, a non-profit organization conducting an impact assessment can keep survey data on file to demonstrate outcomes to funders or to revisit results when designing new programs [16].

In summary, the purpose of surveys and questionnaires is multifaceted, encompassing the collection of standardized and representative data, the measurement of subjective constructs, the exploration of variable relationships, and the provision of evidence for decision-making. They play a central role in both qualitative and qualitative research, offering a practical, scalable, and cost-effective means of gathering information. Whether the aim is to track changes over time, capture individual perspectives, or support strategic planning, surveys and questionnaires remain indispensable tools for researchers, organizations, and policymakers alike. Their continued relevance lies in their adaptability to diverse research contexts and their ability to bridge the gap between data collection and actionable insights [17].

Overall, surveys and questionnaires offer a versatile approach to gathering information about human thought and action, but their effectiveness relies on methodological rigor and cultural sensitivity in question development and administration (Table 3.1).

Application Measuring Attitudes, Opinions, or Behaviours in Research

In research, understanding human perceptions, beliefs, and conduct is fundamental to generating meaningful insights that can inform policy, practice, and theory. Measuring attitudes, opinions, and behaviours plays a critical role in fields such as psychology, sociology, marketing, political science, and public health. These three dimensions—although interconnected—represent distinct yet overlapping constructs

Table 3.1 Comparison table of measuring attitudes, opinions, and behaviours

Aspect	Attitudes	Opinions	Behaviours
Definition	Enduring evaluations or feelings toward an object, idea, or issue	Personal viewpoints or preferences, often influenced by recent experiences	Observable actions or habits, either self-reported or observed
Nature	Deep-seated, relatively stable	More temporary and situational	Action-oriented, can be habitual or occasional
Example questions	"I feel satisfied with my current job role." (Likert scale)	"What is your view on the new company policy?"	"How often do you attend team meetings in a month?"
Measurement tools	Likert scales, semantic differentials, ranking	Multiple-choice, rating scales, open-ended items	Frequency counts, recall-based questions, self-report logs
Data type	Often ordinal or interval	Can be nominal, ordinal, or interval	Often ratio or interval
Usefulness	Understanding underlying motivations and predispositions	Gauging public sentiment and preferences	Tracking actual patterns of activity and habit formation
Potential limitation	May not always predict actual behaviour	Can change rapidly with new information	Self-reports may be inaccurate due to recall bias

that can be systematically examined through various qualitative and qualitative methods, particularly surveys and questionnaires [18, 19].

1. **Attitudes** refer to evaluative tendencies toward people, objects, events, or ideas, often reflecting underlying values and beliefs. They are generally stable over time but can change with new experiences or information. Researchers measure attitudes to determine how individuals feel about specific issues, such as their stance on climate change, their level of trust in institutions, or their perception of workplace culture. By quantifying attitudes, researchers can identify trends, segment populations, and predict future behaviours.
2. **Opinions**, in contrast, are explicit expressions or statements that individuals make about a given topic. While attitudes tend to be deeper and more enduring, opinions are often situational and may shift based on context, current events, or recent experiences. Measuring opinions allows researchers to capture real-time sentiments. For example, political scientists may study voter opinions during election periods to forecast electoral outcomes or evaluate public reception of policy proposals.
3. **Behaviours** involve the observable actions that individuals engage in, which may or may not align with their stated attitudes and opinions. Understanding behaviour is critical in applied research, as actions often have direct implications for policy and intervention. For instance, public health researchers may examine

patterns of exercise, dietary habits, or compliance with vaccination schedules to design targeted health campaigns.

The application of measuring these constructs in research is multifaceted. In market research, companies rely on consumer attitude assessments to tailor products and advertising strategies. Political campaigns use opinion polls to shape messaging and allocate resources effectively. Behavioural studies inform program evaluations, ensuring that initiatives lead to tangible outcomes. In public policy, combining attitude, opinion, and behavioural data allows decision-makers to craft interventions that are both socially acceptable and practically effective. Surveys and questionnaires are the most common tools for capturing these variables. They enable researchers to gather large amounts of data efficiently, apply statistical analysis, and compare results across time and demographics. Well-designed instruments can differentiate between what people say (opinions), how they feel (attitudes), and what they do (behaviours), providing a more complete understanding of the human dimension in research (Table 3.2).

Table 3.2 Comparison table of measuring attitudes, opinions, and behaviours in research

Aspect	Attitudes	Opinions	Behaviours
Definition	Enduring evaluations, feelings, or predispositions toward an object, idea, or person.	Explicit statements or expressed viewpoints on a specific topic or issue.	Observable actions or patterns of activity performed by individuals.
Nature	Often deep-rooted and relatively stable over time.	More flexible and can change quickly depending on circumstances.	Concrete and measurable in real-world settings.
Example	Believing renewable energy is essential for the future.	Saying "I support solar energy projects in my city."	Installing solar panels on one's home.
Measurement methods	Likert scales, semantic differential scales, Thurstone scales.	Structured survey questions, open-ended interviews, polls.	Observation, behavioural logs, self-reported activity questionnaires.
Research application	Predicting long-term tendencies or resistance to change.	Capturing short-term sentiment or reactions to current events.	Evaluating actual outcomes or program impacts.
Potential discrepancy	May not always match behaviours due to external constraints.	May shift depending on context or information received.	May differ from stated attitudes or opinions.

3.2.2 Gathering Descriptive Data

Gathering descriptive data is a fundamental component of both qualitative and qualitative research, aimed at capturing information that paints a detailed picture of the subject, phenomenon, or population under investigation. In research, descriptive data refers to information that explains "what is" without necessarily exploring "why" or "how" it occurs. It focuses on describing the characteristics, patterns, and trends that define a group, behaviour, or environment. By systematically collecting descriptive data, researchers provide a foundation upon which further analytical, interpretive, or experimental investigations can be built. In qualitative research, gathering descriptive data involves recording rich, narrative details about participants' experiences, perceptions, and contexts. The emphasis is on accuracy, completeness, and depth, ensuring that the nuances of human behaviour and social phenomena are preserved. For example, a study on workplace culture might gather descriptive data about the physical office setting, employee demographics, modes of communication, and patterns of social interaction. Such data can be obtained through observation, interviews, focus groups, or document analysis. In these cases, descriptive information is often textual, non-numerical, and designed to capture the complexity of real-life situations [20].

In qualitative research, gathering descriptive data often involves numerical summaries, such as frequencies, percentages, and averages, that help explain the distribution of characteristics within a sample or population. For instance, a researcher examining customer preferences might record age, gender, purchase frequency, and satisfaction ratings. These figures provide an overview of the sample and help identify patterns that warrant deeper analysis. However, even in qualitative contexts, descriptive data remains valuable for contextualizing results and ensuring that findings are interpreted within the right frame of reference. The process of gathering descriptive data typically begins with defining the scope and focus of the research. Clear objectives guide the choice of variables or themes to be described. In qualitative settings, researchers often create observation checklists or interview guides to ensure consistency across data collection sessions. For qualitative approaches, structured questionnaires and data recording templates help ensure uniformity in measurement.

Ethics also play an important role in gathering descriptive data. Researchers must respect participant privacy, obtain informed consent, and ensure that any identifying information is protected. This is especially crucial when descriptive data includes sensitive demographic details or personal narratives. The value of descriptive data lies in its ability to provide context, facilitate understanding, and lay the groundwork for further inquiry. It allows stakeholders, policymakers, or fellow researchers to see the "big picture" and make informed decisions. For example, in public health studies, descriptive data on disease prevalence and demographic patterns can inform targeted interventions. In education research, detailed accounts of classroom dynamics can shape policy and instructional strategies. In summary, gathering descriptive data is not merely about listing facts—it is about creating a detailed, structured, and accurate representation of the research subject. Whether expressed in words, numbers, or a combination of both, descriptive data ensures that research findings are grounded in a well-documented reality. Without it, analysis risks becoming disconnected from the actual conditions being studied, leading to conclusions that lack relevance or applicability [21] (Table 3.3).

This comparison makes clear that while descriptive data gathering exists in both qualitative and qualitative traditions, the epistemological foundations and methodological executions differ significantly. In qualitative studies, the researcher becomes part of the interpretive process, often engaging with participants to co-construct meaning. In qualitative studies, the emphasis is on objectivity, standardization, and replicability, with minimal researcher influence on the responses. By understanding these distinctions, researchers can make more informed methodological choices, ensuring that their approach to descriptive data gathering aligns with the overall objectives of their study and the type of knowledge they aim to produce.

Table 3.3 Comparison of how descriptive data gathering differs in qualitative versus qualitative research

Aspect	Qualitative research	Qualitative research
Primary purpose	To explore and understand meanings, experiences, and perspectives in depth	To measure and quantify variables, relationships, or patterns numerically
Nature of data	Non-numerical, textual, or visual (e.g., interview transcripts, observation notes, photos)	Numerical data (e.g., counts, percentages, statistical scores)
Data collection methods	In-depth interviews, focus groups, participant observation, document analysis	Structured surveys, standardized questionnaires, experiments, existing numerical datasets
Sampling approach	Often non-probability sampling (e.g., purposive, snowball) to select information-rich cases	Typically probability sampling (e.g., random, stratified) to ensure representativeness
Level of structure	Flexible and adaptive; questions can evolve during the study	Highly structured with pre-determined questions and fixed response options
Depth vs. breadth	Emphasizes depth and context of understanding	Emphasizes breadth and generalizability of findings
Type of descriptions	Rich, detailed narratives; thematic patterns	Summaries in the form of charts, tables, and statistical measures
Researcher's role	Researcher is closely involved, often interpreting meanings with participants	Researcher remains detached, focusing on objective measurement
Data analysis	Thematic, narrative, or content analysis	Statistical analysis (descriptive and inferential statistics)
Outcome	Contextualized insights that capture complexity of the subject	Generalizable results that allow prediction or correlation

3.2.3 Testing Hypotheses and Verifying Theories

Surveys and questionnaires play a crucial role in the scientific process, not only in generating data but also in providing evidence to test hypotheses and verify theoretical propositions. In research, a *hypothesis* is a specific, testable prediction about the relationship between variables, while a *theory* is a broader explanatory framework that seeks to describe, predict, and explain phenomena. The application of surveys and questionnaires to this purpose lies in their ability to systematically collect large amounts of structured information from diverse populations, enabling researchers to identify patterns, correlations, and causal relationships [22–24].

In **qualitative research**, testing hypotheses typically involves designing survey instruments with closed-ended questions that operationalize key concepts into measurable variables. For example, a study testing the hypothesis that "job satisfaction positively influences employee productivity" might use a standardized job satisfaction scale combined with self-reported productivity measures. The data collected would then undergo statistical analysis—such as correlation, regression, or ANOVA—to determine whether the predicted relationship holds true within the sample, and by extension, the broader population. Here, questionnaires provide precise, numerical data that allow for hypothesis confirmation or refutation with statistical confidence.

In **qualitative research**, while the concept of "hypothesis testing" is often more flexible, surveys and open-ended questionnaires can still be employed to verify emerging theoretical ideas. This process, sometimes called *theory confirmation*, may involve using qualitative questionnaires to explore whether themes identified in earlier studies are consistently observed across different contexts. For instance, in grounded theory research, after developing a preliminary model of how patients cope with chronic illness, researchers might use semi-structured questionnaires to gather additional perspectives that either reinforce or challenge the proposed model.

A critical advantage of surveys and questionnaires in this domain is their ability to be **replicated** across studies and settings, strengthening the validity of both hypotheses and theories. Replication enables researchers to verify whether findings remain consistent over time, across cultures, or under different conditions. For example, the "Technology Acceptance Model" (TAM) has been tested repeatedly using standardized survey items in multiple industries and countries, reinforcing its theoretical robustness. Moreover, when surveys are carefully designed to minimize bias and maximize validity—through strategies like pilot testing, random sampling, and balanced question phrasing—they produce reliable datasets that can withstand rigorous peer scrutiny. This credibility is essential in hypothesis testing, as unreliable measures can lead to false conclusions, undermining both empirical findings and theoretical frameworks.

In summary, surveys and questionnaires serve as powerful tools for bridging the gap between theoretical abstraction and empirical evidence. In qualitative contexts, they translate theoretical constructs into measurable variables that can be statistically tested. In qualitative contexts, they facilitate the refinement and verification of

Table 3.4 Compares the role and application of hypothesis testing and theory verification

Aspect	Qualitative approach	Qualitative approach
Purpose of hypothesis testing	Hypotheses often emerge during the study; focus on exploring concepts and generating theory	Hypotheses are predetermined; aim to confirm or reject them using statistical tests
Survey/questionnaire design	Open-ended questions allowing participants to elaborate on meanings and experiences	Structured questions with fixed-response options for measurable variables
Data nature	Descriptive, textual, and context-rich narratives	Numerical, quantifiable data amenable to statistical analysis
Theory verification	Theories are adapted, refined, or generated based on findings	Theories are tested for validity against empirical data
Analysis methods	Thematic analysis, content analysis, narrative analysis	Statistical analysis (e.g., t-tests, regression, ANOVA)
Outcome	In-depth understanding and theory development	Evidence for or against a hypothesis, enabling generalization

emerging theories by capturing rich, context-specific data. Ultimately, their value in hypothesis testing and theory verification lies in their adaptability, replicability, and ability to capture both breadth and depth of understanding.

Table 3.4 compares the role and application of hypothesis testing and theory verification when using surveys and questionnaires in qualitative versus qualitative research.

By integrating this comparative understanding, researchers can select the most suitable methodological stance when employing surveys and questionnaires for hypothesis testing. A qualitative design may be most appropriate where the aim is to measure and confirm relationships between variables in a statistically robust manner. Conversely, a qualitative design is ideal when the goal is to uncover nuanced perspectives, generate theoretical insights, and understand phenomena in their social or cultural context. Ultimately, whether adopting a qualitative or qualitative approach, the decision should be informed by the research objectives, the nature of the phenomenon under investigation, and the level of flexibility desired in exploring the underlying theoretical constructs [25].

3.2.4 Ethical Considerations in Survey Research

Survey research, whether conducted through questionnaires, interviews, or online forms, is a valuable method for collecting data in both qualitative and qualitative studies. However, its credibility, validity, and acceptability depend not only on methodological rigor but also on adherence to ethical principles. Ethical considerations in survey research ensure that participants' rights, dignity, and welfare are protected throughout the research process. These considerations address issues such as informed consent, privacy, confidentiality, avoidance of harm, transparency, and responsible data handling. Researchers must comply with ethical guidelines established by institutional review boards (IRBs), professional bodies, and legal frameworks such as data protection laws. One of the most important ethical responsibilities in survey research is obtaining **informed consent**. Participants should be clearly informed about the purpose of the study, the nature of the questions, the expected duration of their involvement, and any potential risks or discomforts. Consent must be given freely without coercion, and participants must understand that they have the right to withdraw at any time without penalty. In the context of online surveys, this often involves including a consent statement on the first page and requiring an affirmative action—such as ticking a box—before proceeding [24].

Closely linked to informed consent is **confidentiality and privacy**. Researchers must ensure that personal identifiers are either removed or securely stored to prevent unauthorized access. When surveys involve sensitive topics—such as political beliefs, health status, or personal experiences—data must be anonymized or pseudonymized to protect respondents' identities. In addition, ethical survey research requires transparency in how data will be stored, used, and shared. For example, specifying that responses will only be used for academic purposes or shared in aggregate form builds trust and aligns with ethical standards. Another key consideration is the **avoidance of harm**. While surveys may seem low risk compared to experimental studies, certain questions can cause emotional distress, embarrassment, or anxiety. Researchers should review their questionnaires carefully to ensure that questions are phrased respectfully, culturally sensitively, and without bias. Providing participants with the option to skip uncomfortable questions or offering contact details for support services can mitigate potential harm.

Ethical practice also involves **honesty and integrity** in reporting and analysis. Researchers must avoid fabricating, falsifying, or selectively reporting data. Additionally, they should acknowledge any limitations of the survey method and ensure that results are presented in a way that avoids misrepresentation. Misleading interpretations can damage public trust and undermine the credibility of the research community. In digital survey research, **data protection and cybersecurity** are increasingly important. With the rise of online survey tools and cloud-based data storage, researchers must comply with relevant data protection regulations such as the EU's General Data Protection Regulation (GDPR) or Nigeria's Data Protection Regulation (NDPR). Measures such as encryption, password protection, and secure servers should be standard practice to safeguard participant data. Finally, ethical considerations extend beyond the data collection phase to the dissemination of findings. Participants should be informed, where possible, about how they can access the study's results. Sharing findings in a transparent, open-access format—while safeguarding privacy—demonstrates respect for participants' contributions and reinforces ethical responsibility. In conclusion, ethical considerations in survey research are not optional formalities; they are integral to ensuring the legitimacy and social value of the study. By prioritizing informed consent, confidentiality, avoidance of harm, integrity, and data security, researchers can protect participants while producing credible and trustworthy results. As survey methods continue to evolve in the digital age, the ethical principles guiding them must remain steadfast, adapting to new challenges in privacy, cultural sensitivity, and responsible use of data [26].

3.3 Types of Survey Instruments

Surveys can be administered using a variety of instruments, each designed to suit specific research objectives, populations, and contexts. The choice of survey instrument influences data quality, respondent engagement, and the feasibility of data collection. In research, survey instruments broadly fall into categories such as structured questionnaires, semi-structured questionnaires, interview schedules, rating scales, and online survey tools. While these instruments share the common aim of eliciting information systematically, they differ in their design, level of flexibility, and mode of administration [27].

1. **Structured Questionnaires** are among the most widely used instruments in survey research, especially in qualitative studies. They consist of pre-determined questions with fixed response options, such as multiple choice, Likert scales, or yes/no answers. The advantage of this format is that it produces standardized data that can be easily coded and analyzed statistically. However, structured questionnaires may restrict respondents from expressing nuanced views, making them less suitable for exploratory qualitative research.
2. **Semi-Structured Questionnaires** offer a balance between structure and flexibility. They contain both closed-ended and open-ended questions, allowing researchers to gather measurable data while also exploring underlying reasons, opinions, and experiences. This hybrid approach is particularly valuable in mixed-method studies, where the goal is to capture both breadth and depth in responses. Semi-structured questionnaires can be administered in person, over the phone, or online, depending on logistical considerations.
3. **Interview Schedules** are commonly used in qualitative research, though they can be adapted for qualitative contexts. Unlike self-administered questionnaires, interview schedules are conducted by a trained interviewer, who asks questions directly and records responses. They can be structured (with fixed wording and order of questions), semi-structured (with flexibility for follow-up probes), or unstructured (guided by broad topics). The interviewer's ability to clarify questions and probe for deeper responses makes this tool particularly effective for studies involving complex or sensitive topics.
4. **Rating Scales** (such as Likert scales, semantic differential scales, or numerical rating scales) are integral to many survey instruments, especially when measuring attitudes, perceptions, or satisfaction levels. These tools enable respondents to indicate the degree to which they agree, approve, or identify with a statement.

While highly compatible with qualitative research, rating scales can also be adapted for qualitative purposes by following up with open-ended prompts that explain the numerical choice.

5. **Online Survey Tools** (e.g., Google Forms, SurveyMonkey, Qualtrics) have transformed data collection in both academic and commercial research. They allow rapid distribution, automated data capture, and integration with data analysis software. Online surveys are cost-effective and scalable, making them ideal for large populations. However, they may exclude individuals without internet access or digital literacy, potentially introducing sampling bias.

When choosing a survey instrument, researchers must weigh factors such as the nature of the research questions, the target population, the required level of standardization, resource availability, and ethical considerations. For instance, a highly structured online questionnaire might be optimal for a large-scale opinion poll, while an in-depth interview schedule may be more appropriate for exploring the lived experiences of a smaller, specialized group (Table 3.5).

This classification underlines that no single survey instrument is universally superior; rather, the optimal choice depends on the research aims, the population being studied, and the balance between depth and breadth of data sought.

Table 3.5 Comparative main survey instrument types and their typical applications in research

Survey instrument	Primary use	Typical research approach	Strengths	Limitations
Structured questionnaire	Collect standardized, quantifiable data	Qualitative	Easy to code and analyze; large-scale applicability	Limited depth; rigid format
Semi-structured questionnaire	Gather both qualitative and qualitative insights	Mixed-method	Flexibility; captures both numbers and narratives	More complex to analyze
Interview schedule	Explore detailed, context-rich responses	Qualitative/ mixed-method	Allows probing and clarification; suited for sensitive topics	Time-consuming; potential interviewer bias
Rating scales	Measure intensity of attitudes, opinions, or satisfaction	Qualitative/ mixed-method	Standardized measurement; easy to compare	Risk of oversimplifying complex views
Online survey tools	Large-scale, rapid data collection	Qualitative/ mixed-method	Cost-effective; automated data handling	Excludes non-digital populations; potential low response rates

3.3.1 Mixed-mode Surveys

Mixed-mode surveys refer to the use of more than one mode of data collection within a single research project to gather information from respondents. Rather than relying solely on one method—such as face-to-face interviews, telephone surveys, or online questionnaires—researchers combine multiple modes to maximize coverage, improve response rates, and reduce potential biases associated with a single method. The increasing diversity of communication channels in the digital era, coupled with variations in respondents' preferences, literacy levels, and technological access, has made mixed-mode surveys an increasingly practical and effective choice in contemporary research. One of the primary reasons for adopting a mixed-mode approach is the recognition that no single survey mode is universally optimal. For example, while online surveys can reach a large number of respondents quickly and at relatively low cost, they may exclude individuals without reliable internet access or those unfamiliar with digital tools. Conversely, in-person interviews offer depth and clarity but are more time-consuming and expensive. By strategically integrating modes, researchers can reduce coverage error and ensure a more representative sample, especially in studies targeting diverse populations. In practical application, mixed-mode surveys can be structured in different ways. A *sequential mixed-mode design* involves using one mode first—such as sending an online questionnaire—and then following up with another mode, like telephone calls, to reach non-respondents. This approach is particularly effective in improving response rates and reducing non-response bias. Alternatively, a *concurrent mixed-mode design* uses multiple modes simultaneously, giving respondents the flexibility to choose their preferred method. For instance, participants may be sent both a web survey link and a paper questionnaire, allowing them to select the format most convenient for them [28].

Mixed-mode surveys are especially valuable when targeting heterogeneous populations. In national or cross-cultural studies, literacy levels, language preferences, and access to technology can vary widely. Combining methods such as online surveys for urban populations and paper-based surveys for rural communities ensures inclusivity. Similarly, in health research, telephone interviews may be ideal for older participants less comfortable with digital tools, while online forms may be more appealing to younger respondents. However, mixed-mode approaches also present challenges. One key issue is *measurement equivalence*. When different modes are used, subtle differences in question presentation, interviewer involvement, or response options can affect how respondents interpret and answer questions, potentially leading to mode effects. For instance, sensitive questions may receive more honest responses in self-administered online formats than in interviewer-led telephone calls. To address this, researchers must carefully standardize question wording and layout across modes to maintain comparability. Cost considerations also play a role. While mixed-mode surveys can improve representativeness, they may require more resources for training, instrument adaptation, and data integration. Additionally, managing and merging data from different modes requires careful planning to ensure consistency in coding and analysis.

In conclusion, mixed-mode surveys offer a flexible and inclusive strategy for data collection, especially in research involving diverse populations or where maximizing response rates is crucial. When designed with careful attention to consistency, accessibility, and measurement equivalence, mixed-mode approaches can strengthen the validity and reliability of survey findings. They represent a pragmatic evolution in research methodology, aligning with the realities of varied respondent preferences and technological accessibility in the modern world [29].

In contrast, single-mode surveys rely on just one method—such as exclusively using online forms, paper questionnaires, or in-person interviews—to collect all data. While this can ensure greater methodological consistency, it may also limit reach or create accessibility barriers for certain respondent groups. The following table compares mixed-mode and single-mode survey approaches, highlighting their respective advantages, disadvantages, and ideal applications (Table 3.6).

This comparison shows that while mixed-mode surveys offer inclusivity and flexibility, they also demand more resources and careful planning. Single-mode surveys, conversely, offer efficiency and simplicity but at the risk of lower coverage and potential bias. The choice between the two depends largely on the study's objectives, target population, budget, and desired level of methodological rigor.

Table 3.6 A comparison mixed-mode and single-mode survey approaches

Aspect	Mixed-mode surveys	Single-mode surveys
Advantages	• Greater coverage of diverse respondent groups • Potential for higher response rates • Reduces mode-specific biases • Flexibility in adapting to participant preferences	• Simpler design and implementation • Lower training and operational complexity • Easier to maintain consistency in data collection • Lower risk of mode-related measurement differences
Disadvantages	• More complex design and management • Higher costs due to multiple modes • Potential for measurement error if questions differ between modes • More complex data integration	• Limited reach to specific respondent groups • Higher non-response risk if chosen mode is inconvenient • Potential for mode-specific bias to go unchecked
Ideal applications	• Studies targeting heterogeneous populations with varying access to technology • National or large-scale surveys where inclusivity is critical • Research aiming to minimize non-response bias	• Small-scale or homogenous population studies • Situations requiring strict methodological consistency • Low-budget projects with limited scope

3.4 Survey Design Principles

Survey research is one of the most widely used methods of collecting data in the social sciences, business, education, and healthcare fields. At the heart of successful survey research lies the design of the survey instrument itself. A survey instrument refers to the tool or set of questions used to gather information from respondents, often in the form of questionnaires or structured interviews. Designing an effective survey instrument requires much more than simply drafting questions; it demands a careful balance between clarity, precision, validity, reliability, and ethical sensitivity. Poorly designed survey instruments risk generating misleading data, low response rates, or biased results, which can compromise the integrity of a study. Conversely, well-designed instruments provide high-quality data that accurately reflects respondents' perspectives and allows researchers to draw valid conclusions. This section explores the principles, processes, and challenges of designing effective survey instruments. It discusses the purpose and role of surveys in research, the importance of reliability and validity, strategies for crafting clear and unbiased questions, decisions around response formats, considerations of layout and administration, pretesting and piloting, cultural sensitivity, ethical concerns, and future directions in digital survey design. By weaving together methodological insights with practical examples, the section underscores that survey instrument design is both a science and an art, requiring methodological rigor as well as creativity in adapting to diverse research contexts [30].

Purpose of Survey Instruments in Research

Survey instruments serve as the primary medium through which researchers access the thoughts, experiences, and behaviours of individuals or groups. The purpose of surveys may range from descriptive to explanatory. In descriptive research, surveys help document demographic characteristics, behaviours, or opinions within a population. For instance, a government survey might measure literacy rates across regions. In explanatory or analytical research, surveys test hypotheses or relationships between variables. For example, an organizational study might use surveys to examine how job satisfaction predicts employee turnover intentions. Survey instruments also play a vital role in applied research and evaluation studies. In healthcare, patient satisfaction surveys inform improvements in service delivery. In education, student feedback surveys guide curriculum development. In marketing, consumer preference surveys shape product design. Across these contexts, the survey instrument functions as a bridge between theory and data, translating abstract concepts such as "satisfaction," "attitude," or "quality of life" into measurable indicators. Thus, the design of the survey instrument directly affects whether a study succeeds in capturing valid evidence for answering its research questions.

Principles of Reliability and Validity

Two key methodological pillars in survey instrument design are reliability and validity. Reliability refers to the consistency of a measurement tool, meaning the extent to which it produces stable results across time, settings, or groups. For example, a reliable job satisfaction scale should yield similar scores for employees with similar levels of satisfaction, regardless of when or how the survey is administered. Techniques such as test–retest reliability, internal consistency (e.g., Cronbach's alpha), and inter-rater reliability are often employed to ensure consistency. Validity, on the other hand, concerns the degree to which a survey instrument measures what it purports to measure. For example, a questionnaire intended to assess stress should genuinely capture stress-related experiences rather than unrelated constructs like fatigue or time pressure. Validity can be examined at different levels: face validity (does the measure appear appropriate?), content validity (does it cover all relevant aspects of the construct?), construct validity (does it align with theoretical expectations?), and criterion-related validity (does it correlate with other established measures?). Without validity, survey findings lack meaning, regardless of how reliable the instrument may be. Effective survey design must balance reliability and validity. An instrument that is consistent but inaccurate provides flawed data, while one that is accurate but inconsistent generates erratic findings. Achieving both requires rigorous testing and thoughtful construction of questions, response scales, and administration methods [31].

Crafting Clear and Unbiased Questions
The wording of survey questions is arguably the most critical component of instrument design. Poorly worded questions can confuse respondents, introduce bias, or lead to inaccurate responses. Clarity should be the guiding principle: questions must be simple, direct, and free from jargon. For instance, instead of asking, "Do you engage in physical activity with a frequency commensurate with recommended health guidelines?" a clearer phrasing would be, "In a typical week, how many days do you exercise for at least 30 minutes?" Avoiding leading or loaded questions is equally essential. Leading questions subtly suggest a desired answer, such as "How much do you agree that our excellent service has improved your experience?" Such phrasing biases respondents toward positive responses. Neutral alternatives like "How would you rate the quality of service you received?" minimize bias. Similarly, double-barreled questions—those that combine two issues in one—should be avoided. A question such as "Do you think the management is supportive and communicates effectively?" is problematic because a respondent may perceive management as supportive but not communicative, or vice versa. Separating them into two questions ensures accuracy. Another important consideration is sensitivity. Questions about income, political beliefs, or personal health may be intrusive. Researchers should justify their inclusion, phrase them respectfully, and provide assurance of confidentiality. Where possible, offering categorical ranges (e.g., income brackets) rather than requiring precise amounts can reduce discomfort [32].

Response Formats and Scaling Decisions
Designing response formats is another crucial aspect of survey instrument development. Response formats determine how data is collected, coded, and analyzed. Closed-ended questions provide fixed response options, such as multiple choice or Likert scales, while open-ended questions allow respondents to answer in their

own words. Each format has advantages and limitations. Closed-ended questions facilitate qualitative analysis and standardization but may oversimplify complex attitudes. Open-ended responses capture richness and nuance but require more time from respondents and more effort from researchers in coding and analysis. Likert scales are especially common in measuring attitudes, opinions, or perceptions. Typically ranging from "strongly disagree" to "strongly agree," these scales provide gradations of intensity. The number of response options (often 5, 7, or 9 points) can influence data sensitivity; more points allow finer distinctions but may overwhelm respondents. Balanced scales should include an equal number of positive and negative options, with or without a neutral midpoint depending on the research aim. Other response formats include ranking questions (ordering preferences), dichotomous items (yes/no), and semantic differential scales (rating between two bipolar adjectives, such as "useful–useless"). The choice depends on the construct being measured and the analytical methods planned. Importantly, response formats should align with the cognitive abilities of the target population. For children or populations with low literacy, visual aids or simplified scales may be more effective [33].

Layout and Administration Considerations

The visual layout and organization of survey instruments significantly affect response rates and data quality. An effective survey should have a logical flow, grouping similar topics together and moving from less sensitive to more sensitive questions. Introductory sections should establish rapport, explain the survey purpose, and provide instructions. Clear formatting, consistent font, adequate spacing, and uncluttered design enhance readability and reduce respondent fatigue. Online and paper-based surveys each present unique challenges. In digital surveys, layout must adapt to different screen sizes, ensure compatibility with mobile devices, and minimize technical glitches. Researchers must guard against "survey drop-off," where respondents abandon the questionnaire midway. Techniques such as progress bars, shorter page lengths, and engaging design features can mitigate this. For paper surveys, clarity of print, appropriate font size, and adequate space for responses are key. In interviewer-administered surveys, interviewer training is critical to ensure standardized delivery and minimize interviewer bias [34].

Pretesting and Piloting

No survey instrument should be deployed without pretesting. Pretesting, also called cognitive interviewing or pilot testing, involves administering the draft instrument to a small sample of respondents to identify potential problems. Pre-tests help reveal issues such as confusing wording, ambiguous response options, or excessive length. Respondents may be asked to "think aloud" as they answer, providing insights into how they interpret questions. Pilot testing, conducted on a larger scale, allows researchers to test reliability, refine scales, and check logistical procedures such as recruitment and data collection. It also provides preliminary data for assessing validity and ensuring the instrument captures the intended constructs. Iterative refinement based on pre-test and pilot results enhances the quality of the final instrument.

Cultural Sensitivity and Contextual Adaptation

Surveys are often administered across diverse populations, making cultural sensitivity a vital consideration in instrument design. Words, phrases, or concepts that are clear in one cultural context may be misunderstood in another. For example, a survey about household roles designed in Western contexts may not translate easily in cultures with different family structures. Similarly, scales that assume linear progression of attitudes may clash with cultural patterns of expression. Translation and back-translation are standard procedures in cross-cultural research to ensure linguistic equivalence. Beyond language, researchers must attend to cultural norms regarding communication, privacy, and authority. For instance, in some societies, respondents may hesitate to disagree openly with authority figures or to disclose personal information. Adapting question formats, providing assurances of confidentiality, and engaging local collaborators can improve cultural appropriateness.

Ethical Considerations in Survey Design

Ethics are central to survey instrument design, especially when sensitive topics are involved. Ethical considerations include informed consent, confidentiality, voluntary participation, and minimizing harm. Consent forms should clearly explain the purpose of the survey, how data will be used, and participants' right to withdraw. Questions that might cause distress should be phrased carefully, and support resources should be provided when necessary. Anonymity and confidentiality are crucial to building trust and ensuring honest responses. For example, anonymous online surveys may encourage greater candor on sensitive issues than face-to-face interviews. However, data security must also be considered, particularly with online surveys where risks of hacking or data breaches exist. Researchers must comply with institutional review board (IRB) guidelines and relevant data protection regulations, such as the General Data Protection Regulation (GDPR) in Europe.

Future Directions in Survey Instrument Design

Technological advances are reshaping survey instrument design. Online platforms such as Qualtrics, Survey Monkey, and Google Forms have made survey creation more accessible and allowed for multimedia integration, adaptive questioning, and real-time data collection. Mobile surveys enable researchers to reach wider audiences but also require optimization for small screens. Artificial intelligence tools are beginning to assist in analyzing open-ended responses, reducing the burden of qualitative coding. Mixed-mode surveys, combining online, telephone, and paper formats, are increasingly used to maximize reach and reduce bias due to non-coverage. However, designing instruments consistent across modes is challenging. Researchers must also contend with survey fatigue in an era of constant digital solicitations. Innovative engagement strategies, shorter instruments, and incentives are being explored to address declining response rates. Looking ahead, survey instrument design must balance methodological rigor with adaptability to changing technologies and societal expectations. Emphasis on inclusivity, accessibility, and user-centered design will be critical in ensuring that surveys remain effective tools for data collection in diverse research contexts.

Thus, designing effective survey instruments is a multifaceted endeavor that combines methodological rigor, clarity of communication, ethical responsibility, and cultural sensitivity. The survey instrument is not merely a set of questions but the foundation upon which reliable and valid data are built. Its effectiveness depends on carefully worded questions, appropriate response formats, logical layout, thorough pretesting, and attention to the needs and characteristics of the target population. Reliability and validity are essential benchmarks, ensuring that instruments provide consistent and accurate measures of intended constructs. Ethical and cultural considerations further ensure that surveys respect participants' rights and contexts. In an era of rapid technological innovation and increasingly diverse research populations, the design of survey instruments continues to evolve. By integrating traditional principles of survey methodology with new digital tools and cross-cultural perspectives, researchers can create instruments that yield high-quality data and support sound decision-making. Ultimately, the success of survey research depends not only on sampling strategies or analytical methods but also—fundamentally—on the design of the survey instruments that mediate the relationship between researchers and participants.

3.4.1 Defining Research Objectives

Defining research objectives is one of the most fundamental and indispensable stages in the research process. Before any data can be gathered, analyzed, or interpreted, a researcher must have a clear and precise understanding of what the study intends to achieve. Research objectives provide the guiding framework that determines the type of data to collect, the methods of analysis, and the eventual conclusions that can be drawn. Without well-formulated objectives, a study risks becoming unfocused, overly broad, or misaligned with the actual problem under investigation. Thus, the act of defining research objectives is not simply an academic exercise; it is a strategic step that influences every subsequent decision within the research design [35].

At its core, a research objective is a clear and specific statement about what a study seeks to accomplish. These objectives are derived from the central research problem or question and are often shaped by existing literature, theoretical frameworks, and practical concerns in the field of study. They help narrow down the scope of research by breaking down broad topics into manageable, measurable, and actionable components. For example, a general research problem such as "understanding the impact of social media on youth behaviour" can be refined into specific objectives such as: "to analyze how social media use influences study habits among university students" or "to examine the role of social media in shaping peer relationships." Each of these objectives provides a sharper focus and points the researcher toward appropriate data collection strategies [36].

The process of defining research objectives begins with identifying the research problem. A well-articulated problem statement highlights the gap in knowledge, unresolved questions, or pressing issues that necessitate investigation. From there, the researcher translates the problem into specific objectives that articulate what exactly needs to be studied. These objectives often fall into two categories: **general objectives** and **specific objectives**. General objectives outline the overall purpose of the research, while specific objectives break the study down into precise, smaller tasks that collectively lead to achieving the general aim. For instance, in a study on customer satisfaction in the banking sector, the general objective might be "to assess customer satisfaction with banking services," while the specific objectives might include "to evaluate customer perceptions of service efficiency," "to examine the impact of digital banking on customer satisfaction," and "to identify areas for improvement in customer service delivery."

An essential characteristic of strong research objectives is that they are **SMART**: Specific, Measurable, Achievable, Relevant, and Time-bound. Specificity ensures that the objective is not vague or overly broad. Measurability allows the researcher to determine whether the objective has been accomplished through empirical evidence. Achievability ensures that the objectives are realistic, considering the researcher's resources, timeframe, and scope. Relevance links the objectives directly to the central problem and the research's contribution to the field. Finally, time-bound objectives establish a temporal scope, clarifying when or within what context the study will take place. Without these features, objectives may remain aspirational rather than actionable.

Defining research objectives also plays a crucial role in determining methodology. The choice between qualitative, qualitative, or mixed-method approaches is strongly influenced by the nature of the objectives. For instance, if an objective seeks to "explore participants' lived experiences of workplace stress," a qualitative approach involving interviews or focus groups would be most appropriate. On the other hand, if the objective is to "measure the correlation between stress levels and productivity," a qualitative approach using surveys and statistical analysis would be more suitable. Thus, objectives serve as the bridge between the research problem and methodological design. Moreover, research objectives influence the sampling strategy and data collection instruments. Objectives help identify the target population, the variables of interest, and the indicators needed to measure those variables. For example, in a health-related study with the objective "to assess the prevalence of diabetes among adults in a rural community," the researcher would need to design a sampling frame of adults within that community and collect relevant health data. Without a clearly defined objective, such decisions risk being arbitrary or misaligned with the ultimate goals of the research [37].

Another critical function of research objectives is their role in guiding data analysis. When data is collected, researchers must decide how to organize, interpret, and present findings. Clearly defined objectives act as a checklist to ensure that the analysis remains focused on answering the research questions. They prevent the study from drifting into unrelated directions, a common issue when researchers encounter unexpected data. For example, if the objective is "to investigate the impact of leadership styles on employee performance," the analysis must remain tightly centered on leadership and performance variables, even if other intriguing patterns (such as generational differences in the workforce) appear in the dataset. In addition to shaping methodology and analysis, research objectives serve a communicative function. They allow researchers to clearly articulate their study's purpose to stakeholders, including funding agencies, supervisors, academic peers, or policy-makers. Well-defined objectives demonstrate that the researcher has a coherent plan and a systematic approach, enhancing the credibility and feasibility of the study. In applied research, objectives are especially critical for decision-makers, as they show how the research findings can be translated into practical interventions, policies, or strategies.

Despite their importance, defining research objectives can be challenging. Researchers often struggle with balancing breadth and depth. Objectives that are too broad risk being unmanageable, while those that are too narrow may overlook important dimensions of the problem. Another challenge is ensuring that objectives are free from bias or assumptions. Poorly defined objectives may inadvertently lead to data collection that confirms preconceived notions rather than objectively investigating the issue. To avoid these pitfalls, researchers are encouraged to conduct thorough literature reviews, consult with experts, and engage in pilot studies before finalizing objectives. In conclusion, defining research objectives is not a mere preliminary task but a cornerstone of the research process. Objectives provide the structure, clarity, and direction necessary for designing, executing, and evaluating a study. They influence every stage of research, from problem formulation and methodology selection to data collection, analysis, and dissemination of findings. By ensuring that objectives are specific, measurable, achievable, relevant, and time-bound, researchers can create a solid foundation for producing meaningful and reliable knowledge. Ultimately, clear objectives not only enhance the quality of the research process but also ensure that the outcomes have practical and theoretical significance [38].

3.4.2 Objectives Process in Qualitative Research

Defining research objectives is a fundamental step in any study, but in qualitative research, the process is often more flexible, iterative, and exploratory compared to qualitative approaches. The objectives in qualitative research are not always rigid hypotheses to be tested; instead, they serve as guiding questions that help the researcher focus on phenomena, behaviours, or social processes of interest. These objectives shape the scope of inquiry while leaving space for discovery and adaptation as new insights emerge during data collection and analysis. The process of setting objectives in qualitative research begins with identifying the research problem or phenomenon that requires in-depth exploration. Unlike qualitative research, which seeks to measure and establish cause-and-effect relationships, qualitative research emphasizes understanding meaning, context, and lived experiences. For example, if a researcher is studying employee motivation, the qualitative objective may be to understand *how employees perceive motivation strategies in the workplace* rather than to measure the extent of motivation statistically [38].

The first step in defining qualitative objectives is to conduct a review of relevant literature and theoretical frameworks. This helps in clarifying gaps in knowledge and shaping broad research questions. For instance, if prior studies have established that leadership style influences employee performance, a qualitative study may aim to explore the lived experiences of employees under different leadership styles. Thus, the objectives may emerge as open-ended aims such as *to explore perceptions of transformational leadership among bank employees* or *to examine how transactional leadership influences workplace relationships.*

Another key element of the process is refining the objectives during the initial stages of fieldwork. Since qualitative research often employs methods like interviews, focus groups, or observations, researchers may discover emerging themes or unexpected issues that necessitate revisiting and adjusting the objectives. For example, a researcher initially aiming to study teacher-student interactions in the classroom may, through early interviews, uncover that cultural factors significantly shape these interactions, prompting an adjustment of objectives to include cultural influences. The iterative nature of qualitative objectives is also reflected in their phrasing. Instead of precise predictions, objectives are expressed as intentions to explore, understand, or interpret. Typical verbs used in qualitative objectives include *explore, examine, understand, interpret, describe,* and *identify.* These reflect the open-ended and interpretive stance of qualitative research.

Ethical considerations are also central in the process of defining qualitative objectives. Because qualitative research often involves sensitive, personal, or contextual data, the objectives must ensure respect for participants' voices and perspectives. Researchers must balance their interests with participants' well-being, ensuring that the objectives are not exploitative but contribute to meaningful understanding. In practice, qualitative objectives provide a foundation for developing research questions, interview guides, and thematic frameworks for analysis. They also serve as a compass throughout the research journey, reminding the researcher to remain aligned with the purpose of inquiry while remaining open to new insights. Ultimately, the process of defining objectives in qualitative research is characterized by its flexibility, adaptability, and focus on meaning. It differs from qualitative research by prioritizing depth over breadth, interpretation over measurement, and context over generalization. This unique process ensures that qualitative studies contribute rich, nuanced, and contextually grounded knowledge to research fields.

3.4.3 Challenges in Defining Research Objectives

Research, at its core, is a systematic process of inquiry designed to generate new knowledge, validate or challenge existing theories, and offer practical solutions to real-world problems. Central to this process is the formulation of research objectives. Research objectives act as a compass: they provide clarity of purpose, define the scope of inquiry, and shape the methodological pathway a study will follow. Without well-defined objectives, research risks becoming vague, unfocused, or irrelevant. While defining research objectives appears to be a straightforward exercise—stating what the researcher intends to achieve—in practice, it is a complex task fraught with challenges. Objectives must strike a balance between being specific enough to guide research and flexible enough to allow adaptation as the inquiry evolves. This challenge becomes even more pronounced across different research paradigms: in qualitative studies, objectives are often narrow, testable, and measurable, while in qualitative studies, they must remain broad, exploratory, and responsive to emerging themes. Complications also arise from contextual and external factors. Multidisciplinary research often requires reconciling conflicting perspectives and methods, while social and cultural sensitivities may restrict how objectives can be articulated. Additionally, pressures from funding bodies, academic institutions, or policymakers may impose priorities that conflict with the researcher's original vision [39].

This section examines the challenges in defining research objectives, beginning with a conceptual overview of their role in research, before analyzing the practical, methodological, ethical, and contextual difficulties researchers face. The discussion also identifies strategies to overcome these obstacles, ultimately underscoring the importance of carefully constructed research objectives in ensuring that studies remain rigorous, ethical, and impactful.

The Conceptual Role of Research Objectives

Research objectives are statements of intent that describe what a study seeks to achieve. They flow from the research problem and often articulate the link between theory and practice. In qualitative research, objectives are frequently tied to hypothesis testing and measurable outcomes. For example, a study may seek to determine the effect of a training program on employee performance, with objectives focusing on measuring variables, establishing correlations, or testing causality. In contrast, qualitative research often adopts broader and more flexible objectives. Here, the emphasis is on exploring experiences, meanings, and social processes. A qualitative study on workplace culture, for instance, may have objectives that focus on "understanding employee perceptions" or "exploring how leadership practices influence collaboration." These objectives are intentionally open-ended, as the researcher acknowledges that new insights may emerge during data collection and analysis. Thus, objectives act as both boundaries and enablers. They narrow the focus of a study to avoid aimless wandering while simultaneously enabling creativity and intellectual curiosity. However, this dual role also creates tension: objectives must be sufficiently defined to maintain clarity yet sufficiently adaptable to accommodate discovery. It is within this tension that many challenges in defining research objectives arise [40, 41].

Challenges in Defining Research Objectives

1. **Conceptual Ambiguity**: One of the foremost challenges in defining research objectives is conceptual ambiguity. Researchers often struggle to translate abstract research problems into clear, actionable objectives. This difficulty is especially pronounced when the research topic is broad or interdisciplinary. For instance, a study on "digital transformation in education" could generate objectives relating to pedagogy, technology adoption, policy, equity, or even psychological impacts on learners. Without a clear framework, researchers risk setting

objectives that are either too scattered or too narrow to capture the complexity of the problem.

2. **Balancing Breadth and Depth**: Defining objectives requires balancing breadth and depth. Overly broad objectives, such as "to study the impact of globalization," provide little guidance for methodological choices and risk producing superficial findings. On the other hand, excessively narrow objectives may fail to capture important dimensions of the problem or limit the study's relevance. For example, an objective such as "to measure the effect of a training manual on employees in one department over two weeks" may be precise but lack broader applicability. Striking this balance is particularly difficult for novice researchers, who may be tempted to either oversimplify or overcomplicate their objectives.
3. **Methodological Alignment**: Objectives must align with methodology, yet this alignment is often challenging to achieve. In qualitative research, poorly formulated objectives can lead to invalid hypotheses, measurement errors, or inappropriate statistical techniques. In qualitative research, objectives that are too rigid can restrict the exploratory nature of the study, preventing researchers from adapting to new insights that emerge in the field. For instance, if the objective is narrowly stated as "to evaluate the effect of leadership training on employee productivity," a qualitative approach may struggle to capture broader contextual themes such as morale or workplace culture that inevitably surface during interviews.
4. **Evolving Nature of Research Questions**: Research is rarely static. Objectives often evolve as the researcher gains more familiarity with the subject through literature reviews, pilot studies, or fieldwork. While such evolution is natural and often beneficial, it creates challenges in maintaining coherence and consistency. A researcher may initially set objectives to "explore student engagement in online learning" but later shift focus toward "investigating the role of digital equity." Such shifts can raise questions about the comparability of objectives, the validity of the research design, and the integrity of the study's outcomes.
5. **External Pressures and Stakeholder Influence**: Research is rarely conducted in isolation; it often involves external stakeholders such as funding bodies, government agencies, NGOs, or private corporations. These stakeholders may impose specific expectations that influence how objectives are defined. For example, a funding agency may prioritize measurable outcomes that demonstrate policy impact, pushing researchers toward qualitative objectives even if the research problem is best explored qualitatively. Similarly, organizational sponsors may discourage objectives that could reveal unfavorable results. This external influence may compromise academic independence, forcing researchers to negotiate between scholarly curiosity and stakeholder demands.
6. **Ethical and Cultural Constraints**: Defining objectives also involves ethical and cultural considerations. Researchers working with vulnerable populations—such as children, patients, or marginalized groups—must carefully frame objectives to avoid exploitation or harm. Objectives that are too intrusive or culturally insensitive risk alienating participants and undermining the validity of findings. For example, a study on family planning in conservative societies may face difficulty

in articulating objectives that balance scientific inquiry with cultural respect. Ethical review boards often play a role in scrutinizing objectives, but researchers themselves must ensure that their goals are respectful, realistic, and justifiable.

7. **Multidisciplinary and Interdisciplinary Complexity**: Modern research increasingly spans multiple disciplines, requiring the integration of diverse theoretical frameworks and methodologies. This creates unique challenges in defining objectives, as different disciplines may have conflicting assumptions, priorities, or epistemological stances. For example, an interdisciplinary study on climate change may involve natural scientists seeking to measure carbon emissions, economists focusing on cost-benefit analyses, and sociologists exploring community adaptation. Defining objectives that satisfy all disciplinary perspectives without diluting the focus becomes a formidable challenge.
8. **Limitations of Time and Resources**: Practical constraints such as limited time, funding, or access to participants often force researchers to adjust their objectives. Ambitious goals may need to be scaled down to match available resources, which can reduce the scope or depth of a study. For instance, a researcher intending to conduct a nationwide survey may be forced by budget limitations to restrict the study to a single region, thereby altering the objectives. Balancing ambition with feasibility remains a persistent challenge.
9. **Novice Researcher Inexperience**: For novice researchers, such as graduate students, defining objectives is particularly daunting. They may lack experience in framing research questions, distinguishing between objectives and hypotheses, or aligning goals with methodology. In many cases, objectives are either too vague (e.g., "to study leadership") or too ambitious (e.g., "to analyze all leadership models in global organizations"). Supervisors often play a critical role in helping novices refine objectives, but the process remains one of the most common stumbling blocks in early research training.

Strategies for Overcoming Challenges

While the challenges in defining research objectives are significant, they are not insurmountable. Researchers can adopt several strategies to improve the clarity, feasibility, and relevance of their objectives [42].

1. **Conducting Thorough Literature Reviews**: A comprehensive review of existing literature helps researchers situate their work within the broader scholarly conversation. It highlights gaps in knowledge, clarifies theoretical frameworks, and provides models of how other researchers have defined objectives. This process reduces conceptual ambiguity and ensures that objectives are both original and grounded in evidence.
2. **Applying the SMART Framework**: The SMART framework—Specific, Measurable, Achievable, Relevant, and Time-bound—provides practical guidance for defining clear and feasible objectives. For instance, instead of stating "to improve employee performance," a SMART objective might be "to measure the impact of a six-month leadership training program on employee productivity scores in Department X." While the SMART framework is more common in

qualitative studies, its principles can also be adapted to qualitative research by ensuring clarity and feasibility without sacrificing flexibility.

3. **Pilot Studies and Pretesting**: Conducting pilot studies or pretests allows researchers to refine their objectives based on preliminary findings. This is particularly useful in qualitative research, where early interviews or focus groups may reveal unanticipated themes that can reshape objectives. In qualitative studies, pilot surveys can test the feasibility of measuring certain variables.
4. **Stakeholder Negotiation**: When external pressures threaten to compromise research independence, researchers must engage in transparent negotiation with stakeholders. This may involve clarifying the scope of objectives, explaining methodological limitations, or setting realistic expectations. Clear communication can help reconcile conflicting priorities while safeguarding academic integrity.
5. **Ethical Reflection and Cultural Sensitivity**: Researchers should critically reflect on the ethical and cultural implications of their objectives. Engaging with community representatives, cultural experts, or ethical review boards can help ensure that objectives are respectful and inclusive. In sensitive contexts, objectives should prioritize minimizing harm, protecting participant autonomy, and promoting mutual benefit.
6. **Collaborative and Interdisciplinary Dialogue**: In interdisciplinary research, open dialogue among collaborators is essential. Joint workshops, planning meetings, and shared conceptual frameworks can help reconcile disciplinary differences and lead to the formulation of coherent objectives. Collaboration encourages the integration of diverse perspectives while maintaining focus.
7. **Supervisory and Peer Support**: Novice researchers benefit greatly from guidance provided by supervisors, mentors, or peers. Constructive feedback helps refine objectives, align them with methodology, and ensure they are realistic. Institutions can also play a role by offering research design workshops and training programs focused on objective formulation.

Defining research objectives is one of the most critical yet challenging stages of the research process. Objectives serve as a roadmap, shaping every subsequent decision from methodology to analysis and interpretation. However, researchers face numerous obstacles: conceptual ambiguity, balancing breadth and depth, methodological alignment, evolving research questions, external stakeholder pressures, ethical and cultural constraints, interdisciplinary complexity, resource limitations, and inexperience among novice scholars. Overcoming these challenges requires deliberate strategies, including comprehensive literature reviews, the application of frameworks like SMART, pilot testing, stakeholder negotiation, ethical reflection, interdisciplinary dialogue, and supervisory support. By adopting such approaches, researchers can craft objectives that are clear, feasible, and ethically sound, while still allowing for creativity and discovery. Ultimately, the process of defining research objectives is not merely a technical exercise; it is a deeply intellectual and ethical endeavor. It requires critical reflection, negotiation of competing interests, and sensitivity to both scientific rigor and human context. Researchers who master the art of defining clear and meaningful objectives are better positioned to produce studies that contribute not only to academic knowledge but also to the broader social good [33].

3.5 Qualitative Research Question Types

Qualitative research has grown into one of the most dynamic and indispensable traditions in contemporary social inquiry. Unlike qualitative methods, which prioritize measurement, statistical testing, and generalization, qualitative research is primarily concerned with exploring meanings, experiences, values, and the complex social realities that shape human life. Central to this process are research questions: the intellectual compass that directs the study's design, methodology, and interpretation of findings. The formulation of qualitative research questions is not only an academic exercise but also a practical necessity that determines the depth and richness of the knowledge generated.

In qualitative research, questions are rarely fixed or narrow at the outset. Instead, they evolve as the researcher engages with participants, observes social contexts, and interprets emerging data. Scholars such as [43, 44] emphasize that research questions in qualitative studies should remain open-ended, flexible, and sensitive to context. Within this tradition, scholars have identified several types of qualitative research questions—exploratory, descriptive, interpretive, explanatory, evaluative, predictive, generative, and critical—each serving distinct epistemological and methodological purposes. This section provides an in-depth exploration of these question types, situating them within real-world case studies that illustrate their application. From education reform in the Global South to gender studies in Western contexts, from healthcare innovations to community-based participatory research, this section highlights how qualitative research questions shape inquiry and contribute to theory, policy, and practice.

1. **Exploratory Questions**

Nature and Purpose
Exploratory questions are concerned with uncovering phenomena that are not well understood. They typically begin with words like "what" or "how," inviting participants to share experiences that have not yet been fully documented or analyzed. These questions are particularly useful in contexts where little prior research exists or where researchers seek to capture emerging social phenomena.

Case Study: Mental Health Narratives among Refugees
A real-life example of exploratory questions can be seen in a study conducted with Syrian refugees in Germany [45]. Researchers sought to understand the lived experiences of trauma and mental health coping strategies among refugees—a context with limited prior research. Exploratory questions such as, "*What challenges have you experienced in accessing healthcare?*" and "*How do you cope with feelings*

of stress or anxiety?" allowed participants to provide rich narratives. The findings revealed previously undocumented barriers such as cultural stigma and language-related misunderstandings. This demonstrates how exploratory questions enable researchers to open the field of inquiry and set the stage for further systematic studies.

2. **Descriptive Questions**

Nature and Purpose
Descriptive questions aim to capture and present a detailed account of a phenomenon, social process, or group experience. They do not attempt to explain causation but instead focus on "painting a picture" of the context under study.

Case Study: Classroom Dynamics in Rural Education
In an ethnographic study of rural schools in India, [46] used descriptive questions to examine classroom dynamics in under-resourced environments. Questions such as "*What teaching strategies do educators employ in large, overcrowded classrooms?*" and "*How do students interact with teachers and peers in these settings?*" helped the researcher construct a detailed account of everyday schooling practices. The findings provided nuanced descriptions of improvisational teaching methods and peer-support networks among students, which were invaluable for policymakers attempting to design interventions for educational equity.

3. **Interpretive Questions**

Nature and Purpose
Interpretive questions seek to understand the meanings individuals or groups assign to their experiences. They go beyond "what is happening" to ask, "what does it mean to those involved?" These questions align closely with hermeneutic traditions of inquiry.

Case Study: Breast Cancer Survivors' Narratives
In a study conducted in the United States, qualitative researchers examined the narratives of breast cancer survivors [47]. The guiding questions included: "*What does survivorship mean to women who have undergone cancer treatment?*" and "*How do survivors interpret changes in their identity and social relationships?*" These interpretive questions revealed that survivors often framed their experiences not only as medical recovery but also as journeys of personal transformation. Such insights illuminated cultural constructions of illness and survival, offering healthcare providers a more empathetic framework for patient support.

4. **Explanatory Questions**

Nature and Purpose
Explanatory questions go a step further than descriptive and interpretive ones. They ask why phenomena occur or what underlying processes or mechanisms account for them.

Case Study: Technology Adoption in Healthcare
In a multi-country study on electronic health record (EHR) adoption, explanatory questions such as, "*Why do some healthcare professionals resist adopting digital technologies despite institutional mandates?*" guided the inquiry [48]. Findings showed that resistance often stemmed from perceived threats to professional autonomy and increased workload, rather than mere lack of training. This explanatory insight helped policymakers design interventions addressing professional identity and organizational culture, rather than simply providing more technical resources.

5. **Evaluative Questions**

Nature and Purpose
Evaluative questions focus on assessing the value, effectiveness, or impact of a program, policy, or intervention. They are particularly common in applied research contexts such as education, healthcare, and social policy.

Case Study: Education Reforms in Kenya
An evaluative case is found in research on Kenya's Free Primary Education (FPE) program, which was introduced to improve access to schooling. Researchers used questions like: "*How effective has the FPE program been in improving student enrolment and retention?*" and "*What are teachers' and parents' perceptions of its success?*" Findings showed that while enrolment surged, issues such as overcrowded classrooms and underfunded schools limited the program's effectiveness. This evaluative questioning provided a nuanced assessment, guiding future reforms to focus not just on access but also on quality.

6. **Predictive Questions**

Nature and Purpose
Predictive questions, though less common in qualitative traditions, aim to anticipate future trends or outcomes based on current experiences and narratives. They are often framed cautiously, since qualitative research is not designed to generalize statistically.

Case Study: Urban Youth Employment in Africa
In a study of urban youth in Lagos, Nigeria, researchers asked predictive questions such as: "*Given current barriers to formal employment, how do young people envision their future livelihoods?*" [49]. Participants predicted that informal economies,

digital entrepreneurship, and migration would become dominant strategies. These predictions helped policymakers anticipate youth labor trends and design proactive vocational training initiatives.

7. **Generative Questions**

Nature and Purpose
Generative questions aim to open new avenues of inquiry, theory, or practice. They are forward-looking and creative, designed to inspire fresh insights.

Case Study: Indigenous Knowledge and Environmental Sustainability
In Canada, a study on Indigenous approaches to environmental sustainability used generative questions such as, "*How can Indigenous ecological knowledge inform contemporary climate change policy?*" [50]. These questions did not simply describe Indigenous practices but generated new frameworks for environmental governance that combined traditional wisdom with modern science. This approach enriched both academic theory and practical policymaking.

8. **Critical Questions**

Nature and Purpose
Critical questions are rooted in critical theory traditions, interrogating issues of power, inequality, and social justice. They seek to expose hidden assumptions and challenge structures of domination.

Case Study: Gender Studies and Workplace Inequality
In a landmark study on gender inequality in corporate organizations, Acker (1990) employed critical questions such as, "*How do organizational practices reinforce gendered hierarchies?*" and "*In what ways do women resist or negotiate these structures?*" The study revealed how supposedly "neutral" organizational practices (e.g., performance evaluations, promotion pathways) were deeply gendered, favoring male employees. These findings influenced later gender equity reforms and demonstrated the transformative potential of critical qualitative questions.

Synthesis and Integration
These case studies illustrate how different types of qualitative research questions serve distinct but complementary purposes. Exploratory and descriptive questions open up new fields of inquiry, while interpretive and explanatory questions deepen understanding. Evaluative and predictive questions connect research to policy and practice, while generative and critical questions push the boundaries of knowledge, inspiring new directions and challenging inequalities. Taken together, these question types reflect the versatility and richness of qualitative inquiry. They also demonstrate that the type of question chosen is never arbitrary: it reflects the researcher's epistemological stance, the nature of the phenomenon under study, and the ultimate

purpose of the research. Qualitative research is built on the foundation of strong, well-formulated questions. Each type of qualitative research question—exploratory, descriptive, interpretive, explanatory, evaluative, predictive, generative, and critical—contributes uniquely to the pursuit of knowledge. Real-life case studies from education, healthcare, gender studies, environmental policy, and refugee experiences demonstrate the applied value of these question types in producing rich, context-sensitive insights. In a world facing complex challenges, the careful formulation of qualitative research questions enables scholars and practitioners to capture human experiences, test social policies, and challenge inequities in ways that qualitative research alone cannot achieve. The examples presented in this section highlight the enduring power of qualitative research to illuminate, critique, and transform human life [1] (Table 3.7).

Table 3.7 Comparative overview table

Type of question	Primary focus	Example	Best suited for
Exploratory	Uncovering new insights, themes	How do first-generation students experience transition to higher education?	Grounded theory
Descriptive	Capturing detailed accounts of experiences	What coping strategies do dementia caregivers use?	Phenomenology
Explanatory	Understanding reasons/ processes	Why do rural women prefer traditional birth attendants?	Case study, ethnography
Interpretive	Meaning-making and sense-making	How do young professionals define success?	Symbolic interactionism
Process-oriented	Examining change and development over time	How do community health initiatives evolve?	Longitudinal studies
Comparative	Exploring similarities and differences	How do male and female teachers perceive digital tools differently?	Case study
Evaluative	Assessing value and effectiveness	What are students' perceptions of blended learning?	Program evaluation
Critical	Examining power, inequality, justice	How do women entrepreneurs face gender bias in finance?	Critical theory, PAR

3.5.1 Qualitative Research Rating Scales

Qualitative research is typically associated with rich, descriptive data that captures experiences, perceptions, and meanings. However, within this paradigm, researchers sometimes employ rating scales to systematize participants' judgments, attitudes, or perceptions while still retaining an interpretive lens. Rating scales in qualitative research differ significantly from those in qualitative contexts: instead of measuring variables numerically for statistical inference, they are often used to support thematic exploration, illustrate intensity of feelings, and provide structured prompts that guide deeper discussion. This section explores the concept of qualitative research rating scales, their types, applications, strengths and limitations, and real-world examples where they have been successfully integrated into qualitative inquiry.

Understanding Rating Scales in Qualitative Research
A rating scale is a tool that allows respondents to assign value, degree, or importance to a phenomenon along a continuum. In qualitative research, scales are commonly used for measurement and statistical comparison. In qualitative research, however, scales serve more as heuristic devices that help researchers understand how participants perceive differences, prioritize choices, or express subjective judgments. For example, in a focus group discussion about healthcare experiences, a researcher might ask participants to rate their satisfaction with hospital services on a 1–5 scale, not to compute averages, but to spark dialogue about why certain services are rated highly or poorly. In this way, the rating scale acts as a springboard for narratives, providing structured entry points into rich qualitative data.

Types of Rating Scales in Qualitative Research

1. **Likert-Type Scales**

A Likert scale is one of the most common forms, asking respondents to indicate their level of agreement or disagreement with a statement (e.g., Strongly Agree to Strongly Disagree). In qualitative research, such scales are not merely analyzed numerically but are accompanied by probing questions that uncover the reasoning behind each choice.

- **Example**: In a study on teacher training, participants might rate their agreement with the statement "The training prepared me to handle classroom diversity." The follow-up discussion would focus on why participants felt prepared or unprepared.

2. **Semantic Differential Scales**

This scale measures attitudes by positioning respondents between two opposite adjectives (e.g., Effective–Ineffective, Helpful–Unhelpful). In qualitative research, the emphasis is on ***why*** respondents locate themselves at a particular point on the continuum.

- **Example**: In marketing research, consumers may rate a product on a scale from "Modern" to "Old-fashioned," followed by interviews probing why they associate the product with certain characteristics.

3. **Ranking and Rating Scales**

Participants may be asked to rank or rate a series of items according to importance, preference, or relevance. This approach is especially useful in participatory qualitative methods such as Delphi studies or focus groups, where collective priorities are explored.

- **Example**: In community development research, residents may rank local needs (e.g., clean water, security, healthcare). Their rankings provide a foundation for discussions about lived priorities and social values.

4. **Visual Analog Scales (VAS)**

A VAS uses a visual line or continuum (e.g., a 10 cm line) where participants indicate intensity of feelings, pain, or satisfaction. In qualitative studies, researchers use VAS to help participants externalize subjective experiences and then probe deeper.

- **Example**: In healthcare research, patients with chronic illness may mark pain levels on a line; interviews then explore how fluctuations in pain affect daily living and identity.

5. **Custom Scales in Participatory Research**

Qualitative studies often design context-specific scales that reflect cultural or situational realities. For instance, a study on rural farmers' perceptions of agricultural programs may use local metaphors (e.g., "a tree with many fruits" to mean highly beneficial). These culturally embedded scales enhance local meaning-making.

Applications of Rating Scales in Qualitative Research

1. **Eliciting Narratives**: Scales act as **conversation starters**, encouraging participants to articulate why they positioned themselves in a particular way.

- *Case Example*: In education research, students rating their level of stress before exams on a 1–10 scale often led to deeper discussions about pressure from parents, institutional policies, and peer competition.

2. **Facilitating Comparisons**: Rating scales allow researchers to compare perceptions across individuals or groups without reducing the inquiry to mere numbers.

- *Case Example*: A gender studies project compared how men and women rated workplace inclusivity on a Likert scale. Follow-up interviews illuminated gendered experiences of macroaggressions and biases.

3. **Structuring Focus Groups**: In group settings, rating exercises help organize discussion and highlight divergences of opinion.

- *Case Example*: In participatory urban planning, community members used ranking scales to prioritize environmental issues. This structured ranking stimulated lively debate about which issues (pollution, traffic, water shortages) mattered most.

4. **Supporting Mixed-Methods Designs**: Rating scales are frequently embedded in qualitative-dominant mixed-methods research, serving as bridges between qualitative depth and qualitative structure.

- *Case Example*: A public health study asked participants to rate satisfaction with healthcare services, then conducted ethnographic interviews. The integration revealed not just satisfaction levels but also cultural narratives of distrust in healthcare systems.

Strengths of Using Rating Scales in Qualitative Research

- **Enhanced Participant Engagement**: Scales give participants a tangible, accessible way to express abstract feelings.
- **Facilitation of Reflection**: Participants reflect on their own choices when asked to justify ratings.
- **Comparability Without Reductionism**: Ratings allow cross-case comparison while still preserving contextual depth.
- **Bridging Tool in Mixed-Methods**: They integrate smoothly into qualitative-qualitative designs.

Limitations and Challenges

Despite their usefulness, rating scales in qualitative research are not without challenges:

- **Risk of Quantification**: Overemphasis on numbers may undermine qualitative richness.
- **Cultural Bias**: Scales designed in one context may not translate meaningfully across cultures.
- **Forced Choices**: Participants may feel constrained by limited scale options.
- **Interpretation Issues**: Different participants may interpret the same scale point differently (e.g., a "5" in satisfaction may mean moderate satisfaction for one and indifference for another).

To mitigate these, researchers often triangulate scale data with open-ended narratives, ensuring the interpretive depth is not lost.

Real-Life Case Study Examples
Case 1: Evaluating Patient Satisfaction in Healthcare

A study in Nigeria asked patients to rate hospital services on a 1–5 Likert scale. Beyond collecting scores, researchers conducted interviews to uncover stories of long waiting times, lack of empathy from nurses, and positive experiences in maternity wards. The scales structured the data, while narratives revealed nuanced healthcare challenges.

Case 2 Education Reform in Finland

During curriculum reform, teachers rated their comfort level with new teaching methods on a visual analog scale. Subsequent focus groups revealed struggles with professional training gaps and institutional resistance, demonstrating how rating scales can both simplify and deepen inquiry.

Case 3: Consumer Behaviour in India

In a branding study, consumers used semantic differential scales to describe mobile phone brands (e.g., "trendy–outdated"). Follow-up interviews contextualized these ratings, linking them to aspirations of social status, youth identity, and peer influence.

Thus, rating scales, though often associated with qualitative research, play a valuable role in qualitative inquiry when thoughtfully adapted. They provide structured entry points into participants' perceptions, stimulate narratives, and facilitate meaningful comparisons without sacrificing contextual richness. Their true strength lies not in the numbers themselves, but in the stories and interpretations that follow the rating exercise. As demonstrated in healthcare, education, and consumer research, qualitative rating scales can enrich analysis, foster participant engagement, and bridge methodological divides. Researchers must, however, remain cautious about over-quantification, cultural misalignment, and interpretive ambiguities. When balanced carefully, rating scales become powerful tools for capturing the subtleties of human experience in qualitative research (Table 3.8).

3.5.2 Strategies of Avoiding Leading or Biased Questions

Qualitative research is grounded in the pursuit of authentic, nuanced, and in-depth understanding of human experiences, behaviours, attitudes, and perspectives. At its core lies the necessity of obtaining data that is as free from distortion as possible. However, one of the persistent challenges researchers face is the tendency to inadvertently (or deliberately) frame questions in ways that introduce bias. Leading or biased questions, whether through wording, tone, or assumptions, can distort participant responses, shape the narratives that emerge, and ultimately compromise the integrity of the research findings. Avoiding such questions is not merely a matter

Table 3.8 Types of qualitative research rating scales

Rating scale type	Description	Advantages	Limitations	Example in qualitative research
Likert scale	A scale where respondents indicate their level of agreement or disagreement with a statement (e.g., strongly agree to strongly disagree).	Simple to design and interpret; captures attitudes and opinions effectively.	May oversimplify complex views; prone to central tendency bias.	Used in education research to measure students' satisfaction with teaching methods.
Semantic differential scale	Measures connotative meaning of objects, concepts, or events along bipolar adjectives (e.g., happy–sad, effective–ineffective).	Provides nuanced understanding of perceptions; captures emotional responses.	Designing appropriate adjective pairs can be challenging; interpretation may vary.	Applied in healthcare research to assess patients' perceptions of hospital services.
Ranking sczale	Respondents order items based on preference or importance.	Highlights priorities and comparative preferences.	Difficult for large lists; may not show the intensity of preference.	Used in organizational studies to rank factors influencing employee motivation.
Rating scale (numerical/ ordinal)	Assigns a number or category to express intensity (e.g., 1–5 for satisfaction).	Easy to administer; provides measurable data that can complement qualitative themes.	Risk of losing context or depth of reasoning behind the number.	Used in community development research to measure perceived effectiveness of programs.
Checklist scale	Respondents select all items that apply from a list of attributes or experiences.	Captures multiple perspectives simultaneously; flexible.	May lead to superficial responses without depth.	Used in gender studies to explore common barriers women face in career advancement.
Visual analogue scale (VAS)	Respondents mark a point on a line between two extremes to represent perception (e.g., "not painful" to "worst pain imaginable").	Captures subtle differences in perception; intuitive for participants.	Data coding can be difficult; less precise if participants cluster near ends.	Applied in psychology research to assess levels of anxiety in therapy sessions.

of methodological rigor but also of ethical responsibility. Biased questioning may silence participant voices, reinforce dominant ideologies, and marginalize already vulnerable groups. In qualitative research traditions such as grounded theory, ethnography, narrative inquiry, and case study designs, the role of open-ended, neutral questioning is particularly emphasized because these approaches rely heavily on participants' unmediated perspectives [51].

This section critically examines the strategies for avoiding leading or biased questions in qualitative research. It blends methodological insights with real-life case studies drawn from education, healthcare, gender studies, and policy research, showing how researchers have grappled with and overcome the challenge of bias. By doing so, it seeks to provide both theoretical and practical guidance to scholars committed to preserving validity, reliability, and ethical accountability in their qualitative work.

Understanding Leading and Biased Questions
Defining Leading Questions

A leading question is one that suggests or nudges respondents toward a particular answer. Instead of allowing participants to articulate their perspectives freely, the question frames the response in advance. For instance, asking "*Don't you think the new teaching methods are effective?*" presupposes effectiveness and subtly pressures the respondent to agree.

Defining Biased Questions
Biased questions embed assumptions, cultural values, or prejudices that influence responses. Bias may emerge through language (e.g., gendered terms), tone (e.g., judgmental inflections), or structure (e.g., double-barreled questions). For example, in healthcare research, asking "*Why do patients fail to follow medical advice?*" frames the issue as patient non-compliance without considering systemic barriers.

Why Avoid Bias?

- **Validity**: Leading and biased questions compromise data accuracy.
- **Ethics**: They may coerce or silence participants, violating informed consent principles.
- **Representation**: They risk misrepresenting marginalized groups, reinforcing dominant stereotypes.
- **Impact**: Biased data leads to flawed policies, interventions, and reforms.

Strategies for Avoiding Leading or Biased Questions

1. **Neutral Wording and Open-Ended Structure**

   Neutrality in wording allows participants to express themselves without being influenced by researcher expectations. Open-ended questions, which encourage expansive answers rather than "yes/no" responses, are crucial here.

Example Strategy: Instead of asking, "*How successful was the policy?*" a neutral phrasing would be "*How do you perceive the outcomes of the policy?*"

Case Study: Education Reform in Nigeria
In a study on the Universal Basic Education (UBE) program, researchers initially used questions like, "*How effective has UBE been in improving literacy rates?*" which assumed improvement. After peer review, they revised the question to "*What changes, if any, have you observed since the introduction of UBE?*" This shift revealed not only areas of improvement but also significant challenges, including overcrowded classrooms and insufficient teacher training, which might have been obscured otherwise.

2. **Avoiding Assumptions in Question Design**

   Questions often embed hidden assumptions. To avoid this, researchers must critically assess whether a question presupposes a particular reality.

   Example Strategy: Replace assumption-laden questions such as "*Why are students uninterested in science?*" with "*What factors influence students' interest or disinterest in science?*"

   Case Study: Healthcare Research in the United Kingdom
   A study investigating patient engagement with digital health records initially framed questions around "*Why don't patients adopt digital health platforms?*"—an assumption that non-adoption was widespread. During pilot testing, patients revealed they *had* adopted such tools but faced usability and privacy concerns. By reframing to "*How do patients experience and use digital health platforms?*" researchers uncovered richer data on both barriers and facilitators.

3. **Pilot Testing and Peer Review of Instruments**

   Pilot studies and expert peer review help identify leading or biased tendencies in questions before full-scale data collection begins.

   Example Strategy: Run a small-scale pilot and ask participants how they interpret each question. Have peer researchers critique the interview guide for neutrality.

   Case Study: Gender Studies in South Asia
   In a feminist study on women's participation in local governance, pilot testing revealed that questions like "*How does your husband support your involvement in politics?*" assumed spousal support. After revision, the question became "*How do family relationships influence your political participation?*" This reframing surfaced stories of resistance, negotiation, and support, enriching the feminist analysis.

4. **Training Interviewers in Reflexivity**

   Research assistants or interviewers may unintentionally introduce bias through tone, body language, or selective probing. Training them in reflexivity—awareness of their positionality, assumptions, and influence—is essential.

   Example Strategy: Encourage interviewers to maintain neutral affect, avoid interrupting participants, and use probes such as "*Can you tell me more about that?*" rather than "*Why do you think that's a problem?*"

   Case Study: Policy Research in the United States
   During a policy evaluation of food security programs, it was found that interviewers unconsciously nodded when participants criticized government programs, subtly encouraging negative responses. After reflexivity training, they adopted neutral responses, leading to a more balanced dataset where both positive and negative experiences were equally represented.

5. **Using Culturally Sensitive and Inclusive Language**

   Language bias arises when terms are culturally inappropriate, stigmatizing, or exclusionary. Researchers must adapt instruments to the linguistic and cultural context of participants.

   Example Strategy: Avoid terms like "*illegal immigrant*" which carry stigma; instead use "*undocumented migrant.*"

   Case Study: Indigenous Education in Australia
   In research with Aboriginal communities, researchers originally used terms like "*school attendance problems.*" Community members highlighted that this phrasing blamed families, whereas structural factors like colonial history and curriculum irrelevance were key. By revising to "*factors influencing attendance patterns,*" researchers aligned the study more closely with community perspectives.
6. **Triangulation of Data Collection Methods**

   Using multiple methods (e.g., interviews, focus groups, observations) reduces reliance on any single potentially biased instrument.

   Example Strategy: Compare interview data with observational field notes to see if responses align.

   Case Study: Healthcare in Sub-Saharan Africa
   A maternal health study combined structured interviews with observations of clinic visits. While some interviews suggested patients were satisfied with services, observations revealed long waiting times and limited resources. This triangulation helped reveal discrepancies that biased questions might have obscured.

7. **Iterative Question Development and Adaptive Interviewing**

   In qualitative research, question design is not static but evolves as data collection proceeds. Researchers should continuously reflect on whether their questions are generating authentic responses or leading participants.

   Example Strategy: Use a "responsive interviewing" model where questions adapt based on participant input, while maintaining neutrality.

   Case Study: Gender-Based Violence Research in Latin America
   Initially, researchers asked survivors, "*How did you cope with the violence?*"—a phrasing that assumed coping. Through iterative reflection, they rephrased to "*What were your experiences following the incident?*" This opened space for participants to discuss not only coping but also systemic failures, retraumatization, and resilience strategies.
8. **Reflexive Journaling by Researchers**

   Maintaining a reflexive journal allows researchers to document their biases, assumptions, and question design choices. This meta-cognitive practice fosters transparency and continuous improvement.

   Example Strategy: After each interview, researchers reflect on how their phrasing or demeanor may have influenced responses.

   Case Study: Education Policy in Kenya
   A researcher studying teachers' responses to curriculum reforms documented how her own background as a teacher made her frame questions around "*implementation challenges.*" Reflexive journaling revealed this bias, prompting her to reframe questions as "*How do teachers experience curriculum reform?*"—a more open inquiry that surfaced unexpected narratives of innovation.

Cross-Case Synthesis: Common Lessons
The above case studies across education, healthcare, gender studies, and policy research highlight common insights:

1. **Bias often originates in assumptions**—about success, compliance, or coping.
2. **Pilot testing and reflexivity** are effective in exposing unintended bias.
3. **Inclusive language** is critical for ethical research.
4. **Iterative adaptation** ensures questions remain responsive to participants.

By applying these strategies systematically, researchers can safeguard both the methodological integrity and ethical standing of their work. The avoidance of leading or biased questions is not a mere technical concern but a cornerstone of credible, ethical, and socially responsible qualitative research. Through neutral wording, avoidance of assumptions, pilot testing, reflexivity training, cultural sensitivity, triangulation, adaptive interviewing, and reflexive journaling, researchers can mitigate bias and ensure that participants' authentic voices are heard. The real-life case studies from education reforms in Nigeria, healthcare innovations in the UK and Sub-Saharan Africa, gender studies in South Asia and Latin America, and policy research in the United States and Kenya demonstrate how these strategies function in practice. Each case illustrates the profound consequences of biased questioning, as well as the transformative potential of methodological rigor. Ultimately, avoiding leading or biased questions is both a methodological discipline and an ethical commitment. It ensures that qualitative research does not simply reproduce dominant narratives but instead provides a platform for diverse, authentic, and contextually grounded human experiences. In a world where research informs policies, reforms, and interventions that shape millions of lives, the stakes could not be higher (Table 3.9).

3.6 Minimizing Survey Bias in Qualitative Research

Qualitative research has become a central approach for exploring human experiences, meanings, and social processes. Unlike qualitative research, which emphasizes measurement and numerical generalization, qualitative research provides deep insights into participants' subjective worlds by employing interviews, focus groups, open-ended questionnaires, and observations. However, one persistent challenge in qualitative inquiry is **survey bias**—the systematic distortion of findings due to flaws in question design, researcher influence, respondent interpretation, or contextual factors. Survey bias not only undermines the validity and reliability of results but also compromises ethical standards by misrepresenting participants' voices [52].

This section critically examines strategies for minimizing survey bias in qualitative research. It is divided into several sections: understanding survey bias, major sources of bias, methodological strategies to address them, and real-life case studies from diverse disciplines. Case examples are drawn from education, healthcare, gender studies, and policy research, showing how scholars and practitioners mitigate bias in practice. Finally, a comparative summary table synthesizes approaches to minimizing bias, offering a practical reference point for researchers.

Understanding Survey Bias in Qualitative Research

Survey bias refers to deviations in the collection or interpretation of qualitative data that systematically favor certain responses or misrepresent participants' true perspectives. Unlike qualitative bias, which often emerges in sampling or statistical errors, qualitative survey bias arises primarily from interpretive processes, language use, power dynamics, and question framing. For example, a researcher asking leading questions during an educational reform study may unintentionally steer participants to confirm their assumptions rather than elicit authentic perspectives. Similarly, in healthcare, vulnerable populations may provide socially desirable answers to avoid stigma, thereby skewing results.

Table 3.9 Comparative summary of strategies for avoiding leading or biased questions in qualitative research

Strategy	Purpose/benefit	Case study example
Use of neutral language	Ensures participants are not influenced by emotionally charged or suggestive wording; encourages authentic responses.	*Education reform research*: In a UK study on inclusive education, instead of asking "how effective has inclusive education been?" researchers asked "what are your experiences with inclusive education policies?" to avoid implying effectiveness.
Avoiding double-barreled questions	Prevents confusion and ensures participants respond to a single, clear issue rather than multiple topics bundled into one question.	*Healthcare research*: In a study on patient satisfaction, instead of asking "how satisfied are you with the hospital facilities and the doctors?" (double-barreled), researchers split it into two distinct questions.
Providing context without suggestion	Clarifies the background of a question while avoiding framing that directs participants toward a particular view.	*Policy research*: In a Nigerian governance survey, researchers framed questions about "access to public services" with factual context rather than suggesting government failure, ensuring unbiased data collection.
Balanced question options	Reduces response bias by offering multiple perspectives and avoiding questions that frame only positive or negative options.	*Gender studies*: In a workplace equality study, instead of asking "how well has the company improved gender equality?" they asked "how has the company addressed gender equality?" allowing for both positive and negative responses.
Avoiding assumptions in questions	Prevents the researcher from embedding unverified beliefs within the question that may alienate or pressure participants.	*Healthcare study*: Instead of asking "how often do you visit your doctor for regular checkups?" (assumes regular visits), researchers asked "can you tell me about your healthcare-seeking practices?"
Pilot testing of questions	Identifies unintended biases or misinterpretations before full data collection, enhancing reliability and neutrality.	*Education pilot study*: A university in South Africa pre-tested its student survey on digital learning. Feedback revealed a leading question that assumed all students had internet access, which was later revised.
Cultural sensitivity in wording	Ensures the questions respect participants' cultural and social contexts, avoiding offense or bias.	*Cross-cultural gender research*: In middle eastern gender studies, researchers rephrased sensitive questions about women's roles using culturally appropriate and non-judgmental language.
Use of open-ended questions	Encourages participants to share in-depth perspectives in their own words rather than being constrained by the researcher assumptions.	*Public Health Research*: During COVID-19, open-ended questions like "what challenges have you experienced accessing healthcare during the pandemic?" revealed diverse experiences beyond structured options.

The key to minimizing bias lies not in eliminating subjectivity—which is inherent in qualitative work—but in acknowledging, reflexively managing, and transparently reporting potential distortions.

Major Sources of Survey Bias

Several types of bias affect qualitative surveys. Recognizing them is the first step toward mitigation.

1. **Question Wording Bias**—Occurs when questions are phrased in ways that lead participants toward particular responses.

   Example: Asking "How beneficial was the new teaching method?" presumes it was beneficial, limiting critical feedback.
2. **Social Desirability Bias**—Participants may answer in ways they think are acceptable rather than expressing true feelings. This is common in sensitive fields such as gender equality, healthcare stigma, or policy corruption studies.
3. **Interviewer Bias**—The researcher's tone, phrasing, or non-verbal cues may shape responses. For instance, nodding approvingly during interviews can unconsciously validate certain answers.
4. **Cultural and Language Bias**—Misinterpretation may occur if survey instruments are not culturally adapted. Words may hold different meanings across contexts.
5. **Recall Bias**—Participants may struggle to remember past events accurately, especially in retrospective studies. For example, patients recalling treatment experiences may unintentionally exaggerate or understate details.
6. **Selection Bias**—When participants are not representative of the population under study, findings may be skewed. In qualitative work, purposive sampling must bc carefully justified.

Strategies to Minimize Survey Bias

Minimizing bias requires methodological rigor, reflexivity, and sensitivity to context. Below are key strategies [52, 53]:

1. **Careful Question Design**

- Use neutral, open-ended questions. Instead of "Did the new curriculum improve your learning?" ask "How did you experience the new curriculum?"
- Pilot-test instruments to identify ambiguous or leading items.
- Avoid jargon or technical terms that participants may not understand.

2. **Triangulation of Methods**

Combining multiple data sources—such as interviews, focus groups, and document analysis—helps cross-check findings and reduce reliance on potentially biased responses.

3. **Reflexivity**

Researchers should critically reflect on their assumptions, positionality, and influence. Keeping reflexive journals ensures that biases introduced by the researcher are documented and addressed.

4. **Ensuring Anonymity and Confidentiality**

Participants are more likely to provide honest responses if they feel secure. Anonymous surveys reduce social desirability bias, especially in sensitive contexts like gender-based violence or political corruption.

5. **Training of Interviewers**

Skilled interviewers minimize unintentional cues and maintain neutrality. Standardized training helps reduce interviewer bias.

6. **Cultural and Linguistic Adaptation**

Translating instruments into local languages and adapting them to cultural contexts avoids misinterpretation. Back-translation and cultural consultation ensure accuracy.

7. **Iterative Data Collection**

In qualitative research, data collection and analysis often proceed simultaneously. Researchers can adjust questions mid-way if they detect potential bias.

Case Studies across Disciplines

1. **Education**: **Avoiding Leading Questions in Curriculum Reform**: In a study examining Nigerian higher education reforms, researchers initially asked, "How has the reform improved student learning outcomes?" This question presupposed improvement, limiting responses. After piloting, they revised it to: "What changes in your learning have you noticed since the reform?" This shift produced richer, more critical feedback, uncovering both strengths and weaknesses of the reform. Minimizing bias allowed a balanced evaluation of reform impacts.

2. **Healthcare**: **Managing Social Desirability Bias in HIV Research**: In sub-Saharan Africa, surveys on HIV treatment adherence often suffer from social desirability bias, as patients hesitate to admit non-adherence. A Ugandan study mitigated this by using anonymous self-administered surveys alongside interviews. This revealed discrepancies: while face-to-face interviews indicated 90% adherence, anonymous surveys showed actual adherence was closer to 60%. By minimizing bias, the study informed targeted interventions for real barriers to treatment.

3. **Gender Studies: Addressing Cultural and Language Bias**: In research on women's empowerment in rural India, cultural and linguistic nuances influenced responses. Directly asking women about "autonomy" yielded little understanding. Researchers instead asked about everyday decision-making practices ("Who decides what to cook?" or "Who controls household money?"). This culturally adapted approach avoided abstract concepts and reduced misinterpretation, leading to more authentic insights into gender dynamics.
4. **Policy Research: Reducing Interviewer Bias in Governance Studies:** A study on corruption in Nigerian local governance faced challenges as participants were reluctant to criticize authorities. Researchers trained interviewers to maintain neutral body language and avoid probing with judgmental tones. They also assured anonymity by using coded responses. This allowed participants to speak more openly about corruption practices, producing policy-relevant insights on governance failures.

Ethical Dimensions of Minimizing Bias

Ethics is central to bias reduction. Leading or misrepresenting participants' voices violates ethical principles of **respect, beneficence, and justice**. In qualitative research, ethical reflexivity ensures that methods protect participants while generating valid insights. For example:

- In healthcare research, ensuring confidentiality is crucial to avoid stigma.
- In gender studies, culturally sensitive phrasing prevents re-traumatization of survivors.
- In education, neutral questions prevent power imbalances between teachers and students from silencing dissent (Table 3.10).

Survey bias is an unavoidable but manageable challenge in qualitative research. By recognizing sources of distortion—whether from question wording, social desirability, interviewer influence, cultural barriers, recall difficulties, or sampling imbalances—researchers can adopt strategies to minimize bias and uphold both methodological rigor and ethical integrity. Real-life examples from education, healthcare, gender studies, and policy research demonstrate that careful design, reflexivity, and cultural sensitivity yield more authentic and trustworthy findings. Ultimately, minimizing survey bias is not about achieving complete neutrality—an impossibility in human-centered inquiry—but about maximizing authenticity and inclusivity in representing participants' voices. In doing so, qualitative research strengthens its role as a vital tool for understanding complex social realities and informing transformative change.

Table 3.10 Comparative summary table: strategies for minimizing survey bias

Type of bias	Source	Mitigation strategy	Case study example
Question wording bias	Leading or presupposing questions	Use open-ended, neutral phrasing; pilot testing	Nigerian education reform surveys revised leading questions
Social desirability bias	Respondents giving "acceptable" answers	Anonymous surveys; confidentiality assurances	HIV adherence research in Uganda
Interviewer bias	Researcher cues, tone, or body language	Interviewer training; reflexivity; standardized protocols	Nigerian governance corruption study
Cultural/language bias	Misunderstanding of terms or abstract concepts	Local adaptation, back-translation, use of culturally relevant examples	Gender empowerment study in rural India
Recall bias	Difficulty remembering past events	Use timelines, diaries, or short recall periods	Healthcare patient recall of treatment experiences
Selection bias	Non-representative participants	Purposive and diverse sampling; transparent reporting	Policy research on marginalized groups in development planning

3.7 Ethical Considerations in Qualitative Research

Ethics are the foundation of qualitative research because of its inherently interactive, interpretive, and often sensitive nature. Unlike qualitative approaches, which tend to rely on numerical abstraction and distance between researcher and subject, qualitative research requires close engagement with people, communities, or institutions in their lived contexts. This intimacy brings opportunities for deeper understanding but also magnifies the risk of harm, exploitation, misrepresentation, or loss of trust. For this reason, ethical considerations must be central to every stage of qualitative research, from design and data collection to analysis, dissemination, and archiving. This section provides a comprehensive discussion of ethical considerations in qualitative research. It begins by exploring the philosophical foundations of research ethics, then addresses core ethical principles such as respect for persons, beneficence, justice, autonomy, and confidentiality. It examines how these principles apply in practice, including issues of informed consent, privacy, data security, researcher reflexivity, and power dynamics. Real-life case studies from education, healthcare, gender studies, and policy research are integrated throughout to demonstrate practical dilemmas and responses. Finally, the section offers strategies for ethically sound qualitative inquiry, concluding with reflections on emerging ethical challenges in the digital era.

Philosophical Foundations of Research Ethics

Qualitative research is guided by moral philosophy as much as by methodological rigor. Several traditions underpin its ethical landscape:

1. **Deontological ethics** (duty-based): Researchers have obligations—such as honesty, respect, and protection of participants—regardless of outcomes. This is reflected in principles like informed consent and confidentiality.
2. **Consequentialism** (outcome-based): The moral worth of an action is judged by its outcomes. For example, in healthcare research, potential benefits to patients may justify certain risks if managed properly.
3. **Virtue ethics**: Focuses on the researcher's character and moral responsibility. Reflexivity and humility are essential, as researchers must constantly evaluate their own biases and motives.
4. **Relational ethics**: Emphasizes relationships and context, recognizing that ethics are not just formal rules but lived practices negotiated between researcher and participant.

Qualitative inquiry, with its reliance on narrative, dialogue, and context, is particularly shaped by relational and virtue ethics. Unlike experimental settings where consent is formalized and interaction minimal, qualitative research often involves prolonged engagement, emergent designs, and evolving researcher–participant relationships.

Core Ethical Principles in Qualitative Research

1. **Respect for Persons and Autonomy**: Respecting participants as autonomous agents means recognizing their right to decide whether to take part in a study, how much to disclose, and how their information is used.

- **Example**: In an education reform study in Nigeria, teachers were invited to share classroom practices. Some initially agreed but later withdrew due to concerns that administrators might misuse the information. Respecting their autonomy required honoring withdrawal requests, even when it complicated data analysis.

2. **Beneficence and Non-Maleficence**: Researchers must maximize potential benefits and minimize harm. While physical risks are rare in qualitative research, emotional, social, or reputational harm can be significant.

- **Example**: A healthcare study on terminal illness narratives risked retraumatizing patients. The research team implemented psychological support referrals to balance the benefits of storytelling with the risk of distress.

3. **Justice**: Justice requires fairness in participant selection and in the distribution of risks and benefits. Vulnerable groups should not be targeted simply for convenience.

- **Example**: In gender studies research on domestic violence, recruiting only women from shelters might stigmatize them. A just approach included engaging policymakers and social workers to contextualize findings without exploiting participants' trauma.

4. **Confidentiality and Privacy**: Maintaining confidentiality is crucial, yet qualitative data (e.g., interviews, observations) are often rich in detail that can inadvertently reveal identities.

- **Example**: In a policy evaluation project, community leaders' statements were so distinctive that anonymization risked erasing their authority. Researchers used pseudonyms but preserved contextual meaning to balance confidentiality with fidelity.

5. **Reflexivity and Power Dynamics:** Qualitative research is never neutral. Researchers' identities, assumptions, and power influence interactions. Reflexivity requires ongoing critical self-examination.

- **Example**: In cross-cultural studies, Western researchers studying rural African communities often carry power imbalances. Ethical reflexivity involves recognizing privilege, co-creating meaning with participants, and avoiding extractive practices.

Informed Consent in Practice

Informed consent is more complex in qualitative research than in clinical trials. Because qualitative studies often evolve, participants may not fully know what will be asked of them.

- **Ongoing Consent**: Consent should be revisited throughout the study. For example, in a longitudinal education study, parents of children gave initial consent but researchers also sought periodic assent from the children as they matured.
- **Collective Consent**: In indigenous research, communities may demand collective decision-making before granting access.
- **Implied Consent**: In ethnographic research in public spaces, obtaining signed forms from every individual may be impractical. Instead, researchers must negotiate what constitutes ethical engagement in context.

Confidentiality and Data Security

Qualitative data—often in the form of recordings, transcripts, and fieldnotes—pose unique challenges for maintaining confidentiality.

- **Anonymization Limits**: Small communities or organizations may be easily identifiable even with pseudonyms.
- **Digital Risks**: Cloud storage and digital transcription tools may expose sensitive data to breaches.

- **Strategies**: Data encryption, secure servers, restricted access, and carefully considered reporting practices are essential.

Case Study (Healthcare): In mental health research, participants expressed fear that recorded interviews might be accessed by employers. The research team reassured participants by storing data offline, demonstrating how ethical practices must adapt to technological realities.

Ethical Issues in Analysis and Interpretation
Beyond data collection, ethical challenges extend to how findings are analyzed and represented.

- **Misrepresentation**: Researchers must avoid distorting participants' voices to fit theoretical frameworks.
- **Ownership of Data**: Who owns narratives—the participant who lived them or the researcher who interprets them? Co-authorship or member checking can resolve disputes.
- **Voice and Empowerment**: Giving participants opportunities to review and comment on interpretations respects their agency.

Case Study (Gender Studies): In research on LGBTQ+ communities, participants challenged the researcher's framing of their narratives as "deviant." Through collaborative workshops, participants reshaped the analysis, transforming the study into an empowering rather than stigmatizing process.

Ethical Considerations in Dissemination
Dissemination carries its own risks, particularly when findings could harm individuals or groups.

- **Risk of Stigmatization**: Publishing findings about high dropout rates in marginalized schools may reinforce negative stereotypes. Researchers must contextualize structural issues rather than blame communities.
- **Accessible Dissemination**: Ethical research requires sharing results with participants and communities, not just academic audiences.
- **Case Study (Policy Research):** In a governance study, findings revealed corruption within local councils. The research team anonymized institutions while sharing results with oversight bodies, striking a balance between transparency and protection.

Ethical Challenges in Cross-Cultural and Online Qualitative Research

1. **Cross-Cultural Research**

- Language barriers, cultural norms, and differing concepts of privacy complicate ethics.

- Example: In rural education research, translating consent forms literally did not capture cultural nuances. Researchers worked with local leaders to redesign consent as oral agreements.

2. **Online Research**

    - The rise of digital ethnography and online surveys creates new dilemmas: Are social media posts "public" or private?
    - Example: In a COVID-19 study of online support groups, researchers debated whether to quote forum posts. They ultimately sought group moderators' permission, respecting collective boundaries.

Strategies for Ethical Qualitative Research

1. **Ethics Committees and Institutional Review Boards (IRBs)**: Provide oversight but may lack sensitivity to qualitative nuances. Researchers must educate boards about context-specific ethics.
2. **Reflexive Journaling**: Helps researchers continuously monitor ethical decisions.
3. **Member Checking**: Involves participants in validating interpretations.
4. **Collaborative Research**: Co-designing studies with communities reduces power imbalances.
5. **Ethical Training**: Continuous researcher education ensures awareness of evolving standards, particularly in digital research (Table 3.11).

Ethical considerations in qualitative research are not peripheral but central, shaping the trustworthiness, integrity, and impact of inquiry. From obtaining consent and safeguarding confidentiality to practicing reflexivity and responsibly disseminating findings, ethical practices determine whether qualitative research serves as a force for empowerment or exploitation. Real-life case studies in education, healthcare, gender studies, and policy research show that ethical dilemmas are context-specific, requiring sensitivity, adaptability, and humility. As qualitative research increasingly engages digital spaces and diverse cultural contexts, new challenges emerge, demanding innovative ethical responses. Ultimately, ethical qualitative research is not only about compliance with rules but about fostering respectful, just, and meaningful relationships with participants and communities. By embedding ethics into every stage of inquiry, researchers can ensure that their work contributes responsibly to knowledge and social transformation.

Table 3.11 Comparative summary table: ethical challenges across contexts

Ethical area	Qualitative research challenges	Case study example	Key strategy
Informed consent	Ongoing, emergent, collective consent issues	Education reform study with parents & children	Use ongoing/assent processes
Confidentiality	Risk of identification in small samples	Mental health interviews in healthcare	Encryption, anonymization, careful reporting
Reflexivity and power	Researcher bias, cultural hierarchies	Cross-cultural rural studies	Reflexive journaling, co-creation
Data interpretation	Misrepresentation of voices	LGBTQ+ gender studies	Member checking, collaborative analysis
Dissemination	Risk of stigmatization or political backlash	Policy corruption study	Contextualize findings, anonymize
Online research	Public versus private data, digital consent	COVID-19 online groups	Moderator permissions, group agreements

References

1. Obi, K.C., Iwuanyanwu, O.: Innovation in the assessment of primary school mathematics: an approach to revamping Nigeria education. J. Teach. Perspect. **18**(1), 51–66 (2023), [Online]. Available: http://repositorio.unan.edu.ni/2986/1/5624.pdf%0Ahttp://fiskal.kemenkeu.go.id/ejournal%0Ahttp://dx.doi.org/10.1016/j.cirp.2016.06.001%0Ahttp://dx.doi.org/10.1016/j.powtec.2016.12.055%0Ahttps://doi.org/10.1016/j.ijfatigue.2019.02.006%0Ahttps://doi.org/10.1
2. Chari, A., et al.: Dynamic capabilities for circular manufacturing supply chains—exploring the role of industry 4.0 and resilience. Bus. Strateg. Environ. **31**(5), 2500–2517 (2022). https://doi.org/10.1002/bse.3040
3. Cluver, L., et al.: A parenting programme to prevent abuse of adolescents in South Africa: study protocol for a randomised controlled trial. Trials. **17**(1), 1–10 (2016). https://doi.org/10.1186/s13063-016-1452-8
4. Cheah, C.W., Low, B., Lee, C.K.C.: Sustainable housing development: the legitimacy-seeking perspective. J. Bus. Ind. Mark. **36**(6), 1027–1041 (2020). https://doi.org/10.1108/JBIM-07-2020-0318
5. Ghavifekr, S., Razak, A., Ghani, M., Ran, N., Meixi, Y., Tengyue, Z.: ICT integration in education: incorporation for teaching and learning improvement. Malaysian Online J. Educ. Technol. **2**(2), 24–45 (2014)
6. Morin, A.: Self-awareness part 1: definition, measures, effects, functions, and antecedents. Soc. Personal. Psychol. Compass. **5**(10), 807–823 (2021). https://doi.org/10.1111/j.1751-9004.2011.00387.x
7. Erdogan, M.R., Camgoz, S.M., Karan, M.B., Berument, M.H.: The switching behaviour of large-scale electricity consumers in the Turkish electricity retail market. Energy Policy. **160**(2020), 112701 (2022). https://doi.org/10.1016/j.enpol.2021.112701

8. Guzman, K., et al.: Acceptability and feasibility of digital adherence technologies for drug-susceptible tuberculosis treatment supervision: a meta-analysis of implementation feedback. PLOS Digit. Heal. **2**(8), e0000322 (2023). https://doi.org/10.1371/journal.pdig.0000322
9. Abdulkarim, S., Mohktar, S., Babaji, N.A., Maisalati, M.S., Musa, M.K., Pwara, K.E.: Roles of road transport policy on economic developmentin federal capital territory Abuja, Nigeria: stakeholders perception. J. Tour. Hosp. Environ. Manag. **7**(27), 179–197 (2022). https://doi.org/10.35631/jthem.727015
10. Owusu-Ansah, A., Soyeh, K.W., Asabere, P.K.: Developer constraints on housing supply in urban Ghana. Int. J. Hous. Mark. Anal. **12**(1), 59–73 (2019). https://doi.org/10.1108/IJHMA-07-2018-0052
11. Levacher, J., Spinath, F.M., Becker, N., Hahn, E.: How did the beginnings of the global COVID-19 pandemic affect mental Well-being? PLoS One. **18**(1), 1–27 (2023). https://doi.org/10.1371/journal.pone.0279753
12. Mertens, A., et al.: Child wasting and concurrent stunting in low- and middle-income countries. Nature. **621** (2023). https://doi.org/10.1038/s41586-023-06480-z
13. Tende, F.B., Ekanem, I.S.: Strategic agility: an intervention prescription to competitive advantage of small businesses in Nigeria. Int. J. Bus. Syst. Econ. **6**(4), 1401–1423 (2018)
14. Pleasence, P., Balmer, N.: Measuring the accessibility and equality of civil justice. **10**(2) (2018). https://doi.org/10.1007/s40803-018-0079-0
15. Kale, E., Aknar, A., Başar, Ö.: Absorptive capacity and firm performance: the mediating role of strategic agility. Int. J. Hosp. Manag. **78**(2018), 276–283 (2019). https://doi.org/10.1016/j.ijhm.2018.09.010
16. Noorzai, E., Gharouni Jafari, K., Moslemi Naeni, L.: Lessons learned on selecting the best mass housing method based on performance evaluation criteria in Iran. Int. J. Constr. Educ. Res. **00**(00), 1–19 (2020). https://doi.org/10.1080/15578771.2020.1867258
17. Oluwatoyin, A., Oluseun, A.: Total quality management: a test of the effect of TQM on performance and stakeholder satisfaction. J. Clin. Nurs. **5**(1), 2007–2012 (2014)
18. Gottman, J.M., et al.: Predicting marital happiness and stability from newlywed interactions published by: national council on family relations predicting marital happiness and stability from newlywed interactions. J. Marriage Fam. **60**(1), 5–22 (2017). https://doi.org/10.1002/job
19. Olajide, O.T.: Employees' attitudes and efficiency of human resource management practices: evidence from Nigeria. Eur. J. Bus. Manag. **6**(31), 68–73 (2014), [Online]. Available: www.iiste.org
20. Schreurs, B., Duff, A., Le Blanc, P.M., Stone, T.H.: Publishing qualitative careers research: challenges and recommendations. Career Dev. Int. **27**(1), 79–98 (2022). https://doi.org/10.1108/CDI-08-2021-0217
21. Olatoye, T.A.: Analysis of modal choice of residents in Lagos state. J. Hum. Ecol. **65**(1–3) (2019). https://doi.org/10.31901/24566608.2019/65.1-3.3157
22. Rehman, S.U., Ashfaq, K., Bresciani, S., Giacosa, E., Mueller, J.: Nexus among intellectual capital, interorganizational learning, industrial internet of things technology and innovation performance: a resource-based perspective. J. Intellect. Cap. **24**(2), 509–534 (2023). https://doi.org/10.1108/JIC-03-2021-0095
23. Ahmadi, H., et al.: Organizational factors affecting the adoption of a technological innovation: a study of the hospital information system (HIS). In: 2019 5th international conference web research ICWR 2019, pp. 240–245 (2019). https://doi.org/10.1109/ICWR.2019.8765270
24. Muhammed, S., Zaim, H.: Peer knowledge sharing and organizational performance: the role of leadership support and knowledge management success. J. Knowl. Manag. **24**(10), 2455–2489 (2020). https://doi.org/10.1108/JKM-03-2020-0227
25. Chierici, R., Mazzucchelli, A., Garcia-Perez, A., Vrontis, D.: Transforming big data into knowledge: the role of knowledge management practice. Manag. Decis. **57**(8), 1902–1922 (2019). https://doi.org/10.1108/MD-07-2018-0834
26. Ali, H.Y., Danish, R.Q., Asrar-ul-Haq, M.: How corporate social responsibility boosts firm financial performance: the mediating role of corporate image and customer satisfaction. Corp. Soc. Responsib. Environ. Manag. **27**(1), 166–177 (2020). https://doi.org/10.1002/csr.1781

27. Galindo-Rueda, F., Millot, V.: Measuring design and its role in innovation. OECD Sci. Technol. Ind. Work. Pap. pp. 1–51 (2015), [Online]. Available: http://www.oecd-ilibrary.org/science-and-technology/measuring-design-and-its-role-in-innovation_5js7p6lj6zq6-en
28. Njonge, T.: Influence of psychological Well-being and school factors on delinquency, during the Covid-19 period among secondary school students in selected schools in Nakuru County: Kenya. **7**(2454), 1175–1189 (2023). https://doi.org/10.47772/IJRISS
29. Ltifi, M., Hichri, A.: The effects of corporate governance on the customer's recommendations: a study of the banking sector at the time of COVID-19. J. Knowl. Manag. **26**(1), 165–191 (2022). https://doi.org/10.1108/JKM-06-2020-0471
30. Nagler, S., Mazumdar, S.: Towards blockchain-IoT based shared mobility: car-sharing and leasing as a case study. J. Netw. Comput. Appl. **200**, 103316 (2022). https://doi.org/10.1016/j.jnca.2021.103316
31. Mursyidto, M.I.: The role of information communication technology in records creation and distribution in Uganda communications commission. Implement. Sci. **39**(1), 1–15 (2014). https://doi.org/10.4324/9781315853178
32. Horvat, J., Sharma, K., Bobek, S.: Key factors for knowledge management: pilot study in IT SMEs. FIIB Bus. Rev. **5**(3), 32–40 (2016). https://doi.org/10.1177/2455265820160306
33. Casey, E.M.D., Mojarrabi, M., Hannan-Jones, M.T., Bogard, J.R.: Measuring dietary intake in low-and middle-income countries: a systematic review of the methods and tools for estimating fish and seafood intake. Nutr. Rev. **00**(0), 1–14 (2023). https://doi.org/10.1093/nutrit/nuad067
34. Oba-Adenuga, O.A., Asikhia, O.U., Akpa, V.O.: Self-management and employee research output in selected private universities in Ogun state, Nigeria. Eur. J. Hum. Resour. Manag. Stud. **5**(4), 180–196 (2022). https://doi.org/10.46827/ejhrms.v5i4.1277
35. Liu, S., Fan, Y., Luh, D., Teng, P.: Applied sciences organizational culture: the key to improving Service Management in Industry 4.0 (2022)
36. Deliu, D.: The intertwining between corporate governance and knowledge Management in the Time of Covid-19-a framework. J. Emerg. Trends Mark. Manag. **1**(1), 93–110 (2020)
37. Siddiqui, F., Akhund, M.A., Ali, T.H., Khahro, S.H., Khoso, A.R., Imad, H.U.: Barriers in adoption of building information Modeling in Pakistans construction industry. Indian J. Sci. Technol. **12**(25), 1–7 (2019). https://doi.org/10.17485/ijst/2019/v12i25/142325
38. Artemova, A.: Engaging generation Z through social media marketing. Tour. Manag. **2**(6), 1–130 (2018), [Online]. Available: https://www.theseus.fi/bitstream/handle/10024/142658/Artemova_Alexandra.pdf?sequence=1&isAllowed=y
39. Philip Onyango, W.: Effects of organization culture on change management: a case of the vocational training Centre for the Blind and Deaf Sikri. Eur. J. Bus. Manag. **6**(34), 2222–2839 (2019)
40. Shafiu, A.M., Manaf, H.A., Muslim, S.: The impact of leadership on organizational performance. Int. J. Recent Technol. Eng. **8**(3), 7573–7576 (2019). https://doi.org/10.35940/ijrte.c6158.098319
41. Balletto, G., Ladu, M., Milesi, A., Camerin, F., Borruso, G.: Walkable City and military enclaves: analysis and decision-making approach to support the proximity connection in urban regeneration. Sustain. **14**(1) (2022). https://doi.org/10.3390/su14010457
42. Alasan, A.I., Eyanuku, J.P.: Human resource management practices and Nigerian banks' performance. LASU J. Employ. Relations Hum. Resour. Manag. **2**(1), 124–140 (2020). https://doi.org/10.36108/ljerhrm/0202.02.0190
43. Yusof, M.M.: An evaluation framework for health information systems: human, organization and technology-fit factors (HOT-fit). Int. J. Med. Inform. **77**(6), 386–398 (2008). https://doi.org/10.1016/j.ijmedinf.2007.08.011
44. Rohan, R., Pal, D., Hautamäki, J., Funilkul, S., Chutimaskul, W., Thapliyal, H.: A systematic literature review of cybersecurity scales assessing information security awareness. Heliyon. **9**(3) (2023). https://doi.org/10.1016/j.heliyon.2023.e14234
45. Reed, P.M., et al.: Multisector dynamics: advancing the science of complex adaptive human-earth systems. Earth's Futur. **10**(3), 1–18 (2022). https://doi.org/10.1029/2021EF002621

46. Pucciarelli, F., Kaplan, A.: Competition and strategy in higher education: managing complexity and uncertainty. Bus. Horiz. **59**(3), 311–320 (2016). https://doi.org/10.1016/j.bushor.2016.01.003
47. Othman, I., Al-Ashmori, Y.Y., Rahmawati, Y., Mugahed Amran, Y.H., Al-Bared, M.A.M.: The level of building information modelling (BIM) implementation in Malaysia. Ain Shams Eng. J. **12**(1), 455–463 (2021). https://doi.org/10.1016/j.asej.2020.04.007
48. Ju, M., Nguyen, N.T.H., Miwa, M., Ananiadou, S.: An ensemble of neural models for nested adverse drug events and medication extraction with subwords. J. Am. Med. Informatics Assoc. **27**(1), 22–30 (2020). https://doi.org/10.1093/jamia/ocz075
49. Adesola, S., den Outer, B., Mueller, S.: New entrepreneurial worlds: can the use of role models in higher education inspire students? The case of Nigeria. J. Entrep. Emerg. Econ. **11**(4), 465–491 (2019). https://doi.org/10.1108/JEEE-08-2018-0076
50. Wei, Q., et al.: A study of deep learning approaches for medication and adverse drug event extraction from clinical text. J. Am. Med. Informatics Assoc. **27**(1), 13–21 (2020). https://doi.org/10.1093/jamia/ocz063
51. Khodeir, L.M., Nabawy, E.M.: Responsive human resource framework for design and building of mega housing development projects in Egypt. Ain Shams Eng. J. **12**(2), 2371–2383 (2021). https://doi.org/10.1016/j.asej.2020.09.025
52. Dube, D.P., Mohanty, R.P.: Application of grounded theory in construction of factors of internal efficiency and external effectiveness of cyber security and developing impact models. Organ. Cybersecurity J. Pract. Process People. **3**(1), 41–70 (2023). https://doi.org/10.1108/ocj-04-2022-0009
53. Elia, N., Von Elm, E., Chatagner, A., Pöpping, D.M., Tramèr, M.R.: How do authors of systematic reviews deal with research malpractice and misconduct in original studies? A crosssectional analysis of systematic reviews and survey of their authors. BMJ Open. **6**(3), 1–10 (2016). https://doi.org/10.1136/bmjopen-2015-010442

Chapter 4
Observational Data Collection for Qualitative Research

4.1 Observational Data Collection

Qualitative research has gained increasing prominence across academic and applied domains because of its ability to capture rich, contextually grounded, and nuanced insights into human behaviour, experiences, and social interactions. Unlike qualitative methods that rely heavily on numerical measurement and statistical generalization, qualitative research seeks to uncover meaning, interpretation, and understanding within specific contexts. At the heart of this paradigm lies observation, one of the most fundamental and versatile data collection methods. Observation allows researchers to engage directly with participants in their natural environments, gathering insights that may not be fully accessible through interviews, surveys, or textual analysis. Observational data collection is particularly significant in qualitative research because it provides a direct lens into "what people do" rather than "what people say they do." While self-reported data is invaluable, it is often constrained by memory, social desirability, or cognitive bias. Observation, in contrast, permits the researcher to witness behaviour in real time, capturing subtleties of interaction, gestures, rituals, and environmental cues that shape meaning. This section explores observational data collection in qualitative research, delving into its types, strengths and weaknesses, ethical implications, and applications in different fields. Real-world case studies will illustrate the practical significance of observational methods, and critical discussions will highlight both methodological challenges and strategies to address them [1].

The Role of Observation in Qualitative Research

Observation has long been regarded as an essential tool in the qualitative research tradition, especially within ethnography, anthropology, sociology, and applied fields such as education, healthcare, and organizational behaviour. Its primary value lies in its ability to provide context-rich data about how people interact with each other and with their environment. Unlike surveys or interviews that rely on participant accounts,

F. A. Alaba et al., *Fundamentals of Qualitative Research*, Studies in Systems, Decision and Control 663, https://doi.org/10.1007/978-3-032-18427-6_4

observation immerses the researcher in the social world under study, allowing them to see behaviour as it unfolds in natural settings. For example, while a teacher may describe their classroom management strategies in an interview, observing them in action can reveal the nuances of how they respond to disruptions, manage transitions, and establish rapport with students. Similarly, in healthcare research, observing nurse-patient interactions can uncover subtle dynamics of care, empathy, or power relations that may not be articulated in formal accounts [2–4].

Observation thus provides:

1. **Descriptive Richness**—Capturing details of actions, setting, and context.
2. **Behavioural Insight**—Going beyond what participants report to what they do.
3. **Contextual Grounding**—Linking individual actions to broader environmental and cultural patterns.

Types of Observation in Qualitative Research

Observation is not monolithic. Different types of observational strategies exist, depending on the level of researcher involvement, structure, and context of the study. Broadly, these can be categorized into participant observation, non-participant observation, structured observation, unstructured observation, overt observation, and covert observation [5].

1. **Participant Observation**: In participant observation, the researcher actively engages in the activities of the group being studied. This approach is common in ethnographic research where immersion in the daily lives of participants is crucial. Participant observation enables researchers to build rapport, gain insider perspectives, and capture meanings from the participant's point of view.

Case Example: Bronisław Malinowski's ethnographic work in the Trobriand Islands is one of the earliest classic examples. By living among the Trobriand Islanders, he observed kinship, trade, and rituals, producing rich ethnographic accounts that shaped modern anthropology. In contemporary education research, participant observation might involve a researcher volunteering as a teaching assistant in a classroom to observe teacher-student interactions while participating in daily routines.

2. **Non-participant Observation**: Here, the researcher observes without direct involvement. This allows for greater objectivity since the researcher is less likely to influence participant behaviour. Non-participant observation is often used in organizational studies or healthcare settings where researchers observe team meetings, patient care, or customer service without active participation.

Case Example: In healthcare, researchers have used non-participant observation to study how surgical teams coordinate during operations. By standing outside the immediate workflow, the observer can document roles, communication, and potential breakdowns in teamwork.

3. **Structured Observation**: Structured observation involves using predefined categories, checklists, or rating scales to systematically record behaviours. While often associated with qualitative methods, structured observation can also

complement qualitative research when specific behavioural aspects need to be consistently captured.

Case Example: In classroom research, structured observation protocols are often used to record the frequency of teacher questioning, student participation, or types of instructional strategies. These observations can then be qualitatively analyzed to interpret teaching styles and classroom dynamics.

4. **Unstructured Observation**: Unstructured observation is open-ended, allowing the researcher to freely record all relevant details without predefined categories. It is especially useful in exploratory research when the researcher seeks to capture the complexity of a social setting.

Case Example: An unstructured ethnographic study in a refugee camp may involve documenting daily life, rituals, and social interactions without fixed categories, allowing the researcher to generate insights grounded in the emergent data.

5. **Overt Versus Covert Observation**

   - **Overt observation** occurs when participants are aware that they are being observed. While this raises issues of reactivity (the Hawthorne effect), it is ethically straightforward since informed consent is obtained.
   - **Covert observation** is conducted without participants' knowledge. This can minimize reactivity but raises ethical challenges, particularly regarding consent and privacy.

Case Example: In workplace studies, overt observation is common, where employees consent to being observed during training. In contrast, covert observation might be employed in public spaces—such as observing social behaviour in a park—where individuals are unaware they are being studied.

Advantages of Observational Data Collection

Observation offers several unique advantages in qualitative research [6, 7]:

1. **Direct Access to Behaviour**: Observation provides first-hand data about actions rather than relying on self-reports, which may be biased or inaccurate.
2. **Contextual Understanding**: By situating behaviour within physical, cultural, and social contexts, researchers can capture a holistic understanding of phenomena.
3. **Nonverbal Cues**: Observation allows researchers to capture gestures, tone, body language, and other subtle indicators often missed in interviews or surveys.
4. **Discovery of the Unexpected**: Open-ended observation often reveals behaviours or patterns that researchers had not initially anticipated.
5. **Rich Descriptive Data**: Observation generates detailed field notes, producing thick description that forms the foundation of qualitative interpretation.

Limitations and Challenges of Observation

Despite its strengths, observational data collection presents several challenges [8]:

1. **Observer Bias**: Researchers may unconsciously interpret behaviour through their own cultural lens or personal expectations.
2. **Reactivity**: Participants may alter their behaviour when aware of being observed (Hawthorne effect).
3. **Time-Intensiveness**: Observational research often requires long periods in the field to generate sufficient data.
4. **Limited Generalizability**: Findings are usually context-specific and may not easily generalize to broader populations.
5. **Ethical Concerns**: Issues of consent, privacy, and deception (in covert studies) pose significant ethical challenges.
6. **Recording Difficulties**: Capturing detailed field notes in real time can be overwhelming, and reliance on memory risks omission.

Applications of Observation in Different Fields

1. **Education**: Observation is a cornerstone of education research. Classroom ethnography, for instance, uses participant and non-participant observation to document teaching methods, classroom cultures, and student learning.

Case Example: A study on "inclusive education" observed classrooms where students with disabilities were integrated into mainstream schools. Through participant observation, researchers documented how teachers adapted lesson plans, the nature of peer interactions, and barriers to inclusion. These insights went beyond what surveys or policy documents could reveal.

2. **Healthcare**: Observation is critical in healthcare research, particularly in studying interactions between healthcare providers and patients. It uncovers both clinical practices and the relational aspects of care.

Case Example: A study in intensive care units used non-participant observation to explore communication breakdowns among healthcare teams. Researchers observed ward rounds, noting subtle nonverbal cues, hierarchical dynamics, and decision-making processes that shaped patient outcomes.

3. **Organizational Studies**: In business and organizational research, observation helps analyze workplace culture, teamwork, leadership, and customer service interactions.

Case Example: Observing employees in open-plan offices provided insights into how physical space influences collaboration, privacy, and productivity. Researchers found that while open offices facilitated interaction, they also increased distractions, leading to mixed outcomes for teamwork.

4. **Gender and Social Studies**: Observation is widely used in social sciences to study issues of identity, power, and inequality.

Case Example: Ethnographic observation in public spaces, such as urban transport systems, has been used to study gendered experiences of safety. Observers documented women's strategies for navigating harassment, revealing the lived realities behind statistical reports on gender-based violence.

Ethical Considerations in Observational Research

Ethics is central to observational research. Key concerns include [9, 10]:

1. **Informed Consent**: In overt observation, participants must be informed about the purpose, procedures, and potential risks of the study.
2. **Privacy**: Observing in public versus private spaces raises distinct issues. While behaviour in public spaces may be observed ethically, care must still be taken to anonymize data.
3. **Covert Observation**: Though sometimes justified (e.g., when studying sensitive topics like workplace discrimination), covert methods must be carefully weighed against participants' rights.
4. **Confidentiality**: Researchers must ensure that detailed field notes do not inadvertently identify participants.
5. **Researcher Safety**: Immersive observation in high-risk contexts (e.g., studying marginalized groups or volatile communities) may place researchers at personal risk.

Strategies for Enhancing Observational Research

To address the challenges outlined, researchers often employ strategies such as [11]:

1. **Triangulation**: Combining observation with interviews, surveys, or document analysis to cross-validate findings.
2. **Reflexivity**: Maintaining reflexive journals where researchers critically reflect on their positionality and potential biases.
3. **Prolonged Engagement**: Spending extended time in the field to build trust and reduce reactivity.
4. **Systematic Recording**: Using audio-visual tools (with consent) or structured note-taking to enhance accuracy.
5. **Member Checking**: Sharing observations with participants to verify accuracy and interpretation.

Observational data collection remains one of the most powerful methods in qualitative research. By directly engaging with social realities, researchers can uncover nuanced, context-rich insights into human behaviour, interaction, and meaning-making. Although challenges such as bias, reactivity, and ethical dilemmas exist, careful design, reflexivity, and methodological triangulation can mitigate these issues. Real-life applications across education, healthcare, organizational studies, and social research demonstrate the enduring relevance of observation in addressing complex research questions. Ultimately, observational data collection reflects the essence of qualitative inquiry: a commitment to understanding human experience as it is lived, performed, and enacted in real-world contexts [12]. In a research landscape increasingly concerned with depth, context, and meaning, observation remains indispensable (Table 4.1).

Table 4.1 Comparative table of observational approaches in qualitative research

Dimension	Structured observation	Unstructured observation	Participant observation	Non-participant observation
Definition	Pre-determined categories and behaviours recorded systematically	Flexible, open-ended, without fixed categories	Researcher actively participates in the setting while observing	Researcher remains detached, observing without involvement
Purpose	To quantify and compare specific behaviours or events	To capture rich, holistic, and contextual data	To gain insider perspectives and understand meanings from within	To minimize researcher influence and maintain objectivity
Degree of control	High—Researcher dictates what is observed and recorded	Low—Observations evolve with the situation	Moderate—Researcher balances participation and observation	High—Researcher maintains distance from subjects
Data type	Qualitative tendencies, often frequency counts	Narrative, descriptive, interpretive data	Rich, contextual insights from both observation and participation	More detached, descriptive accounts
Flexibility	Low—Rigid categories and checklists	High—Flexible, adaptable in real-time	Moderate—Flexible but limited by participant role	Moderate—Structured observation possible, but without direct engagement
Researcher role	Observer as recorder	Observer as explorer	Observer as participant (insider)	Observer as outsider (non-intrusive)
Strengths	Enables systematic comparison			

4.2 Techniques for Collecting Data Through Observation

Observation has long been regarded as one of the cornerstones of qualitative research, offering a direct and often nuanced pathway into understanding human experiences, interactions, and cultural contexts. Unlike qualitative methods, which rely on structured instruments to gather measurable data, observation allows researchers to enter the "lived world" of participants. Through careful, systematic, and ethically guided observation, researchers can document behaviours, settings, and interactions that may otherwise go unnoticed or unarticulated in interviews or surveys. This section explores the major techniques for collecting observational data in qualitative research, their theoretical underpinnings, and real-life case studies across diverse disciplines such as education, healthcare, anthropology, gender studies, and organizational research. By weaving together methodological explanations with concrete examples, the section demonstrates how observation techniques serve as both scientific tools and humanistic practices for uncovering meaning [13, 14].

1. **Structured Observation**

Definition and Characteristics

Structured observation involves systematically recording behaviours or events according to a predefined framework. Researchers enter the field with a clear coding scheme, checklist, or set of categories that guide what is observed. Unlike unstructured approaches, structured observation prioritizes consistency, reliability, and replicability. Structured observation is often used in applied fields where comparability across participants or contexts is crucial, such as education, workplace studies, and healthcare monitoring [15].

Case Study: Classroom Dynamics in Education Reform

A well-known application of structured observation was the Classroom Assessment Scoring System (CLASS) developed by Pianta et al. (2008), which was applied in U.S. schools to evaluate teacher effectiveness. Observers recorded specific teacher behaviours (e.g., instructional support, emotional climate, classroom organization) across multiple classrooms. This structured approach allowed policymakers to evaluate reforms in teaching methods during the "No Child Left Behind" era. The findings revealed that classrooms with higher-quality interactions produced better academic and social outcomes for students, directly informing federal teacher evaluation policies. This case illustrates how structured observation provides consistent, comparable data that can scale to system-wide reforms.

Strengths and Limitations

1. **Strengths**: High reliability, replicability, ease of comparison across contexts, strong for hypothesis testing.
2. **Limitations**: May overlook subtle, emergent, or culturally specific meanings not captured by predefined codes.

2. **Unstructured Observation**

Definition and Characteristics

Unstructured observation is open-ended, flexible, and exploratory. Researchers immerse themselves in natural settings without a rigid checklist, allowing behaviours, interactions, and meanings to emerge organically. This approach is especially valuable when exploring new or poorly understood phenomena [16].

Case Study: Healthcare and Patient-Doctor Interactions

In the 1990s, researchers studying doctor-patient consultations in the UK's NHS employed unstructured observation in clinics to understand how communication affected patient satisfaction. Instead of focusing solely on clinical outcomes, researchers observed the subtleties of tone, eye contact, and waiting room dynamics. These observations revealed that non-verbal cues such as body posture or attentiveness shaped patient trust as much as the medical advice given. This insight led to the development of new training modules in medical schools that emphasise communication and empathy, demonstrating how unstructured observation can reveal latent dimensions of human interaction [17].

Strengths and Limitations

1. **Strengths**: Provides rich, holistic data, captures emergent phenomena, ideal for exploratory studies.
2. **Limitations**: Low comparability, high potential for researcher bias, challenges in replication.

3. **Participant Observation**

Definition and Characteristics

Participant observation requires researchers to immerse themselves in a setting by actively engaging in participants' activities while simultaneously observing them. Popularized by anthropologists such as [18], this method remains central to ethnography.

Case Study: Gender Studies and Everyday Resistance

A powerful example comes from [19] ethnography of Malaysian women factory workers in Export Processing Zones. Ong participated in the daily routines of the women, blending in as a factory worker. Through this immersion, she observed how women subtly resisted exploitative working conditions—sometimes by slowing production or feigning illness—acts that would have been invisible in structured or survey-based methods. The work highlighted the intersection of gender, class, and globalization, reshaping feminist and labor studies.

Strengths and Limitations

1. **Strengths**: Provides insider perspective, captures lived experiences, allows exploration of tacit knowledge.
2. **Limitations**: Risk of "going native" (losing objectivity), ethical complexities in disclosure, time-intensive.

4. **Non-participant Observation**

Definition and Characteristics

In non-participant observation, researchers remain external to the activities of participants. They watch without actively engaging, maintaining an outsider's perspective. This approach is useful when researcher involvement could distort behaviours or when access to direct participation is ethically restricted.

Case Study: Organizational Research in Corporate Settings

An example is found in studies of corporate board meetings (Pettigrew, 1992). Researchers observed high-level meetings without participating in decision-making. This "fly on the wall" approach revealed how informal alliances, humor, and non-verbal signaling influenced strategic decisions—elements not captured in meeting minutes. This case demonstrates how non-participant observation uncovers organizational culture dynamics while minimizing researcher interference.

Strengths and Limitations

1. **Strengths**: Minimizes disruption, preserves natural behaviours, ethically safer in sensitive contexts.
2. **Limitations**: Less access to insider meanings, risk of superficial interpretations, possible observer bias.

5. **Covert Versus Overt Observation**

Definition and Distinction

1. **Covert Observation**: Participants are unaware of being observed, often used when awareness may alter behaviour (Hawthorne effect).
2. **Overt Observation**: Participants know they are being observed, allowing for ethical transparency.

Case Study: Workplace Surveillance and the Hawthorne Studies

The Hawthorne Studies [20, 21] at Western Electric in Chicago exemplify this distinction. Initially overt, workers were aware of researchers observing productivity under changing conditions. However, productivity rose regardless of lighting or other interventions—demonstrating the influence of being observed. Later, covert observation revealed more authentic worker interactions, emphasizing the methodological dilemmas between overt and covert approaches (Table 4.2).

Strengths and Limitations

1. **Covert**: Reduces reactivity but raises ethical concerns.
2. **Overt**: Ethically transparent but risks artificial behaviour.

6. **Recording Techniques in Observation**

Regardless of the observation style, recording methods play a vital role. Techniques include:

Table 4.2 Comparative on observational approaches

Technique	Characteristics	Example case study	Strengths	Limitations
Structured	Predefined categories, systematic recording	CLASS tool in U.S. classrooms	Reliable, comparable	Misses' emergent meanings
Unstructured	Open-ended, flexible	NHS doctor-patient interactions	Rich, exploratory	Hard to replicate, subjective
Participant	Researcher immerses in setting	Ong's Malaysian factory workers	Insider perspective	Risk of bias, time-intensive
Non-participant	Researcher observes externally	Corporate board meeting studies	Minimally disruptive	Limited insider insights
Covert	Participants unaware	Hawthorne Studies	Reduces reactivity	Ethical challenges
Overt	Participants aware	Educational reforms	Ethical transparency	Reactivity risk

- **Field Notes**: Detailed written records of observed behaviours, conversations, and context.
- **Audio/Video Recording**: Captures verbal and non-verbal cues, though raises ethical concerns.
- **Checklists and Rating Scales**: Common in structured observation.
- **Reflective Journals**: Capture the researcher's impressions, reflexivity, and evolving interpretations.

Case Example: Field Notes in Homelessness Research

In a study on homeless youth in Toronto, researchers used daily field notes from both participant and non-participant observation. Notes recorded not only observed behaviours but also the researcher's reflections on ethical dilemmas (e.g., when witnessing illegal activities). This dual-layer recording highlighted the methodological importance of reflexivity.

7. **Triangulation Through Observation**

Observation is rarely used in isolation; it is often triangulated with interviews, surveys, or document analysis to enhance validity.

Case Study: Education Policy in Sub-Saharan Africa

In studies evaluating free primary education policies in Kenya, researchers combined classroom observation with teacher interviews and policy document analysis. While interviews emphasized resource shortages, observations revealed issues like teacher absenteeism and overcrowded classrooms, offering a fuller picture of policy outcomes.

8. **Ethical Considerations**

Observation presents unique ethical dilemmas:

- **Informed Consent**: Especially in overt observation.
- **Privacy**: Risks of recording sensitive interactions.
- **Researcher Safety**: Particularly in covert or participant roles in risky environments.
- **Reflexivity**: Recognizing the observer's influence on data.

Case Example: Observing Domestic Violence Shelters

Researchers studying domestic violence shelters in the U.S. faced ethical challenges in protecting the identities of women and children while documenting shelter dynamics. Strategies included anonymization, non-intrusive observation, and participant-led consent processes.

Observation techniques in qualitative research form a spectrum—from highly structured, systematic checklists to deeply immersive ethnographic participation. Each approach carries distinct methodological, ethical, and practical implications. Structured methods ensure reliability but risk reductionism, while unstructured and participant observation provide richness but challenge replicability. Non-participant and covert observation can uncover behaviours otherwise hidden but raise critical ethical debates. Across diverse fields—education, healthcare, gender studies, anthropology, and organizational research—observation continues to serve as a powerful lens for understanding human behaviour in context. The choice of technique must align with research objectives, ethical responsibilities, and the realities of the field. Ultimately, the value of observation lies in its ability to bridge the gap between lived experience and scholarly interpretation, capturing both the visible and the invisible dimensions of human life.

4.3 Experimental Data Collection

Research, in its broadest sense, is the systematic process of inquiry aimed at generating new knowledge or validating existing knowledge. Within this process, data collection is central because it provides the raw material upon which analysis, interpretation, and conclusions are based. Among the different methods of data collection, experimental data collection is unique in its ability to establish cause-and-effect relationships through controlled manipulation and observation. Unlike observational or survey methods, which often rely on naturally occurring phenomena, experimental methods involve deliberate intervention by the researcher to observe the effect of one or more variables on an outcome of interest. Experimental data collection is the cornerstone of the scientific method. From the controlled laboratory experiments of natural sciences such as physics, chemistry, and biology to the randomized controlled trials (RCTs) in medicine, psychology, and education, the approach provides a rigorous way of testing hypotheses and building evidence. In social sciences, it allows researchers to test interventions and policies in a controlled manner, while in applied fields such as engineering or information technology, experimental methods underpin innovation and technological advancement. The importance of experimental data collection cannot be overstated. It ensures precision, replicability, and objectivity, provided that it is conducted with methodological rigor and ethical care. Nevertheless, it is not without challenges. Ethical considerations, practical limitations, cultural sensitivity, and the risk of bias remain critical issues that must be addressed. Furthermore, in today's digital age, innovations in AI, big data, and IoT devices are reshaping how experiments are conducted and how data is captured in real-time.

This section provides a comprehensive exploration of experimental data collection. It begins with a conceptual framework, before examining different experimental designs, techniques and procedures, ethical issues, challenges, and recent innovations. By integrating theoretical perspectives with practical examples, the discussion highlights the enduring significance and evolving nature of experimental data collection in contemporary research.

Conceptual Framework of Experimental Data Collection

Experimental data collection is grounded in the philosophy of positivism, which emphasizes empirical observation, measurement, and verification. The defining feature of experimental research is its reliance on controlled manipulation of variables. By controlling the environment and isolating the independent variable, researchers can more confidently infer causality [22, 23].

Core Principles of Experimental Data Collection

1. **Manipulation**

   Experimental research involves manipulating one or more independent variables to observe their effect on a dependent variable. For example, in a medical trial, a drug dosage may be manipulated to observe its effect on blood pressure.
2. **Control**

   The researcher seeks to hold constant or neutralize the effects of extraneous variables that might otherwise influence the dependent variable. This is achieved through controlled environments, placebo groups, or statistical techniques.
3. **Randomization**

   Random assignment ensures that participants have an equal chance of being assigned to experimental or control groups, thereby reducing selection bias.
4. **Replication**

   Experiments must be replicable to ensure reliability. Repetition across different samples, contexts, and times enhances the validity of findings.

Experimental Versus Observational Data

It is important to distinguish between experimental and observational data collection. While both involve systematic measurement, observational research does not involve manipulation of variables. For instance, a sociologist observing classroom behaviour is collecting observational data, whereas an educational researcher testing the effect of a new teaching method on student performance is collecting experimental data. The distinction lies in the researcher's intervention.

Qualitative Versus Qualitative Approaches

Although experiments are traditionally associated with qualitative methods, qualitative data can also emerge from experimental designs. For example, in usability testing of software, while performance metrics are qualitatively measured, participant feedback and facial expressions provide qualitative insights. Hence, experimental data collection is not strictly qualitative but may combine mixed methods for richer analysis. The conceptual framework underscores that experimental data collection is not merely a technical process but a methodological philosophy aimed at ensuring rigorous, reliable, and generalizable knowledge.

Types of Experimental Designs and Data Collection Methods

Experimental research can take multiple forms depending on the level of control, the setting, and the purpose. The main designs include true experiments, quasi-experiments, laboratory experiments, and field experiments.

1. **True Experiments (Randomized Controlled Trials)**: True experiments are considered the gold standard because they involve random assignment, control groups, and manipulation of variables. Randomized controlled trials (RCTs) are particularly common in medical research. For example, when testing a new vaccine, participants are randomly assigned to receive either the vaccine or a placebo, ensuring that any observed effect can be attributed to the vaccine rather than extraneous factors. Data collection in such studies involves clinical measurements, blood tests, and self-reported health outcomes.
2. **Quasi-experiments**: Quasi-experiments resemble true experiments but lack random assignment. For instance, an educational researcher may test a new curriculum in one school and compare results with a similar school that uses the old curriculum. While practical in real-world settings, quasi-experiments are vulnerable to selection bias. Data collection tools may include standardized tests, surveys, or classroom observations.
3. **Laboratory Experiments**: Laboratory experiments are conducted in controlled indoor environments, enabling precise manipulation and measurement of variables. For example, a psychology experiment testing the impact of sleep deprivation on cognitive performance may use computerized memory tests in a lab. Laboratory experiments allow high internal validity but may suffer from low external validity (generalizability).
4. **Field Experiments**: Field experiments take place in real-world settings, providing high ecological validity. An economist might run a field experiment to

test whether text-message reminders improve savings behaviour among low-income households. Data collection may involve digital logs from banking systems or participant interviews. Field experiments often strike a balance between control and realism.

5. **Natural Experiments**: Natural experiments occur when external events or policies create conditions resembling an experiment. For instance, if one state enacts a minimum wage increase while another does not, researchers can compare employment data to infer effects. Here, data is collected from administrative records, labor force surveys, and economic databases.

Data Collection Tools in Experimental Research

1. **Surveys and Questionnaires**: Useful in measuring attitudes, perceptions, or self-reported behaviours.
2. **Standardized Tests**: Common in education to measure learning outcomes.
3. **Observation**: Structured observation protocols record participant behaviour during experiments.
4. **Biometric Tools**: Heart rate monitors, EEG machines, and eye trackers provide physiological data.
5. **Digital and Sensor-Based Data**: IoT devices, smartphone apps, and wearable technology enable real-time data collection.
6. **Archival Data**: Administrative and institutional records may supplement primary data.

This classification highlights that different types of experimental designs are suited to different research questions and practical constraints (Table 4.3).

Techniques and Procedures in Experimental Data Collection

To ensure rigor, experimental data collection follows systematic procedures.

1. **Designing Data Collection Instruments:** Instruments must align with research objectives. For example, if the goal is to measure the effect of a teaching intervention, instruments may include pre- and post-tests to capture changes in learning. Instruments must be pilot-tested to ensure clarity and reliability.
2. **Sampling Techniques:** Sampling determines the representativeness of data.

    a. **Random Sampling**: Ensures each member of the population has an equal chance of being included.

Table 4.3 Comparative of laboratory versus field experiments

Feature	Laboratory experiments	Field experiments
Setting	Controlled, artificial environment	Natural, real-world environment
Control of variables	High	Moderate
Ecological validity	Low	High
Cost	Often lower	Often higher
Data accuracy	High (due to instrumentation)	Variable (due to external noise)

b. **Stratified Sampling**: Divides the population into strata (e.g., age, gender) and samples proportionately.
c. **Cluster Sampling**: Selects entire groups (e.g., schools, villages) rather than individuals.
d. **Convenience Sampling**: Used when randomization is not feasible, though it may reduce generalizability.

Measurement Scales

Experimental data may be collected on different scales:

a. **Nominal**: Categories without order (e.g., gender, blood type).
b. **Ordinal**: Ordered categories (e.g., Likert scales).
c. **Interval**: Continuous data without a true zero (e.g., temperature in Celsius).
d. **Ratio**: Continuous data with an absolute zero (e.g., weight, income).

Validity and Reliability

a. **Internal Validity**: Ensures that observed effects are due to the manipulated variable, not confounds.
b. **External Validity**: Ensures generalizability of findings to broader contexts.
c. **Reliability**: Refers to consistency of measurement over time and across instruments.

Data Recording Techniques

a. **Manual Logs**: Field notes or observation sheets.
b. **Software-Assisted Collection**: SPSS, REDCap, or Qualtrics for digital data entry.
c. **Automated IoT Systems**: Sensors and wearables that continuously record physiological or environmental data.

Ethical Considerations in Experimental Data Collection

Ethics are crucial in experimental research because manipulation of variables may expose participants to risks.

1. **Informed Consent**: Participants must understand the nature, purpose, and risks of the study before consenting. For instance, in medical trials, participants receive detailed information sheets.
2. **Confidentiality and Anonymity**: Data must be stored securely, and personal identifiers should be anonymized to protect privacy.
3. **Risk/Benefit Analysis**: Researchers must weigh potential benefits against risks. Experiments that expose participants to harm without adequate justification are unethical.
4. **Deception and Debriefing**: Some experiments, such as Milgram's obedience study, involve deception. Ethical guidelines require that participants are fully debriefed afterward.

5. **Historical Lessons**: The Tuskegee syphilis experiment, where treatment was withheld from African American men, is a cautionary tale of unethical data collection. Today, Institutional Review Boards (IRBs) oversee experimental research to prevent such abuses.

Challenges and Limitations of Experimental Data Collection

Despite its strengths, experimental data collection faces several challenges.

a. **Bias**: Researcher bias, sampling bias, and participant reactivity (Hawthorne effect) may distort results.
b. **Attrition**: Longitudinal experiments often suffer from participant dropout, affecting data integrity.
c. **Practical Constraints**: Experiments can be costly, time-consuming, and logistically complex.
d. **Generalizability Issues**: Findings from artificial lab settings may not generalize to real-world contexts.
e. **Cultural Sensitivity**: Experiments conducted in one cultural context may not apply universally.

Contemporary Trends and Innovations

Technological advancements are reshaping experimental data collection.

1. **Big Data and AI**: Machine learning algorithms enable real-time analysis of experimental data, especially in health and marketing research.
2. **Online Experimental Platforms**: Platforms like Amazon Mechanical Turk (MTurk) and Prolific enable large-scale, low-cost experiments across diverse populations.
3. **Virtual Reality and Simulations**: VR allows for immersive experimental environments, useful in psychology, education, and training simulations.
4. **Wearables and IoT Devices**: Smartwatches, fitness trackers, and biosensors collect continuous, real-time physiological data, enhancing precision.
5. **Remote Data Collection**: Especially during COVID-19, remote experimental methods such as online surveys, remote usability testing, and digital ethnography became popular.

Experimental data collection remains one of the most powerful tools in the researcher's toolkit. By allowing systematic manipulation and control, it enables researchers to establish causal relationships with a degree of confidence unmatched by most other methods. While challenges such as bias, ethical concerns, and generalizability issues persist, careful design and adherence to ethical standards can mitigate many risks. The rise of digital technologies, online platforms, and IoT devices is transforming experimental research, making it more scalable, accessible, and precise. However, as innovations expand possibilities, they also bring new ethical and methodological challenges that must be addressed with care. Ultimately, the strength of experimental data collection lies in its ability to balance rigor with relevance, control with realism, and precision with ethical responsibility. In a world increasingly driven by data, experiments will continue to play a central role in shaping knowledge, informing policy, and driving innovation across disciplines [24].

Case Studies of Experimental Data Collection

Experimental data collection represents one of the most systematic and rigorous methods of inquiry across the sciences, social sciences, education, and applied fields. Unlike purely observational or descriptive approaches, experimental research actively manipulates variables to determine cause-and-effect relationships. It is this characteristic of control and manipulation that distinguishes experimentation from other data collection methods and makes it one of the strongest approaches for establishing internal validity. The evolution of experimental methods has been marked by both innovation and controversy. From early physiological studies in the nineteenth century to complex multi-site randomized controlled trials (RCTs) and technology-driven behavioural experiments in the twenty-first century, the practice of experimental data collection has continuously adapted to new challenges. Case studies such as Milgram's obedience experiments (1960s), Bandura's Bobo doll experiment (1961), the Framingham Heart Study (ongoing since 1948), and modern computer simulations in physics education illustrate both the potential and limitations of experimental methods. It begins with theoretical and philosophical frameworks underpinning experimentation, proceeds to different experimental designs and their applications, surveys data collection techniques, integrates real-world case studies, addresses ethical concerns, and analyzes contemporary challenges and opportunities. Comparative tables will highlight differences across approaches, while examples from psychology, medicine, education, and technology will ground the discussion in practice (Tables 4.4, 4.5, and 4.6).

Philosophical and Theoretical Foundations of Experimental Data Collection

At its core, experimental data collection is based on positivist philosophy: the belief that reality is observable, measurable, and governed by laws that can be uncovered through systematic study. Positivism emphasizes objectivity, replicability, and the testing of hypotheses under controlled conditions. However, over time, post-positivist critiques have challenged the assumption that experiments can ever be entirely free of bias. Post-positivism accepts that while objectivity may be ideal, human inquiry is

Table 4.4 Comparative table: experimental designs

Feature	True experiment	Quasi-experiment	Field/natural experiment
Randomization	Yes	No/Partial	Rarely
Control of variables	High	Moderate	Low
Internal validity	Strong	Moderate	Weak–Moderate
External validity	Often Low (lab setting)	Moderate	High
Example	Salk vaccine trial	Head start evaluation	Card & Krueger minimum wage study

Table 4.5 Comparative of data collection techniques

Technique	Strengths	Limitations	Example case study
Observation	Direct behavioural evidence	Observer bias, limited scope	Bandura (1961)
Surveys	Large-scale, cost-effective	Self-report bias	Milgram (1963)
Physiological data	Objective, precise	Expensive, invasive	Schachter & Singer
Simulations	Scalable, interactive, safe for risky tests	May lack realism	PhET (2000s–)

Table 4.6 Comparative of ethical concerns

Ethical principle	Historical violation example	Modern safeguards
Informed consent	Milgram obedience study	IRB-approved consent forms
Minimizing harm	Stanford prison experiment	Monitoring boards, risk–benefit analysis
Transparency	Tuskegee syphilis study (1932–72)	Mandatory registration of trials

always shaped by context, researcher influence, and participant interpretation. As a result, experimental design must include strategies to minimize error and bias while acknowledging inherent limitations.

The key principles that guide experimental research include:

1. **Manipulation**—Changing one or more independent variables.
2. **Control**—Minimizing extraneous variables that could confound results.
3. **Randomization**—Ensuring participants are assigned to conditions without systematic bias.
4. **Replication**—Repeating experiments to verify findings.
5. **Causality**—Establishing a cause-and-effect link between variables.

These principles are not only theoretical but also practical; they inform how experiments are designed and data is collected across disciplines.

Types of Experimental Designs

Experimental research is not monolithic. Depending on the research objectives, context, and resources, scholars adopt different designs. The three primary categories are true experiments, quasi-experiments, and field/natural experiments.

1. **True Experimental Designs**: True experiments involve **random assignment** of participants to experimental and control groups, manipulation of an independent variable, and measurement of a dependent variable.

    - **Randomized Controlled Trials (RCTs)**: Gold standard in medical research. Example: Clinical drug testing.
    - **Laboratory Experiments**: Conducted in controlled environments (e.g., Skinner's operant conditioning studies).

Case Study: *The Salk Polio Vaccine Trial* (1954) is a classic example of a large-scale RCT. Over 1.8 million children participated in either treatment or control groups, providing robust evidence for the vaccine's effectiveness.

2. **Quasi-experimental Designs**: Quasi-experiments lack full randomization, often due to ethical or practical constraints. They are common in education, social sciences, and policy research.
   - **Non-equivalent Control Group Design**: Pre-existing groups are used.
   - **Interrupted Time-Series Design**: Observations are taken before and after an intervention.

Case Study: *Head Start Program Evaluation* (1965 onward) in the United States examined the impact of preschool programs on low-income children. Full randomization was not possible, but quasi-experimental approaches provided insights into long-term educational outcomes.

3. **Field and Natural Experiments**: These experiments occur in real-world settings rather than controlled labs.
   - **Field Experiments**: Researcher manipulates variables in natural settings (e.g., marketing studies testing and effectiveness).
   - **Natural Experiments**: Variables change due to external events, not researcher intervention (e.g., policy changes).

Case Study: *David Card and Alan Krueger's Minimum Wage Study* (1994) compared employment rates in New Jersey and Pennsylvania after a wage increase, providing critical evidence in economic policy debates.

Data Collection Techniques in Experimental Research

Once the experimental design is established, data collection becomes the focal activity. Several techniques are employed depending on the discipline:

1. **Observation**
   - **Structured Observation**: Predetermined categories, used in behavioural studies.
   - **Unstructured Observation**: Open-ended, exploratory.

Case Study: *Bandura's Bobo Doll Experiment* (1961) used structured observation to record children's aggression levels after exposure to adult models.

2. **Surveys and Questionnaires**

Often used in social psychology and education experiments. They provide large-scale, quantifiable data.

Case Study: *Milgram's Obedience Experiment* (1963) collected post-experiment surveys from participants to assess their perceptions of authority.

3. **Physiological Measures**

Biological data (heart rate, cortisol, brain imaging) enhances objectivity.

Case Study: *Stanley Schachter and Jerome Singer's Two-Factor Theory of Emotion* (1962) used physiological arousal (adrenaline injections) to study how context shapes emotional labeling.

4. **Digital and Simulation-Based Data**

Increasingly, experiments rely on computer simulations, sensors, and AI-driven systems.

Case Study: *PhET Interactive Simulations in Physics Education* (University of Colorado Boulder) use virtual labs to enhance student learning of abstract physics concepts. Data collection involves pre/post-tests, usage logs, and performance analytics.

Ethical Considerations in Experimental Data Collection

Ethics is central to experimental research, especially when human participants are involved. Key issues include **informed consent, deception, harm minimization, confidentiality, and right to withdraw**.

Case Study 1: Milgram's Obedience Experiment (1963)

- Ethical concerns: deception (participants believed they were shocking others), psychological distress.
- Impact: Led to stricter Institutional Review Board (IRB) regulations.

Case Study 2: Stanford Prison Experiment (1971)

- Ethical concerns: participants experienced emotional trauma and abuse.
- Impact: Raised awareness of situational power dynamics; widely criticized for methodological flaws.

Case Study 3: Modern Clinical Trials

- Ethical safeguards include informed consent documents, data monitoring committees, and trial registration to ensure transparency.

Challenges in Experimental Data Collection

Despite its rigor, experimental research faces multiple challenges:

1. **Artificiality Versus Realism**—Lab conditions may lack ecological validity.
2. **Ethical Boundaries**—Some questions cannot be studied experimentally (e.g., long-term harmful effects).
3. **Attrition and Noncompliance**—Participants dropping out or failing to follow instructions threatens validity.
4. **Replication Crisis**—Psychology and other fields struggle with replicability of famous experiments.
5. **Cost and Resource Demands**—Large RCTs require substantial funding.

Case Study: Replication Crisis in Psychology (2010s)

- Many classic experiments, including some priming studies, failed replication.
- Response: Open Science movement emphasizes pre-registration, data sharing, and replication studies.

Contemporary Trends in Experimental Data Collection

The twenty-first century has introduced new technologies and interdisciplinary methods:

1. **Big Data and AI Integration**: Combining experimental control with large-scale digital datasets. Example: Online A/B testing at tech companies (e.g., Google, Facebook).
2. **Virtual Reality and Immersive Experiments**: Allow simulation of complex environments while collecting behavioural and physiological data.
3. **Mixed-Methods Approaches:** Integrating experiments with qualitative interviews to enrich interpretation.
4. **Global and Cross-Cultural Experiments**: Ensuring generalizability beyond Western contexts. Example: Replications of Milgram's obedience paradigm in different cultures.

Experimental data collection remains one of the most powerful methods for uncovering causal relationships across disciplines. Its evolution reflects both the strength of scientific inquiry and the responsibility of ethical practice. Case studies such as Milgram, Bandura, the Salk Vaccine Trial, and PhET simulations reveal both the promise and pitfalls of experimentation. Looking forward, the field must navigate replication crises, ethical boundaries, and new technological landscapes. Digital platforms, VR environments, and AI-driven analytics present both opportunities for innovation and risks of surveillance or bias. Cross-cultural validation will also be critical to ensure findings are not limited to narrow populations. Ultimately, experimental data collection is not merely a methodological choice but a reflection of broader philosophical commitments: a search for truth, balanced with responsibility to participants and society [25].

4.4 Using Secondary Data

The use of secondary data has become a cornerstone of modern research across disciplines, enabling scholars, policymakers, and practitioners to build upon existing knowledge and extract insights without the costly and time-consuming process of collecting new data. Secondary data refers to information that has already been collected, processed, and archived by other researchers, organizations, or institutions for purposes that may or may not align with the new research project at hand. Such data can include official government statistics, published academic works, institutional reports, census data, archival materials, commercial datasets, and increasingly, big data generated through digital platforms. In an era defined by information abundance, the role of secondary data has expanded significantly. Researchers in the social sciences, natural sciences, health sciences, economics, and education have come to rely heavily on secondary sources, not merely as supplements to primary investigations but as rich and sometimes sufficient foundations for independent studies. For example, economists frequently turn to World Bank development indicators to analyze patterns of poverty reduction, while health researchers draw upon the World Health Organization's epidemiological databases to study global disease trends.

Despite these advantages, the use of secondary data is not without challenges. Concerns about relevance, accuracy, comparability, and potential bias require careful scrutiny. Moreover, the ethical dimensions of using data collected for other purposes—particularly personal or sensitive information—demand a nuanced understanding of consent, confidentiality, and data protection laws such as the General Data Protection Regulation (GDPR). This section provides an in-depth exploration of secondary data in research, examining its nature, types, advantages, challenges, sources, methods of analysis, case studies of application, ethical considerations, and comparisons with primary data. It also looks ahead to emerging trends, such as open data initiatives and artificial intelligence-driven analytics, before concluding with reflections on its enduring value to scholarship and practice.

1. **Nature and Types of Secondary Data**: Secondary data broadly refers to data collected by someone else, often for purposes different from the current research project. This distinguishes it from primary data, which is gathered directly by the researcher. Secondary data can be categorized along several dimensions, including publication status, format, and type.
2. **Published Versus Unpublished Secondary Data**: Published data is readily accessible and includes sources such as government reports, census data, international agency databases, academic journals, and statistical yearbooks. For instance, the UK Office for National Statistics regularly publishes household income and labor market data that is widely used in socioeconomic research. Unpublished secondary data, on the other hand, may consist of internal company reports, private organizational records, or raw datasets from past research projects that are archived but not formally disseminated.
3. **Qualitative Versus Qualitative Secondary Data**: Secondary data may be numerical or descriptive. Qualitative secondary data includes time-series statistics, survey results, or financial records, often stored in spreadsheets or statistical databases. Qualitative secondary data, by contrast, may comprise historical documents, interview transcripts, diaries, or newspaper archives. For example, historians often rely on archived letters or oral histories to reconstruct narratives of past events.
4. **Traditional Versus Big Data**: A newer distinction involves traditional structured datasets versus big data. Traditional secondary datasets are structured and relatively small in volume, while big data refers to vast, complex, and often unstructured digital datasets generated through online activities, social media, or mobile applications. Scholars in fields such as computational social science increasingly harness big data to examine trends in communication, consumer behaviour, or political participation.

This typology demonstrates the richness and diversity of secondary data, highlighting its applicability across different research designs.

Advantages of Using Secondary Data

The widespread use of secondary data stems from a number of significant advantages.

1. **Cost and Time Efficiency**: One of the most compelling reasons to use secondary data is its efficiency. Collecting primary data often requires extensive resources, including funding, staff, and time. In contrast, secondary data already exists and can often be accessed quickly and at little or no cost. For example, researchers studying global economic trends can download decades of financial indicators from the International Monetary Fund (IMF) without needing to conduct original surveys across countries.
2. **Broad Scope and Coverage**: Secondary datasets frequently provide access to large-scale information that individual researchers could not feasibly collect. National censuses, international health surveys, and global trade statistics encompass millions of observations across regions and decades, making them indispensable for longitudinal and cross-country analyses.
3. **Reliability and Authority**: Many secondary data sources are produced by reputable organizations with rigorous data collection and verification procedures. Government agencies, central banks, and international organizations often ensure accuracy through professional data collection standards. This lends credibility to analyses based on such data.
4. **Comparative and Longitudinal Analysis**: Secondary data allows researchers to track changes over time and make comparisons across contexts. A sociologist, for instance, may examine trends in urbanization by analyzing census data from multiple decades, while a political scientist might study shifts in voter turnout using election statistics spanning several election cycles.

Limitations and Challenges

Despite its advantages, secondary data is not without shortcomings.

1. **Relevance and Fit**: A major limitation is that secondary data was originally collected for purposes other than the current research question. This can result in mismatches between the available data and the researcher's needs. For example, a public health researcher studying mental health outcomes may find that existing national health surveys do not include adequate measures of psychological well-being.
2. **Accuracy and Reliability Issues**: Not all secondary data is reliable. In some cases, errors may have been introduced during collection, coding, or processing. Moreover, datasets from less developed regions may suffer from incomplete coverage or outdated methods.
3. **Comparability Problems**: Cross-national or cross-temporal research faces comparability issues, as definitions, measurement techniques, and data collection standards may vary. For instance, unemployment statistics may differ across countries depending on how "actively seeking work" is defined.

4. **Data Access Restrictions**: While some datasets are publicly available, others require costly subscriptions, licensing agreements, or special permissions. For example, commercial market research data is often prohibitively expensive for independent researchers.
5. **Ethical and Legal Constraints**: Secondary data often includes personal or sensitive information. Using such data without proper safeguards can lead to privacy breaches or ethical violations, especially in an era of heightened data protection regulations.

Sources of Secondary Data

Secondary data is drawn from a wide range of sources, each offering different strengths.

1. Government sources are among the most common, providing census records, labor statistics, health surveys, and crime data. In the United States, the Census Bureau and the Bureau of Labor Statistics provide foundational datasets for demographic and economic research. International organizations such as the World Bank, IMF, United Nations, and World Health Organization publish extensive datasets on development indicators, health outcomes, and economic performance.
2. Academic publications and repositories, such as JSTOR or institutional archives, provide access to qualitative and qualitative data from prior research. Increasingly, universities mandate that publicly funded research projects deposit their data in open-access repositories.
3. Commercial data providers, including Nielsen and Statista, offer market research and consumer behaviour data, although often at high costs.
4. Finally, digital and online sources such as Google Trends, Twitter archives, and online forums provide vast repositories of big data that can be mined for social science and marketing research.

Methods of Secondary Data Analysis

The analysis of secondary data depends on the type and format of the data in question.

1. Qualitative secondary data often requires statistical analysis, including descriptive statistics, regression models, or econometric techniques. Time-series analysis, for example, is frequently applied to macroeconomic datasets.
2. Qualitative secondary data may be examined using content analysis, discourse analysis, or thematic coding. Historians analyzing archived speeches, for instance, may use qualitative coding to identify recurring themes.
3. Meta-analysis is another method that involves synthesizing findings from multiple published studies to arrive at general conclusions. This is particularly common in medicine and psychology, where evidence-based practice relies heavily on aggregating prior results.
4. The advent of computational methods has expanded the toolkit for secondary data analysis. Machine learning algorithms, natural language processing, and

sentiment analysis now allow researchers to analyze vast digital datasets such as social media content.

Case Studies of Secondary Data Use

Case studies demonstrate the practical significance of secondary data in real-world research.

1. One notable example is the use of World Bank development indicators by economists to study poverty reduction. By analyzing income, education, and health data across decades, researchers can assess the effectiveness of development policies in different countries.
2. The UK Census provides another example, serving as a crucial resource for sociologists and demographers. Scholars have used census data to study migration patterns, housing inequalities, and the changing composition of households over time.
3. During the COVID-19 pandemic, the World Health Organization's epidemiological database became a vital secondary source. Researchers worldwide drew upon daily updates on infection rates, hospitalizations, and vaccinations to model the spread of the virus and evaluate policy responses.
4. Historical research also relies heavily on secondary data, particularly archival documents. For example, education researchers have used archived school records to trace long-term trends in literacy and curriculum development.

These case studies highlight how secondary data enables research that would otherwise be impossible due to constraints of scale, time, or access.

Ethical Considerations in Using Secondary Data

The ethical use of secondary data requires careful attention to issues of privacy, consent, and data protection.

1. One challenge is that data collected in the past may not have anticipated future uses, raising questions about whether participants consented to the current research purpose. Researchers must ensure that their use of secondary data respects the original consent agreements or anonymizes data where necessary.
2. Data protection laws, particularly the GDPR in Europe, impose strict requirements on how personal data may be stored, accessed, and analyzed. Even anonymized datasets must be handled responsibly to avoid re-identification risks.
3. Beyond legal requirements, researchers have an ethical duty to represent secondary data accurately and avoid misinterpretation. This includes acknowledging limitations and contextualizing findings within the constraints of the original data collection.

This comparison underscores the importance of choosing between primary and secondary data—or combining both—based on research objectives, available resources, and the trade-off between specificity and scope (Table 4.7).

Table 4.7 Comparison with primary data

Aspect	Primary data	Secondary data
Collection purpose	Specific to research question	Originally for another purpose
Cost and time	High	Low
Control over quality	High	Variable
Scope and scale	Limited by resources	Often large-scale
Flexibility	High (customizable)	Low (dependent on availability)

Secondary and primary data serve complementary roles, each with distinct strengths and weaknesses

Emerging Trends

Several emerging trends are reshaping the use of secondary data.

1. Open data initiatives, led by governments and international organizations, are making more datasets freely available to researchers and the public. For instance, the European Union's Open Data Portal provides access to thousands of datasets on economics, health, and environment.
2. Artificial intelligence and big data analytics are enabling researchers to extract insights from unstructured digital data, such as social media posts, satellite images, and internet-of-things (IoT) devices.
3. Collaborative data repositories and data-sharing networks are also proliferating, promoting transparency, reproducibility, and cross-disciplinary research.

These developments are expanding both the availability and analytical possibilities of secondary data, solidifying its role as a central pillar of future research.

Secondary data has emerged as one of the most versatile and indispensable resources in research, offering efficiency, scope, and reliability that primary data often cannot match. It has been successfully applied in diverse fields, from economics and public health to history and education, enabling large-scale and longitudinal studies that inform policy and advance knowledge. Nevertheless, secondary data use requires careful attention to issues of relevance, accuracy, comparability, and ethics. Researchers must remain vigilant in assessing the fit between available data and their specific questions, acknowledging limitations, and safeguarding ethical principles of privacy and consent. The future of secondary data is promising, shaped by open data initiatives, big data analytics, and artificial intelligence. As information continues to proliferate in digital form, secondary data will remain at the heart of scholarly inquiry, offering unparalleled opportunities to harness existing knowledge for new insights and solutions.

References

1. Ibrahim, O.M., Ibrahim, R.M., Al Meslamani, A.Z., Al Mazrouei, N.: Role of telepharmacy in pharmacist counselling to coronavirus disease 2019 patients and medication dispensing errors. J. Telemed. Telecare 27272 (2020). https://doi.org/10.1177/1357633X20964347
2. Mulenga, M., Phiri, M., Simukonda, L., Alaba, F.A.: A multistage hybrid deep learning model for enhanced solar tracking. IEEE Access **11**, 129449–129466 (2023). https://doi.org/10.1109/ACCESS.2023.3333895
3. Kacprzyk, J., et al.: Security Framework and Defense Mechanisms for IoT Reactive Jamming Attacks. https://doi.org/10.1007/978-3-031-65929-4
4. Allah, A.N., Akaaba, I.I., Alaba, F.A.: Pilot risk climate and error tendency in Nigeria's aviation industry. Irish Int. J. Eng. Appl. Sci. **9**(2), 73–89 (2025)
5. Alaba, F.A., Madu, I.M., Musa, H.: Attacks, Challenges, and Countermeasures for an Integrated IoT Framework. **12**(2), 28–39 (2024)
6. Omobowale, M.O., Amodu, O.K., Amodu, F.A.: Solving teenage and young mothers' childhood immunization hesitance and non-compliance through mobile immunization friendly service for working mothers in Ibadan, Nigeria—a research note. PLOS Glob. Public Heal. **3**(8), e0002109 (2023). https://doi.org/10.1371/journal.pgph.0002109
7. Alaba, F.A.: Assessment of Community Participation in the Provision of Universal Basic Education in Sanga Local Government, Kaduna State, Nigeria (2025)
8. Trad, A.: The business transformation framework and enterprise architecture framework for managers in business innovation: an applied holistic mathematical model. Int. J. Serv. Sci. Manag. Eng. Technol. **12**(1), 142–181 (2021). https://doi.org/10.4018/IJSSMET.20210101.oa1
9. Kotsev, A., Schade, S., Craglia, M., Gerboles, M., Spinelle, L., Signorini, M.: Next generation air quality platform: openness and interoperability for the internet of things. Sensors (Switzerland) **16**(3), 1–16 (2016). https://doi.org/10.3390/s16030403
10. Nitschke, T., Bour, A., Bergquist, M., Blanchard, M., Molinari, F., Almroth, B.C.: Smokers' behaviour and the toxicity of cigarette filters to aquatic life: a multidisciplinary study. Microplastics Nanoplastics **3**(1), 1–18 (2023). https://doi.org/10.1186/s43591-022-00050-2
11. Schunk, D.H., DiBenedetto, M.K.: Albert Bandura's legacy in education. Theory Pract. **62**(3), 205–206 (2023). https://doi.org/10.1080/00405841.2023.2226560
12. Wark, J.D.: Power up: combining behaviour monitoring software with business intelligence tools to enhance proactive animal welfare reporting. Animals **12**(13) (2022). https://doi.org/10.3390/ani12131606
13. Tajpour, M., Salamzadeh, A., Hosseini, E.: Job satisfaction in IT Department of Mellat Bank: does employer brand matter? Ipsi Bgd Trans. Internet Res. **17**(1 SI), 15–21 (2021)
14. Ramadani, V., Agarwal, S., Caputo, A., Agrawal, V., Dixit, J.K.: Sustainable competencies of social entrepreneurship for sustainable development: exploratory analysis from a developing economy. Bus. Strateg. Environ. **31**(7), 3437–3453 (2022). https://doi.org/10.1002/bse.3093
15. Rady, A., Shokair, M., El-Rabaie, E.L.S.M., Saad, W., Benaya, A.: Energy-efficient routing protocol based on sink mobility for wireless sensor networks. IET Wirel. Sens. Syst. **9**(6), 405–415 (2019). https://doi.org/10.1049/iet-wss.2019.0044
16. Pawlicka, K., Bal, M.: Sustainable supply chain finances implementation model and artificial intelligence for innovative omnichannel logistics. Management **26**(1), 19–35 (2022). https://doi.org/10.2478/manment-2019-0082
17. Bompelli, A., et al.: Social and behavioural determinants of health in the era of artificial intelligence with electronic health records: a scoping review. Heal. Data Sci. **2021**, 1–19 (2021). https://doi.org/10.34133/2021/9759016
18. Bhatt, A.M., Goldberg, A., Srivastava, S.B.: A language-based method for assessing symbolic boundary maintenance between social groups. Sociol. Methods Res. **5**(2), 1–45 (2021) (Online). Available: http://faculty.haas.berkeley.edu/srivastava/papers/SymbolicBoundaries_Language_SMR.pdf

19. Ntuli, L.T., Mahlangu, V.P.: Legal execution of curriculum in primary schools: school management teams' perspectives. Res. Educ. Policy Manag. **5**(2), 177–206 (2023). https://doi.org/10.46303/repam.2023.16
20. Sudirjo, F.: Marketing strategy in improving product competitiveness in the global market. J. Contemp. Adm. Manag. **1**(2), 63–69 (2023). https://doi.org/10.61100/adman.v1i2.24
21. Purnomo, Y.J.: Digital marketing strategy to increase sales conversion on e-commerce platforms. J. Contemp. Adm. Manag. **1**(2), 54–62 (2023). https://doi.org/10.61100/adman.v1i2.23
22. Botes, R., Smit, I.: Taking the mystery out of interpretive research: crossing the infinity approach. JIMS 441–450 (2016)
23. Kim, H.-S.: Oil extraction and the changing dynamics of pastoral conflicts: a conjoint experiment in Turkana, Kenya. J. Mod. Afr. Stud. **61**(1), 23–47 (2023). https://doi.org/10.1017/s0022278x22000398
24. Fu, S., et al.: Automated detection of periprosthetic joint infections and data elements using natural language processing. J. Arthroplasty **36**(2), 688–692 (2021). https://doi.org/10.1016/j.arth.2020.07.076
25. Mohammed, A.A.: Int. J. Educ. Soc. Sci. Res. **2**(05), 35–47 (2019)

Chapter 5
Qualitative Research Measurements Classifications

5.1 Measurement in Qualitative Research

Measurement is traditionally associated with qualitative research, where variables are operationalized and quantified to test hypotheses and establish statistical relationships. However, in qualitative research, measurement takes on a different character—emphasizing meaning, context, and interpretation rather than numerical representation. Instead of seeking generalizability through standardized metrics, qualitative measurement seeks depth, richness, and contextual understanding of phenomena. This distinction highlights the philosophical underpinnings of qualitative inquiry, which is rooted in interpretivism and constructivism, emphasizing subjective experiences, narratives, and meanings constructed by individuals in specific contexts [1].

Conceptualizing Measurement in Qualitative Research

In qualitative research, measurement is not about assigning numerical values but about capturing phenomena in ways that remain faithful to participants lived experiences. According to [2], qualitative measurement involves describing, interpreting, and classifying observations into meaningful categories. For instance, in grounded theory, measurement emerges through coding and constant comparison, where concepts are iteratively refined until theoretical saturation is achieved. Similarly, in phenomenology, measurement occurs through the identification of essential themes that describe the essence of participants' experiences. Thus, the measurement process is embedded within the research design itself, ensuring that it reflects the social realities under investigation [3].

Techniques of Measurement in Qualitative Research

Several strategies allow qualitative researchers to "measure" social phenomena while preserving meaning and authenticity.

F. A. Alaba et al., *Fundamentals of Qualitative Research*, Studies in Systems, Decision and Control 663, https://doi.org/10.1007/978-3-032-18427-6_5

1. **Coding and Categorization**: In grounded theory and thematic analysis, measurement is operationalized through systematic coding. Data is broken into units of meaning, categorized, and interpreted to form themes [4]. This provides a structured way to measure the presence, frequency, or intensity of particular ideas without quantification.
2. **Triangulation**: Measurement is strengthened by using multiple sources of evidence—interviews, observations, documents, or visual data. Triangulation increases credibility and helps confirm whether identified patterns truly reflect the phenomenon [5].
3. **Member Checking**: This involves returning findings or interpretations to participants to confirm accuracy. By doing so, researchers ensure that their measurement of meanings aligns with participants' perspectives [6].
4. **Thick Description**: [7] emphasized that qualitative measurement requires detailed contextual accounts of behaviours and meanings. Such descriptions allow readers to interpret the richness of cultural practices and validate findings through contextual coherence.
5. **Reflexivity**: Researchers measure their own influence on the research process by maintaining reflexive journals. This ensures that interpretations are transparent and that potential biases are acknowledged.

Validity and Reliability in Qualitative Measurement

While qualitative research focuses on statistical reliability and validity, qualitative researchers rely on alternative criteria proposed by [8]:

1. **Credibility**: Equivalent to internal validity, credibility assesses whether findings accurately represent participants' realities. Techniques such as member checking and peer debriefing are used.
2. **Transferability**: Instead of generalizability, qualitative studies emphasize thick description so that readers can judge whether findings apply to other contexts.
3. **Dependability**: Comparable to reliability, this refers to the consistency of findings over time and across researchers, often ensured through audit trails.
4. **Confirmability**: This criterion addresses objectivity by ensuring findings emerge from participants' voices rather than researcher bias.

By applying these standards, qualitative measurement achieves rigor while acknowledging the interpretive nature of the inquiry.

Challenges of Measurement in Qualitative Research

Despite its strengths, qualitative measurement faces several challenges.

First, **subjectivity** can complicate interpretation, as researchers bring their own perspectives, which may shape coding and theme development. Second, **replicability** is difficult because contexts, participants, and interpretations vary widely, making standardized measurement elusive [9]. Third, **ethical issues** such as misrepresenting participants' voices or oversimplifying complex narratives may compromise the integrity of qualitative measurement. Finally, qualitative data often involves large volumes of unstructured information, requiring meticulous organization and coding to ensure meaningful measurement.

The Role of Technology in Qualitative Measurement

Recent advances in digital tools have transformed measurement in qualitative research. Software such as NVivo, ATLAS.ti, and MAXQDA allow researchers to systematically code, categorize, and visualize patterns in textual and multimedia data. These tools support more consistent measurement by providing audit trails, query functions, and visualization options. Furthermore, AI-assisted qualitative analysis is emerging, where natural language processing can assist in coding and identifying themes, though researchers remain responsible for interpretive validity [10].

The Balance Between Measurement and Meaning

A central debate in qualitative research is how much measurement should focus on standardization versus preserving richness and uniqueness. While some argue that qualitative research should incorporate more structured forms of measurement for comparability, others caution that excessive quantification undermines the interpretive goals of qualitative inquiry. The balance lies in adopting systematic approaches to measurement while retaining openness to emergent meanings. Thus, qualitative measurement is less about numerical precision and more about methodological rigor, ethical sensitivity, and interpretive authenticity. Measurement in qualitative research is a complex, interpretive, and context-sensitive process. Unlike qualitative approaches that rely on numbers and statistics, qualitative measurement emphasizes meaning, authenticity, and depth. Through coding, triangulation, thick description, and reflexivity, qualitative researchers capture and represent social phenomena in ways that preserve participants lived realities. While challenges such as subjectivity and replicability persist, qualitative research maintains rigor by adhering to standards of credibility, dependability, transferability, and confirmability. With the advent of digital tools and AI, qualitative measurement continues to evolve, offering researchers new ways to systematically analyze and interpret data. Ultimately, measurement in qualitative research is not about quantifying reality but about authentically representing the complexity of human experience [11] (Fig. 5.1).

5.2 Understanding Variables: Types of Variables

In the field of research, whether qualitative or qualitative, variables play a central role in framing questions, designing studies, and interpreting results. A variable refers to any characteristic, attribute, or factor that can vary or change among individuals, groups, or within an environment. Understanding variables and their types is essential because they form the building blocks of hypotheses, influence the choice of research methods, and determine the statistical analyses used. This section explores the types of variables commonly discussed in research, their significance, and examples from educational and social science contexts.

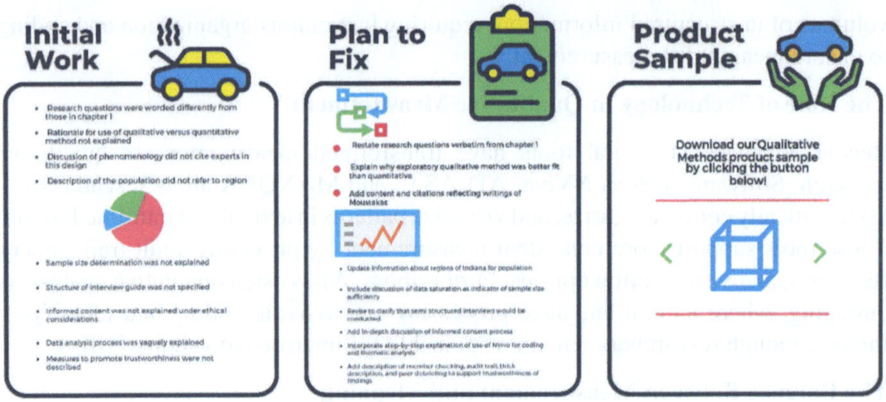

Fig. 5.1 Measurement in qualitative research [12]

5.2.1 *Independent Variable*

In research, particularly within the realms of social science, education, health, and psychology, the independent variable (IV) serves as a cornerstone concept. It is the element that researchers manipulate, categorize, or select to observe its influence on an outcome, known as the dependent variable (DV). By systematically adjusting or identifying the independent variable, researchers can assess cause-and-effect relationships and build theoretical frameworks that explain phenomena. Independent variables are essential to both experimental and non-experimental research because they provide the foundation for hypothesis testing and for drawing valid conclusions. This section explores the meaning, role, types, operationalization, and significance of the independent variable in research, as well as challenges and contemporary applications.

Conceptualizing the Independent Variable

The independent variable is defined as the factor that is manipulated or categorized to observe its effect on the dependent variable. In experimental designs, researchers actively control the independent variable. In contrast, in correlational or observational studies, the independent variable is naturally occurring but still serves as the predictor or explanatory factor [11]. For example, in a study exploring the impact of teaching strategies on student achievement, the teaching strategy (e.g., lecture-based versus flipped classroom) is the independent variable, while student achievement is the dependent variable. By comparing results across conditions, researchers aim to determine whether differences in teaching method cause measurable changes in achievement outcomes.

Types of Independent Variables

Independent variables can be classified into several categories based on their nature and how they are studied.

1. **Experimental Versus Subject Independent Variables**
   - **Experimental IVs**: Variables deliberately manipulated by researchers, such as dosage of a drug in medical research.
   - **Subject IVs**: Naturally occurring characteristics, such as gender, age, or socioeconomic status, which cannot be manipulated but are still used to categorize participants.
2. **Categorical Versus Continuous Independent Variables**
   - **Categorical IVs**: Represent discrete groups (e.g., type of instructional method).
   - **Continuous IVs**: Measured on a scale with a continuum of values (e.g., number of study hours).
3. **Single Versus Multiple Independent Variables**
   - Some studies investigate a single independent variable (e.g., exercise on stress), while others use multiple IVs to examine interaction effects (e.g., exercise and diet on stress reduction).
4. **Moderator as Independent Variable**

Sometimes, variables act as moderators, changing the strength or direction of the IV–DV relationship. For example, stress may predict academic performance differently depending on social support levels. Understanding these types helps researchers design robust studies and select appropriate statistical techniques [13].

Operationalization of Independent Variables

Operationalization refers to the process of defining how an independent variable will be measured or manipulated in a study. It involves moving from an abstract construct to a measurable element. For instance, in studying the impact of physical activity on cognitive performance, researchers must specify what "physical activity" entails—duration, frequency, or intensity. A possible operational definition might be "30 min of moderate aerobic exercise, three times per week, for four weeks." Without operational clarity, the research becomes difficult to replicate and interpret. Operationalization ensures reliability (consistency of measurement) and validity (accuracy in measuring what is intended). It also allows for comparability across studies and contributes to building cumulative scientific knowledge [14].

Role of Independent Variable in Research Design

The independent variable is central to multiple aspects of research design:

1. **Hypothesis Development**: Hypotheses typically predict how changes in the independent variable will affect the dependent variable. For example, "Students taught with project-based learning will score higher on problem-solving assessments than those taught with lectures."
2. **Experimental Control**: Manipulating the independent variable while holding other variables constant allows researchers to infer causality.

3. **Comparative Analysis**: Independent variables enable comparisons across groups, such as treatment versus control groups.
4. **Theory Testing**: By examining the effects of independent variables, researchers validate or refine theoretical models.

Thus, the independent variable provides structure and direction to empirical inquiry [15].

Independent Variable in Qualitative Research

In qualitative research, independent variables are often numerical or categorical factors used in statistical models. Common methods include:

- **Analysis of Variance (ANOVA)**: Compares group means based on categorical IVs.
- **Regression Analysis**: Tests how continuous IVs predict DV outcomes.
- **Factorial Designs**: Examine main effects and interactions when multiple IVs are included.

For example, in health research, an ANOVA might test whether three different diets (IV) produce differences in weight loss (DV). Regression analysis might examine how sleep duration (IV) predicts GPA (DV). Qualitative designs allow for hypothesis testing, generalizability, and statistical inference, relying heavily on the correct identification of independent variables [16].

Independent Variable in Qualitative Research

While independent variables are more central to qualitative research, qualitative studies may also incorporate them conceptually. For example, a phenomenological study exploring the experiences of female versus male teachers in technology integration treats gender as a categorical independent variable. Similarly, in case studies, factors such as school setting (urban vs. rural) may be treated as independent variables shaping participant experiences. However, qualitative research tends to emphasize the complexity of relationships rather than isolating variables. Independent variables in this context are often part of broader themes or contexts rather than strictly manipulated constructs.

Challenges in Using Independent Variables

Despite their importance, independent variables pose several challenges:

1. **Confounding Variables**: Extraneous factors may distort the relationship between IV and DV. For example, studying the effect of exercise on mental health without accounting for diet introduces confounding.
2. **Measurement Issues**: Poor operationalization may lead to unreliable or invalid results.
3. **Ethical Limitations**: Some independent variables (e.g., trauma exposure) cannot ethically be manipulated.
4. **Multicollinearity**: In regression, when IVs are highly correlated with one another, it becomes difficult to assess their independent effects.

5. **Cultural Sensitivity**: Variables may not have the same meaning across contexts, complicating generalizability.

Researchers must anticipate and address these issues to ensure valid results [17].

Independent Variable and Causality

The independent variable is at the heart of causal inference. Establishing causality requires:

1. **Temporal Precedence**: The IV occurs before the DV.
2. **Covariation**: The IV and DV change together.
3. **Control for Confounds**: The relationship is not explained by extraneous variables.

Randomized controlled trials (RCTs) represent the gold standard for demonstrating causality because they involve random assignment to IV conditions, thereby reducing bias.

Contemporary Applications of Independent Variables

1. **Education**: Independent variables such as teaching strategies, technology integration, and curriculum design are widely studied for their impact on learning outcomes.
2. **Health Sciences**: Clinical trials manipulate IVs like drug dosage or therapy type to assess effects on patient outcomes.
3. **Psychology**: Independent variables such as cognitive interventions or stress levels are tested to understand behaviour and cognition.
4. **Business Research**: Marketing strategies, pricing, or leadership styles serve as IVs influencing sales, productivity, or employee satisfaction.
5. **AI and Data Science**: Independent variables include algorithm types, data preprocessing techniques, or system configurations tested for accuracy or efficiency.

These applications highlight the versatility and importance of independent variables across disciplines. The independent variable is a cornerstone of research methodology, serving as the foundation for hypothesis testing, experimental control, and causal inference. By carefully defining, categorizing, and operationalizing independent variables, researchers can design studies that yield valid, reliable, and meaningful insights. Challenges such as confounding variables, measurement issues, and ethical constraints remind us of the complexities inherent in research, yet they also drive methodological innovation. Across disciplines—from education and psychology to health and technology—the independent variable remains central to the pursuit of knowledge. Its proper use not only strengthens research design but also advances theory, practice, and policy [18] (Fig. 5.2).

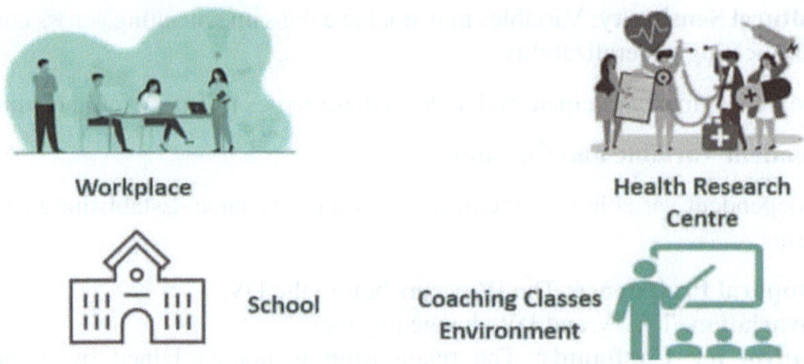

Fig. 5.2 An overview of the independent variable in qualitative research [19]

5.2.2 *Dependent Variable*

Research across disciplines such as education, psychology, health, and the social sciences relies heavily on the concept of variables to examine relationships and test hypotheses. Among these, the dependent variable (DV) holds a unique and critical role as the measurable outcome influenced by changes in other variables, particularly the independent variable (IV). Simply put, the dependent variable is the "effect" or "response" that researchers seek to explain or predict. Understanding the dependent variable is vital because it determines what is being measured, shapes research questions, and influences the choice of statistical analyses. This section explores the meaning, classification, operationalization, challenges, and applications of the dependent variable, highlighting its central role in empirical research.

Conceptualizing the Dependent Variable

A dependent variable is the observed outcome that changes as a function of the independent variable. It is the variable that researchers record, track, or measure to test whether an independent variable exerts an effect. For instance, in an educational study examining whether different teaching methods influence student learning, the **type of teaching method** is the independent variable, while **student performance (e.g., exam scores)** is the dependent variable. In this relationship, performance is "dependent" because its variation is presumed to result from differences in the independent variable. The dependent variable is not only central to hypothesis testing but also represents the phenomenon of interest in most research projects—it is what researchers seek to understand, improve, or explain [20].

Characteristics of the Dependent Variable

The dependent variable has several defining characteristics:

1. **Measurable**: It must be observable and recordable in some form, whether through qualitative metrics (e.g., test scores) or qualitative indicators (e.g., themes in interviews).

2. **Dynamic**: It changes in response to the independent variable.
3. **Central to Hypothesis**: The dependent variable reflects the outcome specified in research questions or hypotheses.
4. **Contextual**: Its meaning and measurement depend on the research design and discipline.

These characteristics distinguish the dependent variable as the endpoint of analysis and evaluation.

Types of Dependent Variables

1. **Qualitative versus Qualitative DVs**

   - **Qualitative DVs**: Numeric outcomes, such as weight, test scores, blood pressure, or income.
   - **Qualitative DVs**: Non-numeric outcomes, such as attitudes, perceptions, or themes derived from interviews.

2. **Continuous versus Categorical DVs**

   - **Continuous DVs**: Can take any value along a scale (e.g., reaction time in seconds, height).
   - **Categorical DVs**: Represent groups or categories (e.g., pass/fail, disease/no disease).

3. **Single versus Multiple DVs**

Some studies focus on a single outcome (e.g., exam scores), while others investigate multiple dependent variables simultaneously (e.g., test scores, motivation, and attendance). Multivariate designs allow for a richer understanding of complex outcomes.

4. **Latent versus Observed DVs**

   - **Observed DVs**: Directly measurable, such as exam grades.
   - **Latent DVs**: Abstract constructs inferred through multiple observed indicators, such as intelligence or job satisfaction.

This classification shows the flexibility of dependent variables across diverse research designs.

Operationalization of Dependent Variables

Operationalization refers to defining how the dependent variable will be measured or assessed in a specific study. Without operational clarity, results cannot be interpreted or replicated.

For example, "academic achievement" as a dependent variable could be operationalized as GPA, standardized test scores, or teacher ratings. Each operational definition yields different insights and implications.

Good operationalization ensures:

- **Reliability**: Consistent measurement across contexts.

- **Validity**: Accurate reflection of the construct being studied.
- **Comparability**: Ability to compare across studies and contexts.

Operationalization is thus crucial to linking theoretical constructs to empirical data.

Role of the Dependent Variable in Research Design

The dependent variable plays a pivotal role in structuring studies:

1. **Framing Research Questions**: Research often begins with identifying the outcome of interest, such as student performance, health improvement, or employee productivity.
2. **Testing Hypotheses**: Hypotheses predict how independent variables will affect dependent variables.
3. **Statistical Analysis**: Choice of tests—t-tests, ANOVA, regression, logistic regression—depends on the nature of the dependent variable (continuous vs. categorical).
4. **Evaluating Effectiveness**: Dependent variables measure whether interventions, treatments, or innovations achieve intended goals.

Thus, the dependent variable is central not only to design but also to interpretation.

Dependent Variable in Qualitative Research

Qualitative research emphasizes the dependent variable as the measurable outcome in statistical analysis. For example:

- In **experimental designs**, researchers manipulate an IV (e.g., drug dosage) and measure its effect on the DV (e.g., blood pressure).
- In **correlational designs**, naturally occurring variables are measured to test associations (e.g., income as IV, life satisfaction as DV).
- In **longitudinal studies**, DVs are tracked over time to assess changes and causal direction.

Statistical methods such as regression, MANOVA, and path analysis often model dependent variables explicitly, making them the core focus of qualitative inquiry [21].

Dependent Variable in Qualitative Research

Although qualitative research does not typically use the term “dependent variable” explicitly, the concept exists as the outcome or central phenomenon being explored. For example:

- In phenomenological studies, the dependent “variable” may be the lived experience of participants.
- In case studies, the dependent outcome might be the effectiveness of a program in a specific setting.
- In grounded theory, the dependent phenomenon is the process or category being developed from participant data.

In this sense, qualitative dependent variables are contextually embedded and are often represented through themes, narratives, or categories rather than numbers.

Challenges in Measuring Dependent Variables

Several challenges affect how dependent variables are defined and assessed:

1. **Complex Constructs**: Many DVs, such as "motivation" or "well-being," are abstract and difficult to measure directly.
2. **Cultural Sensitivity**: Measures may not have the same meaning across different populations.
3. **Measurement Error**: Inaccuracies in instruments (e.g., biased survey items) can distort findings.
4. **Multidimensionality**: Outcomes often consist of multiple dimensions, complicating operationalization.
5. **Ethical Concerns**: Some outcomes, such as trauma or mental health states, may be sensitive and require careful assessment.

Addressing these challenges requires methodological rigor and sensitivity.

Dependent Variables and Causality

Dependent variables are central to causal inference because they represent the outcomes affected by presumed causes. For causality to be established:

1. The independent variable must precede the dependent variable (temporal precedence).
2. There must be covariation between IV and DV.
3. Alternative explanations must be ruled out through control of confounds.

Without a clearly defined dependent variable, causal claims cannot be substantiated.

Contemporary Applications of Dependent Variables

1. **Education**: Student achievement, engagement, and motivation are common dependent variables studied in relation to teaching strategies and curriculum design.
2. **Health Sciences**: Patient recovery rates, treatment adherence, and quality of life are measured as outcomes in clinical trials.
3. **Psychology**: Dependent variables include mental health indicators, cognitive performance, or behavioural changes.
4. **Business**: Customer satisfaction, sales growth, and employee productivity serve as dependent outcomes.
5. **Artificial Intelligence**: Model accuracy, error rates, and user satisfaction often serve as dependent variables in evaluating AI systems.

These applications demonstrate how dependent variables reflect the goals and impact of interventions across fields.

The dependent variable is a fundamental component of research design and analysis, representing the measurable outcome that reflects the effect of independent

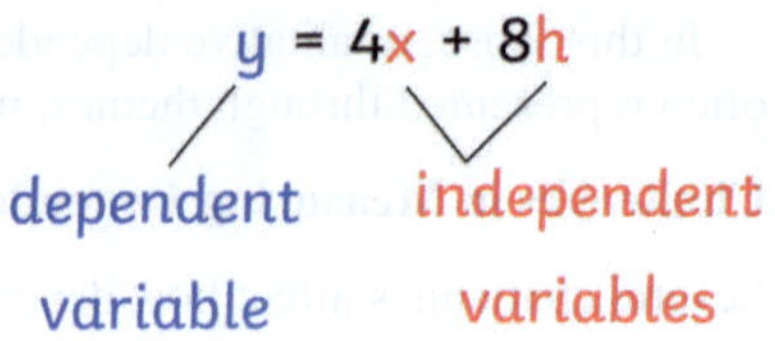

Fig. 5.3 Understanding the dependent variable [22]

variables. By defining, operationalizing, and measuring dependent variables carefully, researchers can test hypotheses, evaluate interventions, and build theoretical frameworks. Despite challenges such as measurement error, abstract constructs, and cultural sensitivity, the dependent variable remains central to advancing knowledge. Across disciplines—from education and health to psychology and AI—dependent variables allow researchers to capture the "effect" in cause-and-effect relationships, making them indispensable to both scientific inquiry and practical applications (Fig. 5.3).

5.2.3 *Control Variable*

In empirical research, understanding relationships between independent and dependent variables is essential for testing hypotheses and building theory. However, many studies are influenced by external factors that can distort or confound the observed relationships. To address this, researchers often include **control variables**—factors that are not the primary focus of investigation but are held constant, statistically adjusted, or accounted for to reduce bias and increase the validity of findings. Control variables play a critical role in ensuring internal validity and enabling more accurate conclusions about causal relationships [1]. This section explores the concept, types, importance, methods of handling, challenges, and applications of control variables across disciplines.

Conceptualizing the Control Variable

A control variable is defined as a variable that could influence the dependent variable but is not the primary focus of the study. Researchers include it to minimize its effect, ensuring that changes in the dependent variable can be attributed more confidently to the independent variable. For example, in a study examining the effect of **study time (IV)** on **academic performance (DV)**, potential control variables might include prior academic ability, socioeconomic background, and access to resources. By holding these factors constant, researchers isolate the effect of study time more accurately. Thus, while independent variables explain variation and dependent variables reflect outcomes, control variables serve as methodological safeguards against confounding.

Characteristics of Control Variables

Control variables share several key features:

1. **Not the Focus of Study**: They are included to account for potential confounds, not to test hypotheses.
2. **Systematically Managed**: Either through experimental design (holding constant) or statistical adjustment.
3. **Potentially Related to DV**: They have the capacity to influence the dependent variable if uncontrolled.
4. **Context-Specific**: The choice of control variables depends on the research question, discipline, and theoretical framework.

By incorporating control variables, researchers strengthen the credibility of causal inferences.

Types of Control Variables

1. **Demographic Controls**: Age, gender, socioeconomic status, and education are frequently controlled to reduce bias in studies involving human participants.
2. **Environmental Controls**: In laboratory research, temperature, lighting, and noise are controlled to ensure uniform experimental conditions.
3. **Temporal Controls**: Time of day, season, or year of data collection can affect results, particularly in longitudinal or field studies.
4. **Procedural Controls**: Instructions, instruments, or testing conditions are standardized to avoid variability introduced by the research process.
5. **Contextual Controls**: In education research, variables such as school location (urban vs. rural) or teacher quality are sometimes controlled to focus on instructional interventions.

These categories illustrate the broad scope of variables that researchers may treat as controls.

Operationalization of Control Variables

Operationalization involves specifying how control variables will be measured or managed.

1. **Experimental Designs**: Researchers hold control variables constant by standardizing procedures (e.g., testing all participants in the same room).
2. **Statistical Designs**: Researchers measure control variables and adjust for them in analyses using regression, ANCOVA, or propensity score matching.

For example, in a study on the effect of exercise on weight loss, diet could be included as a control variable by measuring participants' caloric intake and adjusting for it in statistical analysis [21].

Importance of Control Variables

1. **Increase Internal Validity**: Controlling for extraneous variables ensures that observed effects are due to the IV rather than uncontrolled influences.
2. **Reduce Bias**: Control variables reduce systematic error in data interpretation.

3. **Improve Comparability**: By adjusting for controls, researchers can compare findings across populations and contexts.
4. **Support Causal Inference**: Careful use of control variables helps establish stronger causal claims.

Without control variables, research findings risk being spurious or misleading [23].

Control Variables in Qualitative Research

Control variables are particularly important in qualitative designs:

1. **Regression Analysis**: Includes control variables as covariates to isolate the effects of the independent variable.
2. **ANCOVA**: Adjusts group means for the influence of control variables.
3. **Propensity Score Matching**: Balances control variables across treatment and comparison groups in quasi-experiments.

For example, in economics, income and education are often controlled when studying the relationship between health outcomes and employment. This prevents misattribution of effects.

Control Variables in Qualitative Research

Although control variables are less explicit in qualitative studies, researchers often account for contextual influences:

- In **case studies**, researchers may acknowledge institutional factors (e.g., school policies) as controls.
- In **ethnography**, contextual descriptions serve as controls by situating participant behaviours within broader cultural or environmental settings.
- In **grounded theory**, variation in participant characteristics is often noted to explain differences in emergent themes.

Thus, while not statistically controlled, qualitative research acknowledges and interprets control variables within narratives.

Challenges in Using Control Variables

1. **Over-Control**: Controlling for too many variables can eliminate meaningful variance and obscure real effects.
2. **Collinearity**: Highly correlated control variables can distort regression estimates.
3. **Measurement Error**: Poorly measured control variables can introduce new biases rather than reducing them.
4. **Selection Bias**: In observational studies, identifying appropriate control variables is difficult, and omission can lead to endogeneity.
5. **Context Dependence**: Variables relevant as controls in one setting may be irrelevant in another.

These challenges require researchers to balance methodological rigor with theoretical relevance.

Control Variables and Causality

Causal inference requires accounting for potential confounds. Control variables help rule out alternative explanations by statistically adjusting for factors that could distort the IV–DV relationship. However, control variables alone cannot prove causality. Researchers must combine controls with robust design strategies such as randomization, longitudinal analysis, and replication [24].

Contemporary Applications of Control Variables

1. **Education**: Studies on teaching methods often control for prior academic achievement, socioeconomic background, and school resources.
2. **Health Sciences**: Clinical trials control for age, gender, pre-existing conditions, and lifestyle factors.
3. **Psychology**: Experiments control for variables such as participant mood, fatigue, or environmental distractions.
4. **Business and Economics**: Studies on consumer behaviour often control for demographic variables like income, location, and education.
5. **Artificial Intelligence**: Models often include control variables such as dataset size or user demographics to evaluate algorithm performance.

These examples highlight the broad relevance of control variables across applied research.

Control variables are essential tools for ensuring validity and reliability in research. While independent variables explain and dependent variables measure outcomes, control variables mitigate bias by accounting for external influences. Their role extends across both experimental and non-experimental designs, in both qualitative and qualitative research. Despite challenges such as over-control, collinearity, and measurement error, control variables strengthen causal inference and enhance research credibility. In contemporary research across education, health, psychology, business, and AI, control variables remain indispensable for producing trustworthy and actionable findings (Fig. 5.4).

Fig. 5.4 Control variable [1]

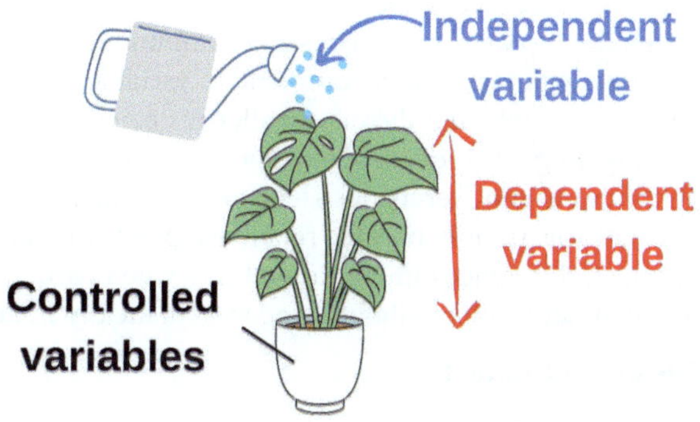

5.2.4 Extraneous Variable

In scientific research, the credibility and accuracy of results depend heavily on the precision of experimental design and the rigorous control of variables. A fundamental principle of experimental methodology is the relationship between the independent variable (IV) and the dependent variable (DV). The independent variable represents the factor deliberately manipulated by the researcher, while the dependent variable reflects the outcome measured. Ideally, the changes observed in the dependent variable should be attributable solely to the manipulation of the independent variable. However, this relationship is rarely so straightforward in practice. Numerous external influences—collectively referred to as extraneous variables—can interfere with this relationship, creating uncertainty and potentially invalidating findings. Extraneous variables are variables other than the independent variable that may affect the dependent variable if they are not properly controlled. Unlike confounding variables, which directly distort the cause-and-effect relationship, extraneous variables represent potential influences that, when unaccounted for, can reduce the internal validity of an experiment. Researchers across disciplines—from psychology and education to medical science and business studies—acknowledge the importance of identifying, understanding, and mitigating extraneous variables to ensure that research findings are reliable, generalizable, and ethically defensible. This section provides a comprehensive analysis of extraneous variables, exploring their definitions, classifications, and theoretical significance. It examines different types of extraneous variables, discusses their impact on research outcomes, and outlines effective strategies for controlling them. Furthermore, the section highlights contemporary challenges and ethical considerations in managing extraneous influences, concluding with reflections on their continued relevance in modern research contexts.

Extraneous variables refer to any factors other than the independent variable that might influence the dependent variable. The classic experimental framework emphasizes the importance of isolating the relationship between cause (IV) and effect (DV). However, in reality, human behaviour, biological processes, and social phenomena are influenced by multiple interacting factors. If these factors are overlooked, they may introduce bias, obscure real relationships, or create spurious associations. The distinction between extraneous and confounding variables is critical. While both can interfere with research, a confounding variable systematically varies with the independent variable and directly challenges causal inference. For example, in a study investigating whether exercise improves mood, diet could act as a confounding variable if those who exercise also tend to eat healthier. In contrast, an extraneous variable may not consistently bias the relationship but still introduces noise into the data. For instance, differences in the time of day when participants complete a mood survey might affect results without being systematically linked to exercise levels [25].

Historical Context

The concept of extraneous variables has been integral to the development of scientific methodology. Early experimental psychologists, such as Wilhelm Wundt and later

B.F. Skinner, emphasized the need to create controlled laboratory environments to eliminate external influences. Similarly, medical researchers designing randomized controlled trials (RCTs) recognized the necessity of randomization and blinding techniques as safeguards against extraneous influences. In contemporary times, with the increasing use of large datasets and machine learning models, extraneous variables remain central concerns, as uncontrolled variables can distort predictive algorithms and mislead decision-making processes.

Types of Extraneous Variables

Extraneous variables can be grouped into several categories, each with distinct implications for research. Understanding these categories allows researchers to design appropriate strategies for minimization and control [26].

1. **Situational Variables**: Situational variables are aspects of the environment that might affect participants' behaviour. Examples include room temperature, noise levels, lighting, time of day, and the presence of other people. For instance, in a cognitive psychology experiment assessing memory performance, participants tested in a quiet room may perform better than those tested in a noisy environment. If such factors are not standardized, they can obscure the true effect of the independent variable.
2. **Participant Variables**: Participant variables are individual differences among participants that can influence outcomes. These include characteristics such as age, gender, intelligence, personality, health status, mood, and prior experience. For example, in a study testing a new teaching method, differences in students' baseline knowledge could affect their learning outcomes independent of the teaching method itself. Random assignment and large sample sizes are often used to balance participant variables across conditions.
3. **Experimenter Variables**: Experimenter variables arise from the influence of the researcher themselves. Subtle cues, such as tone of voice, body language, or expectations, can inadvertently affect participants' behaviour—a phenomenon known as the experimenter expectancy effect. Rosenthal and Jacobson's (1968) study on the "Pygmalion effect" demonstrated how teachers' expectations influenced students' academic performance. Such findings underscore the potential for researcher behaviour to act as an extraneous influence.
4. **Demand Characteristics**: Demand characteristics refer to cues within the experimental setting that lead participants to guess the purpose of the study and adjust their behaviour accordingly. For example, in Milgram's (1963) obedience experiments, some participants speculated about the true purpose of the study and modified their responses to align with perceived expectations. Demand characteristics threaten both the internal and ecological validity of research.
5. **Order Effects**: Order effects occur when the sequence in which participants experience experimental conditions influences their responses. This issue is common in repeated measures designs, where the same participants are exposed to multiple conditions. Practice effects (improved performance due to familiarity) and fatigue effects (declining performance due to tiredness or boredom) can distort results.

For example, in memory studies, participants may recall more words in later trials simply because they become accustomed to the task.

6. **Instrumental or Measurement Variables**: Extraneous influences can also arise from the tools and procedures used to measure outcomes. Faulty equipment, poorly designed questionnaires, or inconsistencies in scoring can all introduce bias. For example, if a blood pressure cuff is not calibrated correctly, the readings may not accurately reflect the participant's actual health status.

Impact of Extraneous Variables on Research

Extraneous variables, if uncontrolled, can have significant consequences for research validity and reliability [27].

Threats to Internal Validity

Internal validity refers to the degree to which an experiment establishes a trustworthy cause-and-effect relationship between variables. Extraneous variables threaten this validity by introducing alternative explanations for observed outcomes. For instance, if a study concludes that a new therapy reduces anxiety, but participants were also exposed to a supportive group environment, the improvement may be due to social support rather than the therapy itself.

Threats to External Validity

Extraneous variables can also limit the generalizability of findings. If a study on workplace productivity is conducted only in one type of office environment with unique conditions, its results may not apply to other workplaces. Situational variables like company culture or working hours may constrain the applicability of findings.

Case Studies

1. **Milgram's Obedience Study**: While groundbreaking, Milgram's experiment was criticized for extraneous influences such as the artificial laboratory setting and the presence of authority figures, which may not reflect real-world obedience.
2. **Clinical Drug Trials**: In pharmaceutical research, extraneous variables such as participants' lifestyle factors (diet, exercise, alcohol use) can influence drug effectiveness. Without randomization and strict control, results could be misleading.
3. **Educational Research**: When evaluating teaching methods, factors such as teacher enthusiasm, class size, and peer influences act as extraneous variables. Without addressing these, it is difficult to isolate the true effect of instructional techniques.

Methods of Controlling Extraneous Variables

Researchers employ a range of techniques to minimize the influence of extraneous variables and protect the validity of their findings.

1. **Randomization**: Randomly assigning participants to conditions helps ensure that participant variables are evenly distributed, reducing systematic bias. Randomization is a cornerstone of experimental design, especially in large-scale clinical trials.
2. **Standardization**: Standardizing procedures ensures that all participants experience identical conditions, minimizing situational variables. For example, using the same instructions, equipment, and testing environments reduces variability.
3. **Counterbalancing**: Counterbalancing is used in repeated measures designs to control for order effects. By varying the sequence of conditions across participants, researchers can balance out practice and fatigue effects.
4. **Blinding Techniques**: Single-blind and double-blind procedures minimize experimenter bias and demand characteristics. In a single-blind study, participants are unaware of which condition they are in, while in a double-blind study, both participants and experimenters are kept unaware. Such methods are standard in clinical trials to reduce placebo effects.
5. **Statistical Controls**: Analytical methods such as ANCOVA (Analysis of Covariance) or multiple regression can statistically control for extraneous variables by accounting for their influence in the analysis. For example, adjusting for age and gender can help isolate the true effect of a treatment.
6. **Sample Size and Diversity**: Larger and more diverse samples increase the likelihood that extraneous variables are evenly distributed, enhancing generalizability and reducing random error.

Contemporary Issues and Advances

In modern research contexts, the challenge of extraneous variables has grown more complex. Big data analysis, machine learning, and social media research introduce new types of uncontrolled influences. Algorithms may pick up on irrelevant correlations (e.g., consumer behaviour linked with weather patterns) that act as extraneous factors, leading to spurious conclusions. Field experiments, while ecologically valid, are particularly vulnerable to extraneous influences such as social context and environmental unpredictability. Balancing ecological validity with experimental control remains a persistent dilemma for researchers. Ethical considerations also arise. Some methods of controlling extraneous variables, such as deception to minimize demand characteristics, must be carefully weighed against participants' rights. Researchers are required to strike a balance between methodological rigor and ethical responsibility, adhering to codes of conduct such as those established by the British Psychological Society (BPS) or the American Psychological Association (APA).

Extraneous variables represent a fundamental challenge in scientific research, as they threaten the clarity and accuracy of the relationship between independent and dependent variables. They can take many forms, including situational, participant, experimenter, and measurement-related influences, each with unique implications. Left uncontrolled, extraneous variables undermine both internal and external validity, jeopardizing the reliability and generalizability of findings. Fortunately, a wide range of methodological tools exist to address these challenges, from randomization and standardization to counterbalancing and statistical adjustment. Nevertheless, contemporary research introduces new complexities, particularly in large-scale data analysis and field settings, where extraneous influences are harder to eliminate. Ultimately, the careful recognition and management of extraneous variables are essential for producing valid, reliable, and ethically sound research. By maintaining awareness of these influences and applying rigorous control techniques, researchers can strengthen the integrity of their findings and contribute meaningfully to knowledge across disciplines [28].

5.3 Scales of Measurement in Qualitative Research

In the field of research, the concept of measurement is central to the process of gathering, analyzing, and interpreting data. While qualitative research is often associated with numerical scales and statistical testing, qualitative research equally relies on careful measurement, though in a different sense. Qualitative inquiry focuses on meanings, interpretations, and experiences rather than numerical values. However, it does not escape the requirement of systematic classification and measurement. Understanding scales of measurement in qualitative research is crucial, because it provides researchers with structured ways to code, categorize, and analyze non-numeric data. The four classical scales of measurement—nominal, ordinal, interval, and ratio—are widely acknowledged in the research community, yet their application in qualitative contexts is more nuanced and interpretive. Exploring how these scales apply to qualitative research enhances the rigor and reliability of findings, ensuring that subjective insights are systematically represented and understood.

The nominal scale is the most fundamental level of measurement and plays a significant role in qualitative research. At this level, variables are categorized without implying any order or hierarchy. Categories are mutually exclusive and exhaustive, serving as labels that distinguish one group from another. For instance, in qualitative studies conducted in healthcare, researchers may classify participants based on their professional roles—support worker, nurse, doctor, or administrator. In the case of an organization such as Padle Healthcare Limited, researchers might categorize support workers according to the type of patients they assist, such as elderly residents, patients with mobility challenges, or individuals with cognitive disorders. These categories do not imply that one is superior or inferior to the other but simply establish distinct labels for analysis. In qualitative analysis, nominal measurement is particularly useful during the coding process. When conducting interviews or focus groups, responses are often grouped into themes or categories, such as "barriers to care," "communication challenges," or "emotional support." The nominal scale allows researchers to treat each theme as a category, thus making qualitative data manageable and analyzable. Although the nominal scale does not provide information about magnitude or direction, it is foundational in ensuring clarity and consistency during qualitative analysis [29].

Moving beyond classification, the ordinal scale introduces order among categories. This level of measurement allows researchers to rank responses or themes according to certain criteria, even though the intervals between ranks are not necessarily equal. In qualitative research, ordinal measurement is often applied in situations where participant perspectives or experiences need to be arranged in terms of intensity, frequency, or importance. For example, in a study exploring workplace stress among support workers at Padle Healthcare Limited, themes that emerge from interviews might include "high workload," "emotional exhaustion," and "lack of resources." While these themes are identified through nominal classification, researchers may rank them according to how frequently participants mentioned them or how strongly they were emphasized. Thus, "high workload" might be placed as the most frequently reported challenge, followed by "emotional exhaustion" and then "lack of resources." Similarly, qualitative researchers may use ordinal scaling when coding participant responses on a Likert-type scale, such as "agree," "somewhat agree," or "disagree." Although qualitative research avoids reducing data purely to numbers, it still benefits from ordinal ranking, because it provides a structured way of representing participant priorities and hierarchies of experience. However, the limitation of the ordinal scale is that it does not capture the precise difference between ranks; for example, the gap between "agree" and "somewhat agree" may not be the same as the gap between "somewhat agree" and "disagree." Nonetheless, ordinal measurement adds depth to qualitative interpretation by allowing researchers to explore gradations in meaning [30].

While interval and ratio scales are typically associated with qualitative research, their application in qualitative contexts should not be overlooked. The interval scale involves ordered categories with equal intervals between them but lacks a true zero point. In qualitative research, interval-level measurement may be applied when researchers use rating systems that capture perceptions, attitudes, or experiences in structured interviews or surveys that complement narrative data. For example, in a qualitative study on patient satisfaction at Padle Healthcare Limited, participants might be asked to rate their satisfaction on a scale of 1 to 5. While the narrative responses provide in-depth understanding of satisfaction drivers, the interval scale adds another layer of interpretation by standardizing responses. The equal intervals between the ratings ensure comparability across participants, even though the data remains embedded in a qualitative framework. Similarly, when participants evaluate aspects of service delivery, such as timeliness of care, communication effectiveness, or emotional support, their ratings can be analyzed as interval-level data alongside thematic findings. This integration strengthens the reliability of qualitative studies by offering consistency in how subjective experiences are measured.

The ratio scale, which possesses all the properties of the interval scale plus a true zero point, is the highest level of measurement. While ratio data is rare in purely qualitative research, certain forms of qualitative inquiry incorporate ratio-level concepts, especially when mixed-methods designs are used. For instance, when exploring the workload of support workers, researchers may document the actual number of patients attended per shift (which has a meaningful zero point). Although this appears qualitative, in qualitative research such numerical data is often contextualized with narratives that explain how workload impacts emotional well-being or patient care quality. Similarly, in educational research examining learning outcomes, the amount of time students spends engaging in active learning activities can be treated as ratio data, while the qualitative narratives explain the significance of these activities in shaping student engagement. The use of ratio scales in qualitative contexts underscores the complementarity of qualitative and qualitative traditions. By embedding ratio-level measures into qualitative designs, researchers enrich the depth of interpretation without undermining the primacy of lived experiences [31].

The application of these scales in qualitative research is not without challenges. One of the central debates in qualitative inquiry revolves around whether it is appropriate to impose numerical scales on non-numeric, narrative data. Critics argue that over-reliance on measurement risks reducing the richness of qualitative data to rigid categories. For example, coding narratives into nominal categories may strip away nuances of meaning, while ranking themes through ordinal measurement may oversimplify the complex interplay of participant experiences. However, proponents contend that scales of measurement, when used thoughtfully, enhance rigor and comparability without compromising depth. The balance lies in treating scales as tools for organizing and interpreting qualitative data rather than as ends in themselves. Researchers must remain reflexive, ensuring that the voices and contexts of participants are not silenced by measurement frameworks.

In addition to the classical scales of measurement, qualitative researchers often employ alternative frameworks tailored to the unique nature of interpretive inquiry. For example, typologies and taxonomies are widely used to classify phenomena in ways that go beyond nominal categorization. A researcher might construct a typology of coping strategies among support workers, distinguishing between "emotional coping," "problem-focused coping," and "avoidant coping." Such typologies reflect both the nominal scale (categorization) and the ordinal scale (hierarchies of preference or effectiveness). Similarly, scaling methods like semantic differential scales and repertory grids, often used in qualitative psychology, allow participants to articulate meanings across a continuum rather than within fixed categories. These approaches demonstrate the flexibility of measurement in qualitative research and highlight how classical scales are adapted to suit the interpretive goals of qualitative inquiry. The importance of measurement scales in qualitative research also lies in their contribution to reliability and validity. Qualitative researchers are often challenged to demonstrate that their findings are credible, dependable, and transferable. By systematically applying measurement scales, they create transparent processes for coding, categorizing, and interpreting data. For instance, when multiple coders classify participant responses into nominal categories, the use of consistent coding schemes enhances inter-coder reliability. Similarly, when ordinal scales are applied to rank themes, researchers can compare findings across different groups or contexts, thereby strengthening the validity of interpretations. Measurement thus provides a structure that prevents qualitative research from devolving into anecdotalist, ensuring that participant voices are represented with rigor and systematicity [32].

Practical examples from healthcare illustrate how scales of measurement function in qualitative studies. In an ethnographic study of patient care at Padle Healthcare

Limited, researchers may use nominal scales to categorize types of care interactions, such as "clinical support," "emotional support," or "administrative support." Ordinal scales might be used to rank which forms of support patients find most meaningful. Interval-level measures, such as patient satisfaction ratings, could complement narrative data to show trends in perceptions. Finally, ratio data, such as the actual number of patient interactions per day, may contextualize narratives about workload. Together, these layers of measurement create a holistic understanding of care delivery, blending narrative richness with structured classification. Despite the utility of these scales, qualitative researchers must always situate measurement within the broader epistemological frameworks guiding their work. Unlike positivist qualitative approaches, qualitative inquiry often draws on constructivist or interpretivist paradigms, which emphasize meaning-making over objectivity. In this sense, scales of measurement in qualitative research serve as heuristic devices rather than rigid frameworks. They guide researchers in organizing data but must remain flexible enough to capture the dynamic, contextual nature of human experience. For example, while ranking coping strategies may help identify common patterns, researchers must also attend to the unique ways individual support workers describe their experiences of stress. This reflexivity ensures that measurement supports, rather than constrains, the interpretive goals of qualitative research.

In conclusion, scales of measurement are integral to qualitative research, even though their application differs from qualitative traditions. The nominal, ordinal, interval, and ratio scales each contribute to organizing, interpreting, and validating non-numeric data in meaningful ways. Nominal scales provide basic categorization, ordinal scales allow for ranking and prioritization, interval scales enable structured comparisons, and ratio scales offer contextual qualitative insights within qualitative frameworks. Together, they create a layered approach to measurement that enhances rigor while preserving the richness of qualitative data. However, their use requires careful reflexivity to avoid reducing complex human experiences to simplistic categories. For organizations such as Padle Healthcare Limited, applying these scales in qualitative research can generate deep insights into workplace experiences, patient care, and service delivery, ultimately improving outcomes for both staff and service users. By embracing the strengths of measurement while respecting the interpretive nature of qualitative inquiry, researchers can ensure that their findings are both rigorous and meaningful, contributing to the advancement of knowledge in diverse fields [32] (Fig. 5.5; Table 5.1).

Importance of Scales of Measurement

Scales of measurement are vital in qualitative research because they provide a structured framework for organizing and interpreting complex, non-numeric data. Unlike qualitative studies, where numerical precision is the central goal, qualitative inquiry emphasizes meaning, interpretation, and context. However, without systematic measurement, qualitative findings risk becoming anecdotal or inconsistent. The use of nominal, ordinal, interval, and ratio scales ensures that researchers can classify, rank, compare, and contextualize data while preserving its interpretive depth. This strengthens the credibility and transparency of the research process. One major

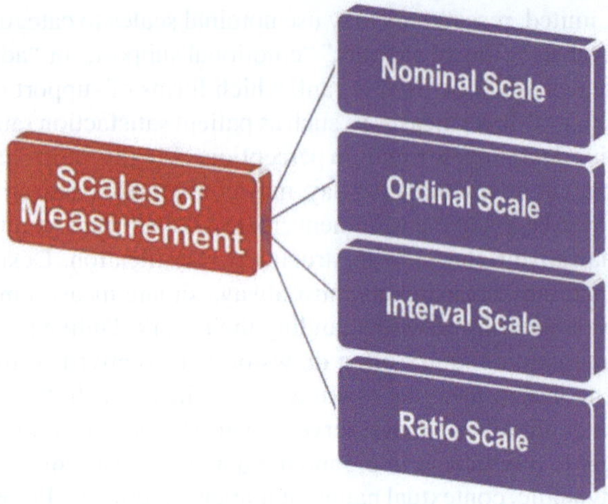

Fig. 5.5 Scales of measurement [33]

Table 5.1 Comparison of scales of measurement in qualitative research

Scale	Definition	Key features	Example in qualitative research
Nominal	Categorizes data into distinct groups without order or ranking	Labels, names, mutually exclusive categories	Coding interview responses into themes such as *"workplace stress," "emotional support,"* or *"training needs"*
Ordinal	Ranks categories in order, but intervals between ranks are not equal	Shows relative position but not precise difference	Ranking coping strategies among support workers as *"highly effective," "moderately effective,"* or *"least effective"*
Interval	Ranks categories with equal intervals but lacks a true zero point	Allows comparison of differences between values; arbitrary zero	Using a 1–5 satisfaction scale in a patient care study to complement narratives
Ratio	Contains all properties of interval data with a true zero point	Absolute comparisons possible; meaningful zero	Recording actual *number of patients supported per shift* alongside narratives about workload

importance of scales of measurement is their role in coding and categorizing data. Nominal scales, for example, allow researchers to group responses into distinct categories, such as themes emerging from interviews or focus groups. This ensures consistency across the analysis and facilitates comparisons between participants. By applying systematic coding frameworks, researchers can demonstrate that their interpretations are not arbitrary but grounded in explicit categories, enhancing reliability [34].

Ordinal scales further enhance qualitative research by providing a means to establish order among responses. In practice, participants' experiences, priorities, or challenges are rarely equal, and ordinal measurement allows researchers to reflect those differences meaningfully. For example, ranking stressors among healthcare support workers in order of importance captures nuances in lived experiences while still preserving their qualitative richness. This contributes to deeper insights that go beyond simple categorization. The importance of interval and ratio scales in qualitative research lies in their ability to bridge qualitative and qualitative traditions. Interval scales allow researchers to capture perceptions or attitudes using standardized measures such as rating scales. Although these are numerical, they add rigor to qualitative narratives by providing a comparable structure across participants. Ratio scales, though less common, contribute by embedding objective measures—such as time spent, number of interactions, or frequency of behaviours—within qualitative inquiry. This enriches interpretation by linking experiential narratives with observable, quantifiable dimensions.

Another critical importance of scales of measurement is their contribution to validity and reliability. By applying clear scales, qualitative researchers can demonstrate that their findings are systematic and replicable. This enhances credibility in contexts where decision-makers rely on qualitative evidence to inform policy or practice. For example, in healthcare research, applying structured measurement allows administrators to compare staff or patient experiences across departments and design interventions grounded in reliable data. Finally, scales of measurement support mixed-methods integration. In modern research, qualitative and qualitative approaches are often combined. Scales provide a common language that allows data from different paradigms to be compared and triangulated, creating more holistic findings. Thus, scales of measurement are not just technical tools but essential frameworks that safeguard rigor, enhance interpretation, and bridge methodological divides in qualitative research [33].

5.3.1 Nominal Measurement

Measurement is central to all forms of systematic research because it provides a structure for classifying, comparing, and analyzing data. Among the four classical scales of measurement identified by Stevens (1946)—nominal, ordinal, interval, and ratio—the nominal scale is the most basic and foundational. Though often regarded as the simplest form of measurement, nominal measurement plays a vital role in both qualitative and qualitative research by providing the categories through which complex phenomena can be understood. This section explores the concept of nominal measurement in depth, its defining characteristics, applications in qualitative and mixed-methods research, advantages and limitations, and its significance in professional practice. Special attention is given to practical examples in healthcare, such as within an organization like Padle Healthcare Limited, to illustrate its application in real-world contexts [35].

Understanding Nominal Measurement

The term "nominal" originates from the Latin word *nomen*, meaning "name." Nominal measurement therefore refers to data that are categorized by name, label, or classification rather than by numerical value. In this scale, variables are divided into mutually exclusive and collectively exhaustive categories. Each category is distinct, and there is no implied order, hierarchy, or magnitude between them. For example, when categorizing individuals based on gender (male, female, non-binary), each classification is separate and equally weighted; one category is not considered higher or lower than the other. Similarly, in healthcare research, patients can be classified according to diagnosis (e.g., diabetes, hypertension, arthritis) without implying that one diagnosis is greater or less than another. Nominal measurement is qualitative by nature because it focuses on identity and distinction rather than measurement of intensity or degree. Its central function is classification, and it underpins all higher forms of measurement. Without categories established at the nominal level, it would be impossible to proceed to ordinal rankings, interval distances, or ratio calculations [36].

Characteristics of Nominal Measurement

There are several key characteristics that distinguish nominal measurement from the other scales:

1. **Categorization without Order**—Categories have no inherent ranking or priority. For example, blood type categories such as A, B, AB, and O are simply distinct labels.
2. **Mutual Exclusivity**—Each unit of analysis can belong to only one category at a time. For instance, in a study on occupational roles at Padle Healthcare Limited, an employee may be categorized as a "support worker" but not simultaneously as a "registered nurse" under the same variable.
3. **Exhaustiveness**—All possible categories must be accounted for so that every case can be classified. For instance, when classifying mental health diagnoses in a study, researchers must include another category to capture diagnoses outside the main list.
4. **Equality of Categories**—Each category is considered equal in value; there is no greater or lesser degree attached to them. For example, in a classification of patient ethnic backgrounds, no category carries a higher weight than the others.
5. **Numerical Coding as Labels Only**—Numbers may be assigned to categories (e.g., 1 = male, 2 = female, 3 = non-binary), but these numbers are merely symbolic identifiers and do not convey magnitude or rank.

Applications of Nominal Measurement in Qualitative Research

In qualitative research, nominal measurement is most visible during the process of data coding and thematic analysis. Interviews, focus groups, and observations often produce rich textual data that need to be categorized systematically. For instance, in a study examining workplace stress among support workers at Padle Healthcare Limited, interview transcripts may be coded into categories such as "emotional exhaustion," "lack of resources," "poor communication," and "workload pressures." These categories are nominal because they are simply distinct themes. None is inherently greater or lesser than the other, they simply represent different aspects of the data. Similarly, in educational research, responses to open-ended questions about teaching strategies may be categorized into nominal groups such as "project-based learning," "lecture-based teaching," "flipped classroom," or "experiential learning." Again, the purpose is classification rather than ranking [37].

Examples of Nominal Measurement

1. **Healthcare**—Classifying patients by medical condition, such as cardiovascular disease, respiratory illness, or musculoskeletal disorder. These categories allow researchers to analyze care needs across different patient groups.
2. **Education**—Categorizing students by their preferred learning style, such as visual, auditory, or kinesthetic. Each style is distinct, and no one style is inherently superior.
3. **Business and Management**—Segmenting customers based on brand loyalty (loyal, occasional, new). This helps companies tailor strategies without implying that one group is more important than another.
4. **Sociological Studies**—Grouping participants according to ethnicity, marital status, or religious affiliation. These categories serve as identifiers that shape the context of analysis.

Advantages of Nominal Measurement

Nominal measurement provides several key advantages that explain its importance in research:

1. **Simplicity**—It is easy to understand and apply, requiring no advanced mathematical calculations.
2. **Foundation for Further Measurement**—All higher levels of measurement build on nominal classification. Without categories, variables could not be ordered or quantified.
3. **Flexibility**—It can be applied across diverse fields, including healthcare, education, psychology, business, and sociology.
4. **Clarity in Data Organization**—Nominal scales help researchers systematically organize large amounts of qualitative data into manageable categories.
5. **Enables Frequency Analysis**—Though nominal data cannot be ranked or measured in magnitude, researchers can still calculate frequencies and percentages for each category, offering useful descriptive insights.

Limitations of Nominal Measurement

Despite its utility, nominal measurement also has limitations:

1. **Lack of Order**—Because categories are unordered, researchers cannot assess intensity, preference, or importance.
2. **Limited Statistical Analysis**—Only basic descriptive statistics (e.g., mode, frequencies) are possible. Advanced statistical techniques cannot be applied directly.
3. **Potential Oversimplification**—Complex phenomena may be reduced to simplistic categories that obscure nuance. For example, categorizing patients as simply "satisfied" or "dissatisfied" may overlook deeper dimensions of their experiences.
4. **Risk of Ambiguity**—Poorly defined categories may lead to overlap or confusion, reducing reliability.

Nominal Measurement in Mixed-Methods Research

Nominal measurement is not confined to purely qualitative designs but is also important in mixed-methods research. For example, in a study examining the impact of training programs at Padle Healthcare Limited, researchers might use nominal categories such as "trained" versus "untrained" staff. They can then compare these categories qualitatively by examining patient outcomes, error rates, or staff retention. Nominal variables thus serve as grouping variables that allow integration of qualitative and qualitative data.

Nominal Measurement and Validity

A crucial consideration in nominal measurement is ensuring validity, that categories accurately represent the phenomenon being studied. This requires clear operational definitions of categories and careful coding procedures. For example, in a study on communication in healthcare, researchers must define what counts as "effective communication" versus "ineffective communication" to ensure consistency among coders. Using coding manuals, inter-coder reliability checks, and pilot testing helps strengthen validity in nominal measurement.

Case Study: Application in Padle Healthcare Limited

At Padle Healthcare Limited, a study might aim to explore the experiences of support workers in caring for patients with different needs. Nominal measurement could be applied to classify workers' responses into categories such as "emotional challenges," "physical demands," "communication issues," and "professional growth opportunities." These categories would allow researchers to examine which issues are most commonly reported. While this approach does not measure the intensity of challenges, it provides an essential foundation for further analysis. Administrators could use this information to identify priority areas for staff training or resource allocation.

Significance of Nominal Measurement

Nominal measurement is often underestimated due to its simplicity, but its significance in research cannot be overstated. It forms the bedrock of data organization and is indispensable for thematic analysis, coding, and classification. In qualitative research, it allows researchers to handle large volumes of textual data by reducing complexity into manageable categories. In mixed-methods research, it enables integration by serving as grouping variables that facilitate qualitative analysis. Beyond methodological significance, nominal measurement has practical implications in healthcare, education, business, and social sciences by providing decision-makers with clear categorical distinctions that inform policy and practice. Nominal measurement, though the most basic scale, is foundational to all research. Its function of categorization without order or magnitude provides the necessary structure for qualitative data analysis and underpins more advanced levels of measurement. By enabling systematic classification, nominal measurement ensures clarity, reliability, and validity in the research process. Its simplicity belies its power, as it allows researchers to capture complex human experiences in structured categories that facilitate interpretation and application. In professional contexts such as healthcare, education, and management, nominal measurement supports evidence-based decision-making by organizing data into actionable categories. While it has limitations in terms of order and statistical power, its strengths make it indispensable. Ultimately, nominal measurement is not merely a preliminary step in research but a vital tool for organizing and understanding the complexity of human experience [38] (Table 5.2).

Applications of Nominal Measurement

Nominal measurement, despite being the most basic level of measurement, plays a critical role in both qualitative and qualitative research. Its primary function of classification and labeling makes it indispensable across diverse academic disciplines and professional fields. By categorizing complex data into distinct, manageable groups, nominal scales provide a foundation for systematic analysis and interpretation. Below, key applications of nominal measurement are discussed across research,

Table 5.2 Nominal versus other scales of measurement

Scale	Definition	Key features	Statistical possibilities	Example
Nominal	Categorizes data into distinct, non-ordered groups	Labels, mutually exclusive categories, no order or magnitude	Frequencies, mode, percentages	Patient diagnosis: *diabetes, hypertension, asthma*
Ordinal	Ranks data into ordered categories without equal intervals	Order matters but distances are not measurable	Median, percentiles, non-parametric tests	Ranking job stress levels: *low, medium, high*
Interval	Ordered categories with equal intervals but no true zero	Allows differences to be measured, but ratios are meaningless	Mean, standard deviation, correlation, regression	Temperature in Celsius: *20, 25, 30 °C*
Ratio	Ordered categories with equal intervals and a true zero	All mathematical operations possible; ratios meaningful	All descriptive and inferential statistics	Patient weight: *0, 50, 70 kg*

healthcare, education, business, and the social sciences, with illustrative examples relevant to practice [38, 39].

1. **Coding and Thematic Analysis in Qualitative Research**: One of the most common applications of nominal measurement is in coding qualitative data. When researchers collect open-ended responses through interviews, focus groups, or written narratives, these responses are typically coded into categories. For instance, in a study on workplace experiences at Padle Healthcare Limited, support workers' responses might be classified into themes such as *"workload challenges," "emotional stress," "team collaboration,"* and *"training needs."* These categories are nominal because they represent distinct themes without inherent ranking. By applying this categorization systematically, researchers can identify recurring patterns, quantify the frequency of themes, and support their interpretive analysis with structured evidence.
2. **Classification of Populations in Healthcare**: In healthcare, nominal measurement is essential for patient classification. Medical professionals and researchers often categorize patients by diagnosis, blood group, or type of treatment. For example, patients may be grouped into categories such as *diabetes, hypertension, asthma, or arthritis.* These categories allow researchers to analyze prevalence rates, compare outcomes across diagnostic groups, and plan interventions. Importantly, the categories themselves carry no hierarchy—hypertension is not "greater" or "lesser" than asthma; they are simply distinct. At Padle Healthcare Limited, this approach enables staff to tailor support services according to specific patient categories, such as mental health needs, physical rehabilitation, or dementia care.
3. **Grouping Students and Learning Styles in Education**: In educational research, nominal measurement is widely used to classify students based on attributes such as learning styles, subject preferences, or participation in teaching methods. For example, students might be grouped into categories like *visual learners, auditory learners,* and *kinesthetic learners*. Similarly, responses to open-ended survey questions about preferred teaching strategies can be categorized into *"project-based learning," "lecture-based teaching," "flipped classroom,"* or *"experiential learning."* These categories help educators and researchers analyze trends in learner needs and design more inclusive teaching practices.
4. **Market Segmentation in Business Research**: In business and management, nominal measurement underpins market segmentation and consumer research. Companies often classify customers into categories such as *loyal customers, occasional customers, and new customers.* Likewise, consumers can be grouped based on product preferences, geographic location, or brand affiliation. These categories provide a structured basis for analyzing consumer behaviour and tailoring marketing strategies. For example, a business may discover that "loyal customers" respond better to reward programs, while "new customers" require introductory offers. Although the categories are not ranked, they are vital for evidence-based decision-making.

5. **Sociological and Demographic Research**: Nominal scales are also central to the study of social structures and demographics. Variables such as ethnicity, religion, marital status, and political affiliation are inherently nominal. They serve to categorize individuals into groups without implying hierarchy or order. For instance, in a community survey, participants may identify their marital status as *single, married, divorced,* or *widowed*. Researchers can then compare how experiences or attitudes differ across these groups. Similarly, political scientists categorize citizens into parties or ideological groups to analyze voting behaviours.
6. **Administrative and Institutional Use**: Nominal classification is critical in administrative contexts. In organizations such as Padle Healthcare Limited, staff roles are classified into categories such as *support worker, registered nurse, physiotherapist,* or *administrator*. These categories allow managers to allocate responsibilities, track staff distribution, and analyze workforce patterns. Similarly, patients may be classified based on *in-patient, out-patient,* or *community care.* This use of nominal measurement supports operational efficiency and organizational planning.
7. **Basis for Frequency and Comparative Analysis**: Although nominal data do not allow ranking or arithmetic operations, they provide a foundation for frequency counts and comparative analysis. For instance, researchers may calculate the number of participants in each category and compare percentages. In a study at Padle Healthcare Limited, researchers might report that 40% of patients present with musculoskeletal conditions, 30% with cardiovascular conditions, and 30% with neurological conditions. Such analysis enables decision-makers to prioritize resources without needing ranked or numerical scales.
8. **Facilitating Mixed-Methods Research**: In mixed-methods designs, nominal measurement plays a crucial role in integrating qualitative and qualitative data. For example, in a study evaluating the impact of staff training, participants may first be categorized into *trained* versus *untrained* groups (nominal). Then, their performance outcomes (qualitative data) can be compared across these groups. Thus, nominal variables serve as grouping variables that link qualitative categories with qualitative measures, allowing a more holistic understanding of phenomena.

The applications of nominal measurement extend far beyond simple classification; they are integral to the design, execution, and interpretation of research across disciplines. In qualitative research, it supports coding and thematic analysis; in healthcare, it enables patient classification and treatment planning; in education, it informs pedagogy; in business, it underpins consumer segmentation; and in sociology, it structures demographic analysis. Furthermore, nominal scales are crucial in administrative management and serve as the foundation for frequency analysis and mixed-methods integration. By organizing data into distinct, meaningful categories, nominal measurement provides the first step in transforming complex realities into structured insights that inform decision-making and practice [40] (Table 5.3).

Table 5.3 Applications of nominal measurement across disciplines

Discipline/field	Application of nominal measurement	Example
Qualitative research	Coding and thematic categorization of interview/focus group data	Categories: "*workload challenges*," "*emotional stress*," "*team collaboration*"
Healthcare	Classification of patients by diagnosis, treatment type, or care needs	*Diabetes, hypertension, dementia care*
Education	Grouping students by learning style, subject preference, or teaching method	*Visual learner, auditory learner, kinesthetic learner*
Business/marketing	Market segmentation and consumer classification	*Loyal, occasional, and new customers*
Sociology/demographics	Categorizing individuals by ethnicity, religion, marital status, or political affiliation	*Married, single, divorced, widowed*
Administration/workforce	Classifying staff roles or organizational units	*Support worker, nurse, physiotherapist, administrator*
Mixed-methods research	Grouping participants into categories for comparative analysis	*Trained versus untrained staff*

5.3.2 Ordinal Measurement

Measurement is one of the central concerns in scientific inquiry because it provides the basis for collecting, organizing, and interpreting data. Among the four classical scales of measurement—nominal, ordinal, interval, and ratio—ordinal measurement occupies a unique position. It extends beyond the basic classification of the nominal scale by introducing order and rank but stops short of the precision offered by interval and ratio levels. Ordinal measurement is crucial in research where variables reflect relative standing, preferences, or rankings but lack measurable distances between categories. This section explores the definition, characteristics, examples, advantages, limitations, and applications of ordinal measurement, with emphasis on its role in qualitative and mixed-methods research. Practical illustrations are drawn from healthcare, education, sociology, and business to highlight its significance in real-world contexts [41].

Understanding Ordinal Measurement

Ordinal measurement refers to a scale in which data are categorized into distinct groups that are arranged in a meaningful order. Unlike nominal measurement, which only labels categories, ordinal measurement introduces hierarchy or ranking. However, the intervals between categories are not equal or known, meaning that the degree of difference between ranks cannot be measured. For instance, in a survey on patient satisfaction at Padle Healthcare Limited, respondents might rate their experience as *"very satisfied," "satisfied," "neutral," "dissatisfied,"* or *"very dissatisfied."* These categories clearly indicate order, but the distance between them is unknown: the difference between *"very satisfied"* and *"satisfied"* may not be the same as between *"neutral"* and *"dissatisfied."* Ordinal measurement, therefore, captures relative position but not the magnitude of difference. It is especially useful in research involving perceptions, attitudes, or preferences, where variables are inherently subjective and cannot be measured with absolute precision [42].

Characteristics of Ordinal Measurement

1. **Order and Rank**—Data categories have a clear sequence (e.g., low → medium → high).
2. **No Equal Intervals**—The "distance" between categories is unknown or inconsistent.
3. **Relative Comparison**—It is possible to say that one category is higher or lower than another, but not by how much.
4. **Mutually Exclusive and Exhaustive Categories**—Each observation falls into one category, and all relevant categories are included.
5. **Statistical Operations**—Median, mode, and non-parametric tests (e.g., Mann–Whitney U, Kruskal–Wallis) are appropriate, but mean and standard deviation are not meaningful.

Examples of Ordinal Measurement

1. **Healthcare**—Severity of illness categorized as *mild, moderate, severe.*
2. **Education**—Ranking students as *first, second, third* in class performance.
3. **Business**—Customer satisfaction surveys with options from *"very dissatisfied" to "very satisfied."*
4. **Sociology**—Social class categories such as *lower class, middle class, upper class.*
5. **Psychology**—Levels of anxiety categorized as *low, moderate, high.*

Applications of Ordinal Measurement

1. **Patient Satisfaction and Healthcare Outcomes**: In healthcare research, ordinal scales are widely used to assess patient satisfaction, quality of life, or symptom severity. At Padle Healthcare Limited, a survey might ask patients to rank their level of pain as *no pain, mild pain, moderate pain, or severe pain.* While these categories cannot quantify the exact degree of pain, they provide a meaningful ordering that guides treatment decisions.
2. **Education and Learning Outcomes**: Ordinal measurement is crucial in evaluating student performance and perceptions. For instance, learners may be asked to rate their agreement with statements such as *"The flipped classroom improves my learning"* on a scale from *strongly disagree to strongly agree.* While ordinal, such responses provide insights into the effectiveness of pedagogical strategies.
3. **Market Research and Consumer Preferences**: Businesses often rely on ordinal scales in customer satisfaction surveys, product ratings, or brand preferences. For example, a consumer might rank three smartphone brands as first, second, and

third choice. The ranking reveals preference order but does not indicate the size of the gap between them.

4. **Social Sciences**: Sociologists use ordinal scales to measure phenomena like socioeconomic status (low, middle, high), political attitudes (conservative, moderate, liberal), or quality of community life. These categories enable comparative analysis of social groups and policy implications.
5. **Workplace Research**: In organizational contexts, ordinal scales are used to assess employee satisfaction, engagement, or stress levels. At Padle Healthcare Limited, support workers' stress might be categorized into *low, medium, high,* which helps administrators design interventions.

Advantages of Ordinal Measurement

1. **Captures Order and Hierarchy**—Enables researchers to assess preferences, attitudes, and rankings.
2. **Simple to Use**—Easy for participants to understand and respond to.
3. **Widely Applicable**—Useful across healthcare, education, business, and social sciences.
4. **Allows Non-Parametric Analysis**—Facilitates analysis using tests that do not assume normal distribution.
5. **Bridges Qualitative and Qualitative Approaches**—Ordinal data often emerge from qualitative responses but can be analyzed qualitatively.

Limitations of Ordinal Measurement

1. **Lack of Equal Intervals**—Cannot measure the exact distance between categories.
2. **Limited Statistical Power**—Mean and standard deviation are inappropriate; only median and mode apply.
3. **Subjectivity**—Categories may reflect perceptions rather than objective realities.
4. **Risk of Oversimplification**—Complex experiences may be reduced to simplistic rankings.
5. **Cross-Study Comparisons Difficult**—Differences in scale construction (e.g., 5-point vs. 7-point Likert scales) make direct comparison challenging.

Ordinal Measurement in Mixed-Methods Research

In mixed-methods designs, ordinal variables serve as a bridge between qualitative insights and qualitative analysis. For example, after interviewing staff at Padle Healthcare Limited about job stress, researchers may develop a survey asking participants to rank stress levels on a 5-point Likert scale. This ordinal data can then be analyzed alongside narrative accounts, offering a richer picture of workplace challenges [43].

Case Study: Ordinal Measurement in Healthcare Research

Consider a study conducted at Padle Healthcare Limited on patient experiences with mental health care. Researchers might develop a survey with the following ordinal categories:

- Quality of care: *very poor, poor, fair, good, excellent*
- Comfort with staff: *very uncomfortable, uncomfortable, neutral, comfortable, very comfortable*
- Recovery progress: *no improvement, slight improvement, moderate improvement, significant improvement*

Though not precise, these categories allow administrators to identify patterns and prioritize areas for improvement. They also allow for frequency analysis and comparisons across demographic groups.

Significance of Ordinal Measurement

Ordinal measurement holds a significant place in research because it balances simplicity and analytical usefulness. While it does not provide precise measurements, it offers valuable insights into relative positioning and preferences. In healthcare, it helps clinicians assess symptom severity; in education, it supports evaluation of teaching effectiveness; in business, it measures customer loyalty; and in sociology, it captures social class and attitudes. Its ability to transform subjective perceptions into structured, analyzable data makes it indispensable. Ordinal measurement is a vital scale of research measurement that extends beyond the nominal level by incorporating order and rank. While it cannot measure exact intervals or magnitudes, it provides meaningful insights into attitudes, perceptions, and rankings that shape decision-making in healthcare, education, business, and social sciences. Its advantages include simplicity, wide applicability, and the ability to support non-parametric analysis, while its limitations include lack of precision and statistical power. Nonetheless, ordinal measurement serves as a critical bridge between qualitative and qualitative traditions, making it a cornerstone of modern research practice (Table 5.4).

Table 5.4 Ordinal versus nominal, interval, and ratio scales

Scale	Definition	Order present?	Equal intervals?	True zero?	Example
Nominal	Categorizes data into distinct groups without ranking	✕ No	✕ No	✕ No	Types of mental illness: *anxiety, depression, psychosis*
Ordinal	Categorizes data with a meaningful order but unequal intervals	☑ Yes	✕ No	✕ No	Patient satisfaction: *very dissatisfied → very satisfied*
Interval	Ordered categories with equal intervals but no true zero	☑ Yes	☑ Yes	✕ No	Temperature in Celsius: *20, 30, 40 °C*
Ratio	Ordered categories with equal intervals and an absolute zero	☑ Yes	☑ Yes	☑ Yes	Weight: *0, 50, 70 kg*

5.3.3 *Interval Measurement*

Interval measurement is a unit of measurement that quantifies variables with equal intervals between values, allowing for meaningful comparisons of differences. It is different from nominal and ordinal scales as it lacks a true zero point, meaning zero does not indicate the complete absence of the variable being measured. Interval data supports many mathematical operations, including addition and subtraction, averages, and standard deviation [44]. Examples of interval measurement across fields include education, healthcare, business, environmental science, social sciences, and political science. Interval data can be analysed using descriptive statistics such as mean, median, and mode, as well as inferential statistics like parametric tests and correlation and regression analyses. Its strengths include precision, analytical flexibility, and applicability in science, education, and social research. However, it has limitations such as no true zero, arbitrary scales, and potential misuse, such as treating Likert scales as interval data [45].

In summary, interval measurement is a valuable tool for analysing variables across various fields, such as education, healthcare, business, environmental science, social sciences, and political alignment. Its limitations include the lack of a true zero-point, arbitrary scales, and potential misuse of Likert scales (Table 5.5).

Applications of Interval Measurement

1. **Education**: Comparing student performance over time using standardized tests.
2. **Healthcare**: Monitoring changes in patient health metrics, such as temperature or psychological scores.
3. **Business**: Evaluating customer satisfaction trends through interval-based sentiment scores.
4. **Environmental Science**: Assessing variations in air quality indices or pollutant levels.

Real-World Example: Interval Data in Action

Study: Investigating the Impact of Workplace Temperature on Productivity

- **Independent Variable**: Room temperature (measured in Celsius).
- **Dependent Variable**: Employee productivity scores.

Table 5.5 Interval measurement versus other scales

Feature	Nominal	Ordinal	Interval	Ratio
Categories	Yes	Yes	Yes	Yes
Order/rank	No	Yes	Yes	Yes
Equal intervals	No	No	Yes	Yes
True zero point	No	No	No	Yes

- **Analysis**: Calculate mean productivity for each temperature setting, and use ANOVA to determine if differences in temperature significantly impact productivity.

Interval measurement is an essential scale that provides a higher degree of precision than nominal and ordinal scales. By offering consistent intervals between values, it enables robust statistical analysis, though its lack of a true zero limits certain interpretations. Researchers in various fields rely on interval data for meaningful insights into trends, relationships, and effects. Understanding its strengths and limitations ensures its effective use in research and data analysis.

5.3.4 Ratio Scales Measurement

Measurement is central to research because it allows investigators to capture, quantify, and analyze phenomena systematically. Among the four classical scales of measurement—nominal, ordinal, interval, and ratio—interval measurement represents a critical step toward precision. It moves beyond the categorical ordering of nominal and ordinal scales by introducing equal intervals between categories, while still lacking an absolute zero. This makes interval measurement highly versatile in capturing variables like attitudes, perceptions, and psychological constructs that cannot be measured with ratio precision.

This section explores the concept of interval measurement, outlining its definition, characteristics, examples, strengths, limitations, and applications. It also highlights its role in bridging qualitative and qualitative research, with practical illustrations drawn from healthcare, education, business, and the social sciences.

Understanding Interval Measurement

Interval measurement refers to a scale where values are ordered, and the intervals between them are equal and meaningful, but there is no true zero point. In other words, the differences between values can be measured consistently, but the ratios are meaningless because the zero point is arbitrary. For example, temperature in Celsius is an interval scale: the difference between 20 and 30 °C is the same as between 30 and 40 °C. However, zero degrees does not indicate the absence of temperature—it is merely a point on the scale. Thus, while differences can be compared, ratios (e.g., "40 °C is twice as hot as 20 °C") are invalid.

In research contexts, interval measurement is particularly relevant for psychological scales, attitude surveys, and educational assessments, where constructs are quantified but cannot be expressed in absolute terms.

Characteristics of Interval Measurement

1. **Order**—Data are ranked in meaningful sequence.
2. **Equal Intervals**—The difference between values is consistent and measurable.
3. **No Absolute Zero**—Zero does not indicate a complete absence of the variable.

4. **Statistical Operations**—Mean, median, mode, variance, correlation, and regression are meaningful.
5. **Arbitrary Units**—Units of measurement may depend on convention (e.g., Celsius vs. Fahrenheit).

Examples of Interval Measurement

1. **Temperature Scales**—Celsius and Fahrenheit (but not Kelvin, which is ratio).
2. **Psychological Testing**—IQ scores (e.g., difference between IQ 100 and IQ 110 equals difference between 110 and 120).
3. **Attitude Surveys**—Likert scales (e.g., 1 = strongly disagree, 5 = strongly agree).
4. **Educational Testing**—Standardized test scores (e.g., SAT, GRE).
5. **Calendar Dates**—The difference between years (2000–2010) is measurable, but "year zero" is arbitrary.

Applications of Interval Measurement

1. **Attitude and Perception Research**: Interval scales are widely used in Likert-type surveys, which measure attitudes toward policies, products, or teaching strategies. For example, in evaluating support workers at Padle Healthcare Limited, researchers may ask participants to rate agreement with statements such as *"I feel supported by my supervisor"* on a 1–5 scale. The intervals are equal, allowing meaningful analysis of trends.
2. **Educational Assessment**: Interval measurement underpins many standardized tests. For instance, the difference between a student scoring 500 and 600 on the SAT is equivalent to the difference between 600 and 700, but the zero point is arbitrary. Similarly, in classrooms, student evaluations often use rating scales that assume equal intervals.
3. **Healthcare Research**: Patient satisfaction surveys and quality-of-life measures are often structured as interval scales. For example, patients may be asked to rate pain intensity on a 1–10 scale. Although subjective, the equal spacing of the categories allows healthcare providers at Padle Healthcare Limited to track changes in patient perceptions across time and treatments.
4. **Market Research**: Businesses often rely on interval measurement to gauge consumer preferences. Product satisfaction ratings (e.g., 1–7 scale) allow marketers to assess trends and design interventions. While ratios are invalid, averages and correlations across products or demographics can be calculated.
5. **Social Science Research**: Sociologists and political scientists use interval scales to study opinions, attitudes, and psychological traits. For example, measuring "political ideology" on a scale from 1 (very conservative) to 7 (very liberal) provides equal intervals, even though the zero point is arbitrary.

Advantages of Interval Measurement

1. **Equal Intervals**—Allows precise comparison of differences between values.
2. **Robust Statistical Analysis**—Supports a wide range of parametric tests.
3. **Versatility**—Useful in psychology, education, healthcare, business, and social sciences.

4. **Bridges Subjective and Objective**—Transforms perceptions into analyzable data.
5. **Facilitates Longitudinal Studies**—Enables tracking changes over time.

Limitations of Interval Measurement

1. **No True Zero**—Ratios and proportions are meaningless.
2. **Assumption of Equal Intervals**—Especially in Likert scales, equal spacing between categories is assumed but may not reflect reality.
3. **Potential for Misinterpretation**—Researchers may mistakenly treat interval data as ratio data.
4. **Arbitrary Scale Units**—Different conventions (Celsius vs. Fahrenheit) can complicate cross-study comparisons.
5. **Subjectivity**—In attitude scales, participants may interpret scale points differently.

Interval Measurement in Mixed-Methods Research

Interval measurement is particularly important in mixed-methods research because it allows qualitative constructs to be quantified. For instance, interviews with healthcare staff may reveal themes of *job stress* and *workload management*. Researchers can then design an interval scale (e.g., 1–10) to measure the degree of stress across participants. This enables both thematic analysis (qualitative) and statistical comparison (qualitative).

Case Study: Interval Measurement in Healthcare

At Padle Healthcare Limited, researchers might investigate staff well-being using a Likert-style survey:

- *"I feel my workload is manageable."* (1 = strongly disagree; 5 = strongly agree)
- *"I receive adequate emotional support from management."* (1 = strongly disagree; 5 = strongly agree)
- *"My work environment promotes mental well-being."* (1 = strongly disagree; 5 = strongly agree)

Because intervals are equal, administrators can calculate averages and correlations, allowing them to compare staff satisfaction across departments. While ratios (e.g., "twice as satisfied") are meaningless, interval data provide actionable insights into organizational improvement.

Significance of Interval Measurement

Interval measurement holds significant importance in research because it provides the precision needed for advanced statistical analysis without requiring a true zero. It allows researchers to compare differences, calculate averages, measure correlations, and test hypotheses across diverse contexts. Its versatility makes it essential in psychology, healthcare, education, and business, where many variables cannot be

Table 5.6 Interval versus nominal, ordinal, and ratio scales

Scale	Definition	Order present?	Equal intervals?	True zero?	Example
Nominal	Categorizes data into distinct, non-ordered groups	✕ No	✕ No	✕ No	Types of diagnosis: *anxiety, depression, schizophrenia*
Ordinal	Groups data into categories with meaningful order but unknown intervals	☑ Yes	✕ No	✕ No	Patient satisfaction: *low, medium, high*
Interval	Orders data with equal intervals but no absolute zero	☑ Yes	☑ Yes	✕ No	Temperature in Celsius; Likert scales (1–5 agreement)
Ratio	Orders data with equal intervals and an absolute zero	☑ Yes	☑ Yes	☑ Yes	Weight (*0, 50, 70 kg*); income (*$0, $100, $200*)

measured with ratio-level precision. By translating subjective perceptions into structured, analyzable formats, interval measurement enriches the scientific understanding of human behaviour and social phenomena. Interval measurement represents a pivotal stage in the hierarchy of measurement scales. By combining order with equal intervals, it offers researchers a level of precision that ordinal and nominal scales cannot achieve, while falling short of the absolute precision of ratio measurement. Despite its limitations, particularly the lack of a true zero, interval measurement is indispensable in fields that deal with psychological, perceptual, or social constructs. Whether used in patient satisfaction surveys at Padle Healthcare Limited, standardized educational testing, or business market research, interval measurement transforms subjective experiences into structured evidence that informs decision-making (Table 5.6).

5.4 Reliability and Validity

Measurement is the backbone of scientific inquiry. It allows researchers to quantify abstract constructs, compare findings across studies, and generate evidence-based conclusions. However, for measurement to be meaningful, it must be both **reliable** and **valid**. These two concepts are central to research design, data collection, and analysis. Reliability ensures that measurements are consistent and stable over time, while validity ensures that the measurements accurately reflect the construct they are intended to measure. This section provides a detailed discussion of **reliability**

and validity, including definitions, types, examples, strengths, limitations, and applications in healthcare, education, business, and the social sciences. The discussion highlights how both concepts complement each other and why they are indispensable in producing credible and generalizable research findings.

1. **Understanding Reliability**

Reliability refers to the consistency or stability of a measurement instrument. A reliable instrument yields the same results when applied repeatedly under similar conditions. For example, if a patient satisfaction survey at Padle Healthcare Limited is administered today and again next week under the same circumstances, the responses should be consistent if the instrument is reliable.

Types of Reliability

1. **Test–Retest Reliability**: Measures stability of results over time. Example: A stress questionnaire given to support workers at Padle Healthcare Limited should produce similar scores if repeated within a short period, assuming no real change in stress levels.
2. **Inter-Rater Reliability**: Consistency across different observers or raters. Example: Two nurses rating a patient's pain on a scale should assign similar scores if the tool is reliable.
3. **Internal Consistency Reliability**: Consistency of responses across items within the same instrument. Example: In a survey measuring job satisfaction, all items related to workplace support should yield correlated responses. Cronbach's alpha is commonly used to assess this.
4. **Parallel-Forms Reliability**: Consistency between two equivalent versions of a test. Example: Two versions of a cognitive test given to patients should yield similar results if they measure the same construct.

2. **Understanding Validity**

Validity refers to the extent to which an instrument measures what it is intended to measure. A tool can be reliable but not valid—for instance, a faulty thermometer may consistently give the same reading (reliable) but not reflect the actual temperature (not valid).

Types of Validity

1. **Face Validity**: The degree to which a test "appears" to measure what it claims. Example: A depression scale asking about sadness, sleep, and appetite seems valid on the surface.
2. **Content Validity**: Extent to which an instrument covers all relevant aspects of a construct. Example: A quality-of-life scale for patients should include physical, emotional, and social dimensions.
3. **Construct Validity**: Degree to which a test actually measures the theoretical construct. Example: A stress scale should correlate with known indicators of stress such as cortisol levels or absenteeism.
4. **Criterion Validity** Correlation of a measure with an external criterion. Two types:

- **Concurrent validity**—correlation with current outcomes (e.g., a new pain scale correlating with existing pain assessments).
- **Predictive validity**—ability to forecast future outcomes (e.g., an entrance exam predicting academic success).

3. **Relationship Between Reliability and Validity**

While related, reliability and validity are distinct. Reliability is necessary but not sufficient for validity. An instrument cannot be valid if it is unreliable, but it can be reliable without being valid.

- Example: A broken scale that always shows 5 kg too high is reliable (consistent) but not valid (inaccurate).
- Conversely, a scale that gives fluctuating readings is neither reliable nor valid.

4. **Applications in Different Fields**

   1. **Healthcare**: In patient care, reliable and valid tools ensure accurate diagnosis, treatment planning, and evaluation of outcomes. At Padle Healthcare Limited, reliable mental health assessments are essential to avoid misdiagnosis and ensure appropriate intervention.
   2. **Education**: Reliable and valid assessments allow educators to measure student performance accurately. For example, standardized tests must consistently evaluate knowledge (reliability) and align with curriculum objectives (validity).
   3. **Business and Marketing**: Customer satisfaction surveys require reliability and validity to ensure that results reflect actual consumer preferences rather than random variation or poorly designed questions.
   4. **Social Sciences**: In sociology and psychology, constructs like anxiety, motivation, or job satisfaction are abstract. Reliable and valid scales enable researchers to capture these constructs systematically, facilitating generalizable conclusions.

5. **Methods for Enhancing Reliability and Validity**

   1. **Clear instrument design**—Questions should be unambiguous and culturally sensitive.
   2. **Pilot testing**—Helps identify and correct weaknesses.
   3. **Training raters**—Improves inter-rater reliability.
   4. **Using established tools**—Enhances both reliability and validity through prior validation.
   5. **Triangulation**—Using multiple methods (e.g., surveys, interviews, observations) to enhance validity.

6. **Challenges in Ensuring Reliability and Validity**

   1. **Subjectivity of Respondents**—Attitude surveys may be influenced by mood, culture, or social desirability.

2. **Contextual Factors**—Results may vary depending on time, place, or situation.
3. **Resource Limitations**—Validity testing (e.g., predictive studies) may require large samples and long follow-ups.
4. **Cross-Cultural Issues**—Tools validated in one culture may lack validity in another.

7. **Case Example: Reliability and Validity at Padle Healthcare Limited**

A support worker survey is designed to measure workplace stress.

1. **Reliability check**: Internal consistency analysis (Cronbach's alpha) shows strong correlation among stress-related items.
2. **Validity check**: The scale is correlated with absenteeism records and patient complaints, confirming predictive validity.

This ensures that management decisions are based on credible evidence rather than flawed instruments.

Reliability and validity are the cornerstones of scientific measurement. Reliability ensures consistency, while validity ensures accuracy. Together, they provide confidence that research findings reflect reality and can inform policy, practice, and further inquiry. In fields such as healthcare, education, business, and social sciences, their importance cannot be overstated. At Padle Healthcare Limited, ensuring that surveys, diagnostic tools, and assessments are both reliable and valid directly affects the quality of care and staff well-being. Ultimately, rigorous attention to reliability and validity strengthens the integrity of research and its contributions to knowledge and practice (Table 5.7).

Table 5.7 Comparison of reliability and validity

Aspect	Reliability	Validity
Definition	Consistency of measurement; produces stable results over time	Accuracy of measurement; assesses whether the tool measures what it intends to
Key question	"Does the instrument give the same result every time?"	"Does the instrument measure the right construct?"
Types	Test–retest, inter-rater, internal consistency, parallel-forms	Face, content, construct, criterion (concurrent and predictive)
Requirement	Necessary but not sufficient for validity	Requires reliability but goes beyond it
Example (healthcare)	Pain scale that always gives the same reading regardless of user	Pain scale that accurately reflects actual patient pain levels
Example (education)	Student test that yields consistent scores over repeated administrations	Test that truly measures subject knowledge rather than memory tricks

5.5 Operationalization of Concepts

In research, particularly within the social sciences, education, business, and healthcare, scholars frequently grapple with abstract concepts such as *stress*, *motivation*, *well-being*, or *job satisfaction*. These constructs are intangible and cannot be measured directly, yet they are central to theory building and empirical investigation. To study such phenomena systematically, researchers must make them measurable, a process referred to as operationalization. Operationalization is the bridge between abstract theorization and empirical analysis; it ensures that concepts are transformed into observable indicators that can be tested, compared, and analyzed. This section explores the meaning, importance, processes, challenges, and examples of operationalization, with practical illustrations from healthcare and education.

1. **Understanding Operationalization**

Operationalization of concepts refers to the process of defining abstract variables in measurable terms by specifying the procedures, instruments, or indicators that will be used to capture them. For example, while "mental well-being" is an abstract construct, it can be operationalized through self-report surveys (e.g., WHO-5 Well-Being Index), frequency of healthcare visits, or observed behaviours such as absenteeism. Operationalization therefore converts theoretical language into empirical tools. It ensures that research is not confined to abstract theorizing but instead produces data that can be analyzed qualitatively or qualitatively.

2. **Importance of Operationalization**

    1. **Clarity and Precision**: Operationalization prevents ambiguity by defining exactly what is meant by a concept. For instance, "stress" can be ambiguous; operationalization may define it as elevated cortisol levels, perceived workload, or self-reported anxiety on a scale.
    2. **Replicability**: Clearly defined operational variables allow future researchers to replicate studies, thereby strengthening reliability and generalizability.
    3. **Linking Theory and Practice**: Theories remain abstract until operationalized. For instance, theories of job satisfaction in healthcare become actionable when measured through employee surveys and performance data.
    4. **Facilitating Measurement and Analysis**: Operationalization enables the use of statistical tools and qualitative coding frameworks to analyze constructs systematically.
    5. **Improving Validity and Reliability**: Operational definitions align measurement tools with research objectives, enhancing both accuracy (validity) and consistency (reliability).

3. **Steps in Operationalizing Concepts**

Step 1: Conceptualization

The process begins with a clear definition of the concept at a theoretical level. For example, *"job satisfaction"* may be defined as the extent to which employees feel content with their roles, working environment, and organizational support.

Step 2: Identification of Dimensions

Most concepts are multidimensional. Job satisfaction, for instance, may include dimensions such as pay, working conditions, interpersonal relationships, and career advancement.

Step 3: Selection of Indicators

Each dimension is then linked to measurable indicators. For example, *pay satisfaction* might be measured by agreement with the statement: *"My salary reflects the effort I put into my work."*

Step 4: Measurement Design

The researcher then designs tools such as surveys, interviews, observational protocols, or physiological tests.

Step 5: Data Collection

The defined indicators are applied in real-world settings. For instance, healthcare workers at **Padle Healthcare Limited** may complete a standardized job satisfaction questionnaire.

Step 6: Analysis and Interpretation

The data collected are analyzed using appropriate statistical or qualitative techniques to test hypotheses or answer research questions.

4. **Examples of Operationalization**

Healthcare

- **Concept**: Patient well-being.
- **Operationalization**: Measured through patient self-reports, number of hospital visits, pain scores, or WHO-5 Well-Being Index.

Education

- **Concept**: Academic motivation.
- **Operationalization**: Assessed through attendance records, assignment completion rates, or survey items like *"I enjoy learning new things."*

Business

- **Concept**: Customer loyalty.
- **Operationalization**: Frequency of repeat purchases, membership duration, or Net Promoter Scores.

Social Sciences

- **Concept**: Political participation.
- **Operationalization**: Voter turnout, frequency of political discussions, or participation in protests.

5. **Challenges in Operationalization**

    1. **Abstractness of Concepts**: Some constructs, like *happiness* or *trust*, are inherently difficult to quantify.
    2. **Cultural Sensitivity**: Indicators may vary across cultural contexts. For example, job satisfaction in Nigeria may be measured differently from in Europe due to differences in workplace values.
    3. **Subjectivity in Self-Reports**: Respondents may overstate or understate their experiences, affecting validity.
    4. **Multidimensionality**: Many concepts have multiple facets, making it difficult to decide which indicators best capture the construct.
    5. **Dynamic Nature of Concepts**: Constructs such as *stress* or *mental health* change over time and context, complicating operationalization.

6. **Operationalization in Qualitative Versus Qualitative Research**

    - **Qualitative Research**: Operationalization often involves transforming variables into numerical scales. For instance, stress may be measured using a 1–10 scale.
    - **Qualitative Research**: Operationalization may focus on coding textual or observational data into categories. For example, interview transcripts about workplace challenges can be coded into themes such as *"workload pressure"* or *"support from management."*

In mixed-methods research, both strategies are combined to provide a comprehensive understanding of the concept.

7. **Operationalization and Measurement Scales**

Operationalization is closely linked to **scales of measurement**:

- **Nominal**—Categorizing (e.g., gender, job title).
- **Ordinal**—Ranking (e.g., satisfaction: low, medium, high).
- **Interval**—Equal spacing without true zero (e.g., Likert scale responses).
- **Ratio**—Equal spacing with true zero (e.g., income, age).

Choosing the right measurement scale is critical in operationalization because it determines the type of statistical analysis that can be applied.

8. **Case Example: Operationalization at Padle Healthcare Limited**

At **Padle Healthcare Limited**, a study is conducted to assess support workers' **stress levels**.

- **Conceptualization**: Stress is defined as the psychological and physical response to work-related demands.
- **Dimensions**: Emotional stress, workload pressure, physical fatigue, and social support.
- **Indicators**:

1. Emotional stress: Measured by self-report items on anxiety and irritability.
2. Workload: Number of patients per worker per shift.
3. Fatigue: Hours of sleep reported per night.
4. Social support: Likert scale responses to *"I feel supported by my colleagues and supervisors."*

Through operationalization, the abstract idea of "stress" becomes measurable, allowing management to identify areas for intervention.

9. **Significance of Operationalization in Research**

    1. **Bridges Theory and Practice**—It allows abstract theories to be tested empirically.
    2. **Enhances Communication**—Provides a common language among researchers.
    3. **Supports Policy and Decision-Making**—Data derived from operationalized concepts inform interventions.
    4. **Promotes Accuracy and Objectivity**—Reduces ambiguity by linking concepts to observable indicators.
    5. **Facilitates Cross-Disciplinary Research**—Enables psychologists, sociologists, educators, and healthcare workers to study similar constructs systematically.

Operationalization of concepts is a cornerstone of empirical research. Without operationalization, abstract constructs such as *stress*, *job satisfaction*, *loyalty*, or *well-being* would remain unmeasurable and, therefore, untestable. By defining concepts in terms of measurable indicators, researchers can generate data that are reliable, valid, and comparable across studies. While operationalization presents challenges—particularly with abstract, multidimensional, or culturally variable constructs—it remains essential for advancing knowledge and applying research findings to practice. At Padle Healthcare Limited, operationalization allows staff well-being and patient outcomes to be assessed systematically, guiding interventions and improving care quality. Ultimately, operationalization ensures that research moves beyond abstract theorizing and contributes to tangible improvements in human lives (Table 5.8).

Table 5.8 Examples of operationalization of common concepts

Concept	Theoretical definition	Operational definition (indicators/measures)	Example tool/method
Stress	Psychological and physical response to demands exceeding coping ability	• Self-reported anxiety scores • Number of patients per shift • Sleep hours per night	Perceived stress scale (PSS), workload records
Job satisfaction	Degree of contentment with job roles and work environment	• Satisfaction with pay • Quality of supervisor support • Career advancement opportunities	Job Satisfaction Survey (JSS)
Well-being	A holistic state of physical, mental, and social health	• WHO-5 Well-Being Index scores • Frequency of healthcare visits • Social interaction levels	WHO-5, patient interviews
Motivation	Internal drive to initiate and sustain goal-directed behaviour	• Attendance rates • Assignment completion • Self-reported interest in tasks	Academic motivation scale (AMS)
Loyalty (business)	Commitment to a product, service, or organization	• Frequency of repeat purchases • Duration of membership • Likelihood to recommend (NPS)	Customer loyalty surveys
Political participation	Engagement in activities influencing governance	• Voting frequency • Attendance at rallies • Social media advocacy	Election data, surveys

References

1. Obi, K.C., Iwuanyanwu, O.: Innovation in the assessment of primary school mathematics: an approach to revamping Nigeria education. J. Teach. Perspect. **18**(1), 51–66, 2023. http://repositorio.unan.edu.ni/2986/1/5624.pdf%0A, http://fiskal.kemenkeu.go.id/ejournal%0A, https://doi.org/10.1016/j.cirp.2016.06.001%0A, https://doi.org/10.1016/j.powtec.2016.12.055%0A, https://doi.org/10.1016/j.ijfatigue.2019.02.006%0A, https://doi.org/10.1
2. Mate, S.E., McDonald, M., Do, T.: The barriers and enablers to career and leadership development: an exploration of women's stories in two work cultures. Int. J. Organ. Anal. **27**(4), 857–874 (2019). https://doi.org/10.1108/IJOA-07-2018-1475
3. Intarapanich, C.: Teaching methods, approaches and strategies found in EFL classrooms: a case study in Lao PDR. Procedia Soc. Behav. Sci. **88**, 306–311 (2013). https://doi.org/10.1016/j.sbspro.2013.08.510

4. Cucca, R., Friesenecker, M.: Potential and limitations of innovative housing solutions in planning for degrowth: the case of Vienna. Local Environ. **0**(0), 1–15 (2021). https://doi.org/10.1080/13549839.2021.1872513
5. Hitt, L.M., Wu, D.J., Zhou, X.: Investment in enterprise resource planning: business impact and productivity measures. J. Manag. Inf. Syst. **19**(1), 71–98 (2002). https://doi.org/10.1080/07421222.2002.11045716
6. Aldaihani, F.M.: Justification for adopting qualitative research method, research approaches, sampling strategy, sample size, interview method, saturation, and data analysis. J. Int. Bus. Manag. January (2021). https://doi.org/10.37227/jibm-2021-09-1494.
7. Lohmer, J., da Silva, E.R., Lasch, R.: Blockchain technology in operations & supply chain management: a content analysis. Sustainability **14**(10), 0–83 (2022). https://doi.org/10.3390/su14106192
8. Joshi, S., Sharma, M.: Impact of sustainable supply chain management on performance of SMEs amidst COVID-19 pandemic: an Indian perspective. Int. J. Logist. Econ. Glob. **9**(3), 248 (2022). https://doi.org/10.1504/ijleg.2022.120811
9. Lin, Y.-H., Tseng, M.-L.: Assessing the competitive priorities within sustainable supply chain management under uncertainty. J. Clean. Prod. **112**, 2133–2144 (2016). https://doi.org/10.1016/j.jclepro.2014.07.012
10. Talukder, M.A., et al.: A dependable hybrid machine learning model for network intrusion detection. (2022). http://arxiv.org/abs/2212.04546
11. Kochovski, P., Stankovski, V.: Supporting smart construction with dependable edge computing infrastructures and applications. Autom. Constr. **85**(October 2017), 182–192 (2018). https://doi.org/10.1016/j.autcon.2017.10.008.
12. Oke et al. (2022)."An empirical study on challenges to the adoption of the Internet of Things in the Nigerian construction industry (2022). https://doi.org/10.1080/20421338.2020.1819117
13. Dr. Chinnaiah, P.M.: Impact of dividend payout on the firm value: a study of firms listed on national stock exchange. EPRA Int. J. Environ. Econ. Commer. Educ. Manag. **1**(1), 14–19 (2020). https://doi.org/10.36713/epra5211
14. Alayemi, S., Akintoye, R.: Strategic management of growth in manufacturing companies in Sub-Saharan Africa: a case study of Nigeria. Br. J. Econ. Manag. Trade **6**(2), 151–160 (2015). https://doi.org/10.9734/bjemt/2015/13796
15. Asif, M.K., Nambiar, P., Ibrahim, N., Al-Amery, S.M., Khan, I.M.: Three-dimensional image analysis of developing mandibular third molars apices for age estimation: a study using CBCT data enhanced with Mimics & 3-Matics software. Leg. Med. **39**(May), 9–14 (2019). https://doi.org/10.1016/j.legalmed.2019.05.003
16. Herison, R., Sahabuddin, R., Azis, M., Azis, F.: The effect of working capital turnover, accounts receivable turnover and inventory turnover on profitability levels on the Indonesia stock exchange 2015–2019. Psychol. Educ. **59**(1), 385–396 (2022). https://journal.unismuh.ac.id/index.php/jeb/article/view/2184
17. Kebede, K., Yadete, F.D., Kant, S.: Is paradigm shift from traditional marketing mix to digital marketing mix effects the organizational profitability in Ethiopia? A multivariate analysis. Partners Univ. Int. Res. J, March, 122–134 (2023). https://doi.org/10.5281/zenodo.7772438
18. Titah, R., Shuraida, S., Rekik, Y.: Integration breach: investigating the effect of internal and external information sharing and coordination on firm profit. Int. J. Prod. Econ. **181**, 34–47 (2016). https://doi.org/10.1016/j.ijpe.2016.01.002
19. Kathleen, S., et al.: Wardah marketing communication strategy to improve brand image during Covid-19 pandemic. A Soc. Sci. Entrep. J. **1**(1), 29–34 (2021)
20. Joremi, L., Muda, R., Wan Zuraida Wan Mohd Zain, N.A.R.: The challenges of Asnaf entrepreneurs in forming entrepreneurial resilience. Inf. Manag. Bus. Rev. **15**(3), 242–252 (2023)
21. Galindo-Rueda, F., Millot, V.: Measuring design and its role in innovation. In: OECD Science, Technology and Industry Working Papers, pp. 1–51 (2015). http://www.oecd-ilibrary.org/science-and-technology/measuring-design-and-its-role-in-innovation_5js7p6lj6zq6-en

22. Aswir, Misbah, H.: Smartphones as personal digital archives? Recentering migrant authority as curating and storytelling subjects. Photosynthetica **2**(1), 1–13 (2018). http://link.springer.com/10.1007/978-3-319-76887-8
23. Polit, D.F., et al.: Focus on research methods: Is the CVI an acceptable indicator of content validity? Appraisal and recommendations. Res. Nurs. Heal. **30**(4), 459–467 (2007). https://doi.org/10.1002/nur.20199
24. Shamout, M., Ben-Abdallah, R., Alshurideh, M., Alzoubi, H., Al Kurdi, B., Hamadneh, S.: A conceptual model for the adoption of autonomous robots in supply chain and logistics industry. Uncertain Supply Chain Manag. **10**(2), 577–592 (2022). https://doi.org/10.5267/j.uscm.2021.11.006
25. Meneses Silva, C.V., Silva Fontes, R., Colaço Júnior, M.: Intelligent fake news detection: a systematic mapping. J. Appl. Secur. Res. **16**(2), 168–189 (2021). https://doi.org/10.1080/19361610.2020.1761224
26. Bhattacharya, S., Kumar, P., Maddikunta, R., Pham, Q.: Deep learning and medical image processing for coronavirus (COVID-19) pandemic : a survey. Sustain. Cities Soc. **65**(November 2020), 102589 (2021). https://doi.org/10.1016/j.scs.2020.102589
27. Almuiet, M.Z., Zawaideh, F.: Intelligent agent framework for knowledge acquisition in supply chain management. Int. J. Sci. Technol. Res. **8**(9), 1984–1990 (2019)
28. Cluver, L., et al.: Reducing child abuse amongst adolescents in low- and middle-income countries: a pre-post trial in South Africa. BMC Public Health **16**(1), 1–11 (2016). https://doi.org/10.1186/s12889-016-3262-z
29. Mazhar, N., Salleh, R., Zeeshan, M., Hameed, M.M.: Role of device identification and manufacturer usage description in IoT security: a survey. IEEE Access **9**, 41757–41786 (2021). https://doi.org/10.1109/ACCESS.2021.3065123
30. Zaman, U., Imran, Mehmood, F., Iqbal, N., Kim, J., Ibrahim, M.: Towards secure and intelligent internet of health things: a survey of enabling technologies and applications. Electronics **11**(12), 1893 (2022). https://doi.org/10.3390/electronics11121893
31. Khan, M.A., Salah, K.: IoT security: review, blockchain solutions, and open challenges. Futur. Gener. Comput. Syst. **82**, 395–411 (2018). https://doi.org/10.1016/j.future.2017.11.022
32. Hammi, B., Zeadally, S., Khatoun, R., Nebhen, J.: Survey on smart homes: vulnerabilities, risks, and countermeasures. Comput. Secur. **117**, 1–24 (2022). https://doi.org/10.1016/j.cose.2022.102677
33. Subburaj, A., Sriram, V.P., Mehrolia, S.: Effects of supply chain integration on firm's performance: a study on micro, small and medium enterprises in India. Uncertain Supply Chain Manag. **8**(1), 231–240 (2020). https://doi.org/10.5267/j.uscm.2019.7.001
34. Mathuramaytha, C.: Supply chain collaboration—What's an outcome? : A theoretical model. In: International Conference on Financial Management and Economics, vol. 11, pp. 102–108 (2011). http://www.ipedr.com/vol11/20-R10023.pdf
35. Mursyidto, M.I.: THE role of information communication technology in records creation and distribution in Uganda Communications Commission. Implement. Sci. **39**(1), 1–15 (2014). https://doi.org/10.4324/9781315853178
36. Muniz, E.C.L., Dandolini, G.A., Biz, A.A., Ribeiro, A.C.: Customer knowledge management and smart tourism destinations: a framework for the smart management of the tourist experience—SMARTUR. J. Knowl. Manag. **25**(5), 1336–1361 (2020). https://doi.org/10.1108/JKM-07-2020-0529
37. Cerchione, R., Centobelli, P., Zerbino, P., Anand, A.: Back to the future of knowledge management systems off the beaten paths. Manag. Decis. **58**(9), 1953–1984 (2020). https://doi.org/10.1108/MD-11-2019-1601
38. Adekunle, S.A., Aigbavboa, C., Ejohwomu, O., Thwala, W.D., Aliu, J.: Leaders of tomorrow: status and outlook of BIM implementation among young quantity surveyors. In: Proceedings of International Structural Engineering and Construction, vol. 9, no. 1 (2022). https://doi.org/10.14455/10.14455/isec.2022.9(1).aae-15
39. Camilleri, M.A., Troise, C., Strazzullo, S., Bresciani, S.: Creating shared value through open innovation approaches: opportunities and challenges for corporate sustainability. Bus. Strateg. Environ. **32**(7), 4485–4502 (2023). https://doi.org/10.1002/bse.3377

40. Barnett, P., et al.: Interventions to improve social circumstances of people with mental health conditions: a rapid evidence synthesis. BMC Psychiatry **22**(1), 1–68 (2022). https://doi.org/10.1186/s12888-022-03864-9
41. Dada, E.G., Oyewola, D.O., Yakubu, J.H., Fadele, A.A.: Predicting protein secondary structure based on ensemble Neural Network. ITEGAM J. Eng. Technol. Ind. Appl. **MI**, 49–56 (2021). https://doi.org/10.5935/jetia.v7i27.732
42. Alaba, F.A., Othman, M., Abaker, I., Hashem, T., Alotaibi, F.: Author's accepted manuscript internet of things security: a survey. J. Netw. Comput. Appl. **88**(December 2016), 10–28 (2017). https://doi.org/10.1016/j.jnca.2017.04.002
43. Fadele, A.A., et al.: A novel countermeasure technique for reactive jamming attack in internet of things. Multimed. Tools Appl. **23**(34), 23–41 (2018). https://doi.org/10.1007/s11042-018-6684-z
44. Riyadi, S.: The effects of image, brand and quality on customer loyalty of Sharia Banking. J. Asian Financ. Econ. Bus. **8**(3), 1315–1325 (2021). https://doi.org/10.13106/jafeb.2021.vol8.no3.1315
45. Ekeh, M.C., Ramsaroop, S.: Mathematics subject supervisors' role in ensuring quality teaching in preprimary and primary schools. South African J. Child. Educ. **12**(1), 1–9 (2022). https://doi.org/10.4102/sajce.v12i1.1220

Chapter 6
Qualitative Research Data Preparation and Cleaning Process

6.1 Data Preparation and Cleaning

Qualitative research is a systematic approach to exploring complex human experiences, behaviours, and social phenomena through the collection of non-numeric data such as interviews, focus groups, observations, and documents. Unlike qualitative research, which relies heavily on statistical analysis and numerical precision, qualitative research emphasizes depth, meaning, and context. The credibility of any qualitative study depends significantly on the quality of data collected and how rigorously it is prepared and cleaned before analysis. Data preparation and cleaning represent foundational stages in the research process because they ensure that the information being analysed is both trustworthy and usable. Without careful attention to these stages, findings can be distorted, misinterpreted, or compromised by inconsistencies and errors. This section critically examines the processes of qualitative data preparation and cleaning, highlighting key steps such as transcription, data organisation, coding readiness, verification of accuracy, and ethical considerations. It also discusses the challenges researchers encounter during these stages, strategies for overcoming them, and the broader implications of proper data preparation for the integrity and validity of qualitative inquiry [1].

Importance of Data Preparation and Cleaning in Qualitative Research

The preparation and cleaning of qualitative data serve as the bridge between raw information collected in the field and meaningful insights derived through analysis. In qualitative research, data often come in unstructured forms such as audio recordings, handwritten notes, or digital transcripts. Unlike qualitative datasets, which can be cleaned using automated statistical tools, qualitative data require a more nuanced and human-centred approach. Researchers must interpret, transform, and organise this data into formats that can be systematically analysed while preserving the richness and contextual depth inherent in participants' voices. Data preparation and

F. A. Alaba et al., *Fundamentals of Qualitative Research*, Studies in Systems, Decision and Control 663, https://doi.org/10.1007/978-3-032-18427-6_6

cleaning are also crucial for ensuring research credibility. Fadele et al. [2] introduce a framework for trustworthiness, credibility, dependability, confirmability, and transferability, relies heavily on the accuracy and authenticity of data handling. If raw data are not properly transcribed, anonymised, or verified, the results may lack dependability. Furthermore, errors introduced during preparation may undermine the researcher's ability to provide thick description and produce findings that resonate with real-world contexts.

1. ***Transcription as the First Stage of Data Preparation***: Transcription is the process of converting spoken words or observed behaviours into written text. It is often the most time-consuming aspect of qualitative data preparation, but also one of the most essential. Recordings of interviews and focus groups are transcribed verbatim to capture the exact words, pauses, and non-verbal cues such as laughter or hesitation. There are multiple approaches to transcription, ranging from verbatim transcription to more refined versions such as "intelligent verbatim" or "clean transcription." Verbatim transcription is typically preferred for rigorous qualitative analysis because it preserves all participant expressions, including informal speech and filler words. However, in cases where clarity and readability are priorities, researchers may adopt a cleaner version while ensuring that meaning is not lost. The accuracy of transcription is fundamental for analysis. Misheard words, omitted details, or misattributed speaker turns can distort findings. To mitigate these risks, many researchers adopt a two-step process: initial transcription followed by verification through re-listening to recordings or involving a second coder to check for accuracy. Advances in digital transcription software and speech-to-text tools have sped up the process, but human oversight remains indispensable due to nuances such as accents, emotions, and contextual references.
2. ***Organisation and Structuring of Data***: Once transcribed, qualitative data must be organised into formats that facilitate systematic analysis. Organisation often involves assigning identifiers such as participant codes, dates, and contextual details to each transcript. This ensures that data can be traced back to its source while maintaining anonymity. For example, participants in a study might be coded as P1, P2, or P3 rather than by name. Data organisation also involves creating a logical storage system. Researchers commonly employ qualitative data analysis software (QDAS) such as NVivo, Atlas.ti, or MAXQDA to store, manage, and retrieve transcripts efficiently. These platforms allow researchers to categorise documents, annotate sections, and eventually apply coding schemes. Proper organisation not only enhances efficiency but also supports transparency and reproducibility by enabling other researchers to follow the data trail. Another important aspect of organisation is segmentation. Large blocks of transcribed text are often divided into smaller units such as sentences, paragraphs, or thematic chunks. This segmentation prepares the data for coding, making it easier for researchers to link specific parts of the text to emerging themes.
3. ***Cleaning and Refining the Data***: Data cleaning in qualitative research is not about removing anomalies as in qualitative analysis but about ensuring clarity,

consistency, and relevance. Cleaning involves checking transcripts for accuracy, standardising formats, and addressing errors introduced during collection or transcription. For instance, background noises, inaudible segments, or overlapping speech may be clarified with notations such as "[inaudible]" or "[overlapping voices]" to maintain transparency. Researchers must also clean data by removing irrelevant content that does not contribute to answering the research question. For example, small talk at the beginning of interviews may be excluded unless it provides valuable context. However, caution is necessary to avoid discarding data that may later reveal unexpected insights. Another cleaning task is ensuring consistency in spelling, grammar, and formatting across transcripts. While spoken language often contains grammatical errors, researchers may choose to standardise certain aspects for clarity while keeping the authenticity of participants' voices intact. Cleaning decisions must always be documented in reflexive notes to maintain transparency and allow future verification.

4. ***Verification and Quality Checks***: Quality assurance is a critical part of data preparation and cleaning. Verification involves checking transcripts against original recordings, cross-referencing field notes, and ensuring that all data has been accurately represented. In some cases, researchers conduct participant validation or "member checking," whereby participants review their transcripts or summaries to confirm accuracy. This not only enhances credibility but also fosters trust between the researcher and participants. Triangulation also plays a role in verifying data integrity. By comparing multiple sources such as interviews, observations, and documents, researchers can identify inconsistencies and strengthen the reliability of their findings. Peer debriefing, where colleagues review transcripts and coding, adds another layer of quality control and helps identify overlooked details or biases [3].

Ethical Considerations in Data Preparation and Cleaning

1. Ethics underpin every stage of qualitative research, particularly during data preparation and cleaning. Participants often disclose sensitive, personal, or vulnerable information that must be handled with utmost confidentiality. Anonymisation of transcripts is therefore an essential step. Identifiable information such as names, addresses, or workplace details must be removed or replaced with pseudonyms to protect participants' privacy.
2. Secure storage of transcripts is also critical. Digital files should be password-protected and stored on encrypted devices or secure servers, while physical notes should be locked away. Researchers must comply with ethical guidelines established by institutional review boards (IRBs) and data protection laws such as the General Data Protection Regulation (GDPR).
3. Ethical responsibility also extends to representation. Cleaning decisions must not distort participants' intended meanings or silence marginalised voices. Researchers must remain reflexive about their role in shaping data through transcription and cleaning processes, acknowledging how their choices influence interpretation.

Challenges in Data Preparation and Cleaning

1. Despite its importance, data preparation and cleaning in qualitative research are fraught with challenges. One common difficulty is the sheer volume of data. Long interviews or multiple focus groups can produce hundreds of pages of transcripts, requiring extensive time and resources for transcription and organisation.
2. Another challenge is balancing accuracy with readability. Transcribing every filler word and hesitation may preserve authenticity, but can make transcripts cumbersome for analysis. Conversely, over-editing risks stripping data of its natural richness. Researchers must make informed choices about transcription styles depending on research aims.
3. Technical issues also present obstacles. Poor-quality audio recordings, background noise, and heavy accents can compromise transcription accuracy. Although digital tools assist in transcription, they struggle with contextual nuances and may introduce errors that require painstaking correction.
4. Emotional challenges can also arise. Researchers transcribing sensitive interviews, such as those involving trauma or abuse, may experience emotional fatigue. This underscores the importance of researcher well-being and the need for adequate support systems during data preparation.

Strategies for Effective Data Preparation and Cleaning

To address these challenges, researchers adopt several strategies. Planning by allocating sufficient time and resources for transcription is crucial. Employing professional transcribers with training in research contexts can enhance accuracy, though researchers must still verify transcripts for consistency. Using qualitative data analysis software not only aids organisation but also allows for systematic cleaning by highlighting inconsistencies and tracking changes. Reflexive journaling is another valuable strategy, helping researchers document decisions and maintain transparency about how data were prepared and cleaned. Training in ethical handling of data ensures that researchers remain aware of their responsibilities regarding confidentiality and representation. Furthermore, collaboration among research teams—through double coding, peer reviews, and regular discussions—helps ensure quality and reduces bias during preparation and cleaning stages [4].

Implications for Qualitative Research

The quality of qualitative research findings is directly tied to the robustness of data preparation and cleaning. Well-prepared and cleaned data enhances the credibility of analysis, supports the generation of meaningful themes, and ensures that participants' voices are authentically represented. By contrast, poorly prepared data can lead to misinterpretations, unreliable conclusions, and diminished trustworthiness of the study. Moreover, rigorous data preparation strengthens the transferability of findings. When researchers provide clear, transparent accounts of how data were transcribed, cleaned, and organised, others can better understand the context and judge whether findings apply to different settings. This enhances the broader impact of qualitative research in informing practice, policy, and theory. Data preparation and cleaning form the backbone of qualitative research, ensuring that raw information collected from participants is transformed into accurate, trustworthy, and analysable formats. Processes such as transcription, organisation, cleaning, verification, and ethical handling are not merely technical tasks but deeply interpretive and ethical practices that shape the quality of research outcomes. While challenges such as time demands, technical limitations, and ethical dilemmas persist, effective strategies, including careful planning, reflexive documentation, and collaborative verification, can mitigate these issues. Ultimately, rigorous data preparation and cleaning safeguard the credibility of qualitative inquiry, allowing researchers to generate insights that truly reflect participants' lived experiences and contribute meaningfully to knowledge [5] (Fig. 6.1).

Fig. 6.1 Steps in data preparation and cleaning [6]

6.1.1 Preparing Data for Analysis

Data preparation is one of the most crucial stages in the research process because it determines the quality, accuracy, and reliability of findings. Whether in qualitative or qualitative studies, raw data collected during fieldwork cannot be immediately subjected to analysis. Instead, it requires systematic organisation, cleaning, verification, and transformation to make it suitable for meaningful interpretation. Preparing data for analysis is, therefore, not a mere technical exercise but an integral part of research integrity. Poorly prepared data can distort results, misrepresent participants' experiences, and compromise the credibility of a study, while properly prepared data ensures transparency, accuracy, and robustness. This section explores the process of preparing data for analysis with a particular focus on qualitative research. It examines the stages of transcription, organisation, cleaning, verification, and ethical management, while also reflecting on challenges and strategies that can enhance reliability. The section argues that preparing data is not only a practical task but also a methodological and ethical commitment to protecting the trustworthiness of research [7]

Data preparation is essential because raw data is often unstructured, inconsistent, or incomplete. For instance, interviews may include long pauses, background noise, or tangential conversations, while survey datasets may contain missing values, outliers, or errors. Without systematic preparation, such imperfections can lead to misinterpretation or invalid conclusions. Preparing data ensures that information is usable, consistent with research objectives, and aligned with methodological frameworks. In qualitative research, preparation allows researchers to immerse themselves in the data and begin identifying patterns, meanings, and emerging themes. In qualitative contexts, preparation enables the use of statistical techniques by ensuring that datasets are free from coding errors or inconsistencies. Thus, regardless of methodology, preparation enhances the integrity of analysis and lays the foundation for drawing credible inferences [8].

For qualitative studies, transcription is often the first step in preparing data for analysis. Interviews, focus groups, or recorded observations must be converted into written text to allow systematic examination. Transcription is more than mechanical typing; it requires interpretive judgment about what to include and how to represent speech accurately. Researchers may choose verbatim transcription, which includes every word, pause, and filler, or a more polished approach such as "clean verbatim," which focuses on meaning while omitting non-essential sounds. The choice of transcription style depends on the research purpose. For studies emphasising discourse analysis, every linguistic detail matters, whereas for thematic analysis, cleaned transcripts may be more practical. Regardless of approach, accuracy is essential. Verification through repeated listening or peer review ensures that transcripts faithfully represent participants' voices. Transcription also provides the first stage of data familiarisation, enabling researchers to reflect on nuances and initial themes even before formal coding begins.

Once transcripts or other raw materials are available, organisation becomes vital. Data organisation involves labelling, cataloguing, and storing material systematically. Researchers typically assign unique identifiers to each transcript, such as codes representing participants, dates, or interview settings. This step preserves anonymity while allowing traceability. Technological tools play a central role in organisation. Qualitative data analysis software such as NVivo, Atlas.ti, or MAXQDA allows researchers to store transcripts, attach contextual metadata, and prepare material for coding. For qualitative studies, statistical software like SPSS, Stata, or R requires datasets to be formatted with variables clearly defined and consistently labelled. Organisation not only makes the researcher's work more efficient but also enhances transparency, enabling others to audit or replicate the process.

Data cleaning involves checking for accuracy, consistency, and completeness. In qualitative contexts, cleaning may mean clarifying inaudible segments, correcting typographical errors in transcripts, or removing irrelevant material such as off-topic small talk. The aim is not to erase the authentic voice of participants but to ensure that data is clear and reliable for analysis. Researchers may annotate transcripts with notes such as "[inaudible]" or "[overlapping voices]" to indicate areas of uncertainty rather than guessing or omitting content. In qualitative studies, cleaning typically involves handling missing data, identifying outliers, and correcting entry errors. Researchers may use imputation methods to address missing values or apply statistical tests to decide whether outliers should be retained or excluded. Regardless of method, cleaning must be carefully documented to maintain transparency.

Preparation does not end with transcription and cleaning; verification is essential to ensure accuracy. Verification may include cross-checking transcripts against recordings, comparing multiple data sources, or conducting member checking by asking participants to review their transcripts. Peer debriefing, where colleagues examine prepared data, also strengthens credibility by highlighting overlooked errors or biases. Quality assurance is equally critical in qualitative research. Techniques such as double data entry, consistency checks, and automated error detection reduce the likelihood of mistakes. Verification ensures that data is not only accurate but also dependable, providing a solid foundation for trustworthy analysis.

Ethics is embedded in every aspect of data preparation. Participants often disclose sensitive or personal information that must be handled with confidentiality and care. Preparing data involves anonymising transcripts by removing names, locations, or other identifying details and replacing them with pseudonyms. Researchers also have a responsibility to secure data storage, using encryption, password protection, or locked storage systems in compliance with regulations such as the GDPR. Ethical data preparation also means respecting participants' voices. Cleaning and organising should not distort meaning or silence marginalised perspectives. Reflexivity is key here, as researchers must acknowledge how their decisions during preparation influence what becomes visible in analysis. Transparent documentation of preparation decisions therefore, ensures both ethical accountability and research integrity. Preparing data is rarely straightforward. One challenge is the volume of material. Qualitative interviews can generate hundreds of pages of transcripts, while surveys may produce large datasets requiring extensive cleaning. Another challenge is balancing accuracy with readability. Capturing every hesitation and filler word may make transcripts unwieldy, but removing them risks losing important nuances. Technical limitations also arise. Poor audio quality, background noise, or heavy accents can compromise transcription accuracy. Similarly, in qualitative research, incomplete questionnaires or misrecorded responses complicate cleaning. Emotional challenges may also surface, especially when dealing with sensitive or traumatic material. Researchers transcribing distressing narratives must navigate emotional fatigue and ethical responsibility [9].

Effective preparation requires careful planning and systematic approaches. Allocating sufficient time for transcription and cleaning, employing transcription software with human verification, and using qualitative data management tools all improve efficiency and accuracy. Reflexive journaling allows researchers to document decisions about data preparation, enhancing transparency. Collaboration also strengthens preparation. Double coding, peer checking, and regular supervision ensure that data is consistently prepared and free from unnoticed errors. Training in the ethical handling of data and familiarity with relevant software further enhance quality. In qualitative research, statistical training ensures that missing data and outliers are addressed systematically rather than arbitrarily. Preparing data is not an isolated step but a foundation for analysis. For qualitative research, well-prepared transcripts make coding easier, facilitate the identification of themes, and support robust interpretation. For qualitative studies, properly cleaned datasets allow for valid statistical testing and model building. Ultimately, the accuracy and credibility of research findings depend directly on the rigour of data preparation. Preparing data for analysis is a critical stage in the research process, encompassing transcription, organisation, cleaning, verification, and ethical management. It ensures that raw material collected during fieldwork is transformed into usable, accurate, and trustworthy data that truly reflects participants' experiences or accurately represents measured variables. Despite challenges such as time constraints, technical limitations, and ethical dilemmas, effective strategies, including careful planning, collaboration, and reflexive documentation, can overcome these difficulties. Ultimately, data preparation is more than a technical necessity; it is an ethical and methodological commitment to integrity. By preparing data rigorously, researchers safeguard the credibility of their findings and contribute to the production of knowledge that is both reliable and meaningful [10].

6.1.2 Handling Missing Data

Missing data is a common yet critical challenge in research, affecting both qualitative and quantitative studies. It arises when information intended for collection is incomplete or unavailable, leaving gaps in the dataset. Missing responses in survey questions, incomplete transcripts in qualitative interviews, or absent measurements in clinical trials all present difficulties that must be addressed. If ignored, missing data can undermine validity, distort findings, and weaken the credibility of conclusions. Scholars such as [11–13] have demonstrated that the manner in which missing data is handled has significant implications for statistical inference and the overall quality of research. This section provides a critical discussion of missing data, examining its causes, types, and consequences before exploring various methods for handling it. It highlights both traditional and modern approaches, comparing their strengths and limitations, while also considering ethical responsibilities and practical challenges. Ultimately, the section argues that effective management of missing data is not only a statistical necessity but also an ethical obligation for safeguarding research integrity.

Missing data can result from numerous sources, many of which depend on study design, context, and participant behaviour. In survey-based research, participants may skip sensitive or confusing questions, producing gaps. In longitudinal studies, attrition—where participants drop out before completion—creates incomplete datasets. Technical errors, such as recording malfunctions during interviews or system failures in online surveys, also contribute. In health research, missing clinical measurements may occur if patients refuse invasive tests or fail to attend follow-up appointments. Understanding the causes of missing data is crucial because it shapes the strategies used to manage it. If missingness is systematic rather than random, conventional methods of handling data may produce biased results. For this reason, researchers must examine the underlying mechanisms that generate missingness before selecting an approach to address it.

Types of Missing Data

Statistical theory distinguishes three main mechanisms of missing data, as outlined by [14]:

1. **Missing Completely at Random (MCAR)**: Data is missing in a way unrelated to either observed or unobserved values. For example, if a survey response is missing because a page was accidentally torn out, the missingness is random. MCAR does not bias results but reduces statistical power.
2. **Missing at Random (MAR)**: Missingness depends on observed data but not on unobserved values. For instance, in a study on income, higher non-response might be linked to participants' level of education (an observed variable) but not to the actual income (an unobserved variable). MAR introduces challenges but can be addressed using appropriate statistical methods.
3. **Missing Not at Random (MNAR)**: Missingness depends on the unobserved value itself. For example, if people with very high incomes are less likely to disclose them, the missingness is systematically related to the unreported values. MNAR is the most difficult type to handle, requiring advanced modelling or strong assumptions.

This classification is central because it informs methodological choices. Treating MNAR as if it were MCAR, for instance, can result in severely biased estimates.

The consequences of missing data are both statistical and interpretive. Statistically, missing data reduces sample size, which diminishes power and increases standard errors. If the missingness is not random, it introduces bias, producing misleading parameter estimates and faulty inferences. For example, if a public health survey disproportionately loses responses from minority groups, the findings may inaccurately represent population health disparities. In qualitative research, missing data compromises the completeness of narratives. Missing interview segments due to recorder malfunction or participant withdrawal can distort thematic analysis, particularly if absent voices belong to marginalised populations. Beyond technical consequences, missing data also raises ethical concerns. Failure to handle missingness transparently can undermine participants' contributions and weaken trust in research findings.

Traditional Methods of Handling Missing Data

Historically, researchers relied on relatively simple approaches to deal with missingness, though many are now criticised for their limitations.

1. **Listwise Deletion**: The most common method is listwise deletion, where any case with missing data is excluded from analysis. While easy to apply, it reduces sample size and can produce biased results if the data are not MCAR. For example, if lower-income respondents disproportionately omit income questions, deleting their responses skews results toward wealthier groups.
2. **Pairwise Deletion**: Pairwise deletion uses all available data by excluding only the specific variables with missingness for each calculation. While it preserves more data than listwise deletion, it creates inconsistencies in sample sizes across analyses and complicates interpretation.
3. **Mean Substitution**: Another approach is to replace missing values with the mean of observed responses. This method retains sample size but underestimates variability and biases correlations, since it assumes homogeneity where variability exists.

Although these methods are simple and still occasionally used, they are increasingly regarded as inadequate because they distort data distributions and threaten validity. Contemporary approaches to missing data management have advanced considerably, offering more accurate and theoretically grounded alternatives.

1. **Maximum Likelihood Estimation** (**MLE**): MLE incorporates missing data directly into model estimation by maximising the likelihood function based on available data. It assumes MAR and produces unbiased parameter estimates under this condition. MLE is widely applied in structural equation modelling and multilevel analysis.
2. **Multiple Imputation** (**MI**): MI is one of the most robust methods for addressing missing data. It involves creating several plausible datasets where missing values are replaced by estimated values based on observed data and random variation. Each dataset is analysed separately, and results are pooled to produce final estimates. MI accounts for uncertainty in the imputed values and produces more reliable inferences than single imputation.
3. **Bayesian Approaches**: Bayesian methods integrate prior information with observed data to model missingness. They are especially useful for handling

MNAR data, as they allow researchers to specify assumptions about unobserved mechanisms. However, they require careful modelling and computational resources.

These modern methods represent significant progress, but they require advanced statistical skills and careful justification of assumptions. Without thoughtful application, they can still produce misleading results.

While qualitative strategies dominate the literature, missing data is also a challenge in qualitative research. Unlike surveys, qualitative studies cannot rely on imputation because data are context-rich and unique. Instead, qualitative researchers adopt different strategies:

1. **Triangulation**: Using multiple data sources (e.g., combining interviews, observations, and documents) reduces the impact of missing segments from any single source.
2. **Member Checking**: Returning to participants for clarification or follow-up interviews can fill in missing information or validate incomplete accounts.
3. **Reflexive Documentation**: Researchers explicitly record gaps and reflect on how they may influence interpretation. Rather than concealing missingness, qualitative traditions often acknowledge it as part of the research narrative.
4. **Contextual Interpretation**: In some cases, the absence of data itself becomes analytically meaningful. For example, silence in response to sensitive questions may reveal discomfort or resistance.

Thus, qualitative research treats missing data less as a statistical nuisance and more as a reflexive issue tied to meaning and interpretation.

Handling missing data is not only a technical task but also an ethical responsibility. Researchers must ensure that decisions about missingness do not misrepresent participants or silence marginalised voices. In qualitative contexts, discarding incomplete cases without reflection may disproportionately exclude vulnerable groups, reinforcing inequalities. In qualitative contexts, failure to acknowledge missing segments can present an artificially coherent narrative that erases participant ambivalence or disengagement. Transparency is therefore essential. Researchers should report the extent of missing data, the strategies employed to address it, and the assumptions underlying chosen methods. Ethical guidelines from organisations such as the American Psychological Association (APA, 2020) stress that transparency about data handling is integral to research integrity.

Practical Challenges and Strategies

Practical challenges in handling missing data include limited statistical expertise, time constraints, and software availability. Advanced methods like multiple imputation require specialised knowledge and computational tools, which may be inaccessible to some researchers. Another challenge is diagnosing the mechanism of missingness, since distinguishing MAR from MNAR is often difficult without strong assumptions. To address these challenges, researchers should incorporate missing data strategies into study design. This includes planning for potential attrition, designing user-friendly surveys to reduce non-response, and ensuring robust data collection systems. Training in advanced statistical techniques and collaboration with methodologists also improve handling practices. Furthermore, reporting guidelines such as CONSORT for clinical trials encourage transparent documentation of missing data and its management.

Implications for Research Integrity

How researchers handle missing data has far-reaching implications for the validity and credibility of research. Poorly managed missingness can lead to biased results, reduced generalisability, and ethical lapses. Conversely, rigorous handling enhances the robustness of findings and supports evidence-based decision-making in fields such as healthcare, education, and social policy. Moreover, handling missing data reflects broader commitments to transparency and reflexivity. In both qualitative and qualitative paradigms, acknowledging missingness and applying appropriate strategies safeguards not only methodological rigour but also respect for participants' contributions. Missing data is an inevitable aspect of research, but how it is handled determines whether findings remain credible or become compromised. Causes of missingness vary widely, from participant non-response to technical errors, and mechanisms such as MCAR, MAR, and MNAR shape the severity of the problem. Traditional methods like listwise deletion or mean substitution, while simple, are increasingly inadequate. Modern approaches such as maximum likelihood estimation, multiple imputation, and Bayesian modelling offer more reliable solutions, though they require careful application. In qualitative research, handling missing data involves reflexive, interpretive strategies such as triangulation, member checking, and acknowledging gaps in narratives. Across methodologies, ethical responsibility and transparency are central. Missing data management is therefore not a marginal technical step but a core aspect of research integrity, influencing trustworthiness, validity, and the ultimate impact of scholarly work [15].

6.1.3 Data Transformation

Data transformation is a central stage in the research process, bridging raw data collection and meaningful analysis. It refers to the systematic conversion of raw, unprocessed information into structured formats that are compatible with analytical techniques and aligned with research objectives. In both qualitative and qualitative research traditions, transformation is not merely a technical adjustment but a methodological step that ensures data integrity, enhances interpretability, and supports the credibility of findings. When poorly executed, data transformation can distort meaning or introduce bias, but when carried out with rigor, it enables researchers to unlock insights that would otherwise remain hidden. This section explores the concept and practice of data transformation, considering its role in research, its application in both qualitative and qualitative paradigms, the methods employed, and the ethical and practical challenges it raises. The section argues that transformation is more than a mechanical stage; it is a process deeply tied to interpretation, representation, and research integrity.

Raw data rarely exist in a format suitable for immediate analysis. Qualitative surveys may contain inconsistent coding, categorical responses, or skewed values, while qualitative interviews yield unstructured transcripts with contextual nuances. Data transformation is necessary to prepare such material for systematic interpretation. In qualitative research, transformation allows datasets to be standardised, normalised, or aggregated for statistical modelling. For example, a researcher may transform categorical responses such as "strongly agree" or "agree" into numerical codes, or apply logarithmic transformations to address skewed distributions. In qualitative research, transformation involves converting raw transcripts, field notes, or multimedia material into coded segments, thematic clusters, or narrative structures that can be systematically analysed. Beyond technical necessity, transformation is central to research validity. Without appropriate transformations, statistical tests may produce spurious results, or thematic interpretations may overlook important nuances. Thus, transformation safeguards the reliability and credibility of research while enabling meaningful conclusions [4].

In qualitative research, transformation primarily refers to mathematical and statistical adjustments that ensure compatibility with analytical techniques. Several common practices illustrate this process:

1. **Coding and Recoding Variables**: Raw categorical data, such as gender, ethnicity, or income groups, must often be numerically coded for statistical software. Recoding may also involve collapsing multiple categories into broader groups to simplify analysis.
2. **Normalisation and Standardisation**: Qualitative variables are frequently transformed to a common scale, particularly when combining indicators into composite scores. Standardisation (z-scores) ensures comparability, while normalisation scales data to a defined range (e.g., 0–1).
3. **Logarithmic and Other Mathematical Transformations**: Skewed distributions can violate assumptions of parametric tests. Log, square root, or inverse transformations are applied to reduce skewness, stabilise variance, and approximate normality.
4. **Aggregation**: Data collected at individual levels may be aggregated to group or organisational levels to facilitate multilevel analysis. This requires careful justification to ensure that aggregation does not obscure important within-group variation.
5. **Dummy Variable Creation**: For regression models, categorical variables must often be transformed into binary dummy variables. This transformation enables statistical models to capture group differences effectively.

These transformations not only make datasets compatible with statistical models but also ensure that findings are valid and generalisable. However, they require careful application, as inappropriate transformations may distort relationships or obscure meaningful variation.

In qualitative traditions, data transformation takes a different form, as researchers deal with rich, unstructured, and context-dependent material. Transformation here involves interpretive processes that render textual, visual, or auditory data analysable while retaining its depth.

1. **Transcription**: The initial transformation often involves converting spoken words into written text. This process is interpretive because researchers must decide whether to capture verbatim detail, including pauses and filler words, or to produce a cleaner transcript.
2. **Coding**: Raw transcripts or notes are segmented into meaningful units and assigned codes that capture key ideas, concepts, or actions. Coding transforms unstructured material into organised categories that can be compared across cases.
3. **Thematic Development**: Codes are further transformed into broader themes, patterns, or narratives. This process represents an interpretive transformation that connects individual fragments to larger conceptual frameworks.
4. **Quantitising Qualitative Data**: In mixed-methods research, qualitative material may be transformed into numerical form, for example, by counting the frequency of specific codes or themes. This enables integration with statistical analyses while risking loss of nuance.
5. **Visual or Narrative Transformation**: Qualitative findings are often transformed into visual models, diagrams, or case narratives that synthesise complex data into accessible formats.

Unlike qualitative transformation, qualitative transformation is not primarily mathematical but interpretive. It requires reflexivity to ensure that the researcher's choices do not distort participants' voices but instead illuminate their experiences.

Data Transformation in Mixed-Methods Research

Mixed-methods research provides a unique context where data transformation facilitates integration between qualitative and qualitative strands. This often involves data conversion, where one form of data is transformed into another. For example, researchers may transform survey responses into open-ended qualitative narratives or convert qualitative codes into frequency counts. Such transformations enable triangulation, where findings from different methods are compared and integrated. However, they raise methodological tensions: while quantifying qualitative data may enhance comparability, it risks stripping context and richness. Conversely, qualitative transformation of numeric patterns may oversimplify statistical relationships. Thus, mixed-methods transformation requires careful balance to respect the integrity of both paradigms. Data transformation is not a neutral or purely technical process; it is deeply tied to ethical representation. Decisions about how to code, aggregate, or recategorise data shape what is visible and what is silenced. In qualitative research, collapsing categories (e.g., merging ethnic groups for statistical simplicity) can obscure diversity and perpetuate the invisibility of minority voices. Similarly, transforming sensitive data without careful anonymisation risks breaching confidentiality. In qualitative research, choices about transcription detail or thematic aggregation may unintentionally distort meaning or strip away the subtleties of participants' narratives. Researchers must remain reflexive, documenting decisions and acknowledging their influence on interpretation. Ethical data transformation requires transparency, respect for participants, and alignment with research aims [16, 17].

Challenges in Data Transformation

Several challenges complicate data transformation [18, 19].

1. First, technical challenges arise from the complexity of modern datasets. Large-scale qualitative studies may involve thousands of variables requiring careful coding and transformation. Similarly, qualitative datasets with hundreds of pages of transcripts demand rigorous organisation to avoid loss of meaning.
2. Second, methodological challenges emerge when transformations risk distorting assumptions. For instance, inappropriate statistical transformations may produce misleading results, while thematic aggregation in qualitative research may oversimplify complex realities.
3. Third, ethical challenges persist, particularly when transformations obscure minority perspectives or misrepresent participants' voices. Researchers must balance analytical efficiency with ethical responsibility.
4. Finally, practical challenges such as time, resource constraints, and limited expertise can impede rigorous transformation. Advanced statistical transformations require specialised training, while qualitative transformation demands reflexive interpretation skills that may take years to develop.

Furthermore, researchers employ several strategies to ensure effective data transformation.

1. **Planning**: Anticipating transformation needs during research design prevents unnecessary complexity. For example, designing surveys with consistent coding frameworks reduces later recoding.
2. **Reflexivity**: In qualitative research, maintaining reflexive journals documents decisions about transcription, coding, and thematic development, ensuring transparency.
3. **Collaboration**: Peer debriefing and double-coding strengthen transformation by reducing individual bias and enhancing credibility.
4. **Software Tools**: Technological tools such as NVivo, Atlas.ti, SPSS, or R enhance accuracy and efficiency in transformation. They also provide audit trails that increase transparency.
5. **Documentation**: Recording all transformation steps supports reproducibility and accountability, enabling others to trace and evaluate methodological choices.
6. **Ethical Oversight**: Consulting ethical guidelines and review boards ensures that transformation respects confidentiality, anonymity, and representation.

Through these strategies, researchers can transform data rigorously while safeguarding both methodological integrity and participant trust.

Data transformation directly impacts the credibility, reliability, and transferability of findings. Poorly executed transformations undermine validity, while rigorous and transparent processes strengthen trustworthiness. In qualitative paradigms, transformation ensures that statistical assumptions are met and results are unbiased. In qualitative traditions, transformation enables researchers to move from raw narratives to meaningful interpretations that resonate with lived experience. In mixed-methods studies, transformation enables integration and triangulation across paradigms. Ultimately, data transformation is integral to research integrity. It is not a mechanical step but a reflective and ethical practice that shapes how raw information becomes evidence and how evidence contributes to knowledge. Data transformation is a crucial stage in the research process, encompassing both technical adjustments and interpretive processes. Qualitative research involves coding, standardisation, mathematical transformations, and aggregation. In qualitative research, it entails transcription, coding, thematic development, and narrative construction. In mixed-methods designs, transformation enables integration across paradigms but raises unique challenges. Despite technical, methodological, and ethical challenges, careful planning, reflexive practice, and transparent documentation ensure that transformation enhances rather than distorts data. By treating transformation not as a marginal technical step but as a core methodological practice, researchers safeguard integrity, respect participants' contributions, and produce findings that are both credible and meaningful [20–22].

6.2 Descriptive Statistics

Descriptive statistics constitute one of the most fundamental aspects of qualitative research and data analysis. They are concerned with summarising, organising, and presenting data in ways that enable researchers to identify patterns, describe distributions, and communicate findings effectively. Unlike inferential statistics, which aim to conclude populations based on samples, descriptive statistics focus on making raw data more comprehensible by reducing it to manageable forms. They provide the initial step in almost all empirical investigations, forming the foundation for subsequent analytical processes. Without descriptive statistics, researchers would struggle to interpret large volumes of data, and the communication of results would be difficult for both academic and non-academic audiences. This section critically examines descriptive statistics, beginning with their purpose and importance in research. It explores the different measures of central tendency, dispersion, and distribution, while also considering graphical methods of representation. The section further discusses the role of descriptive statistics in research design, their applications in real-world contexts, and the ethical and interpretive issues associated with their use. It concludes by emphasising that while descriptive statistics are limited in their ability to infer causality or generalise beyond observed data, they are indispensable tools for data organisation, clarity, and initial analysis [23, 24].

The primary function of descriptive statistics is to simplify complex data. In many research studies, especially those involving large samples, raw data consists of extensive lists of numbers or responses that are not immediately interpretable. Descriptive statistics condense such information into meaningful summaries that highlight key trends. For example, in a study of healthcare service satisfaction, descriptive statistics can be used to summarise the average level of satisfaction, the range of responses, and the percentage of respondents falling into various categories of satisfaction. Beyond simplification, descriptive statistics also play a communicative role. Policymakers, educators, or healthcare practitioners may not have the expertise or time to engage with raw data or complex inferential models. Descriptive statistics present findings in accessible forms, whether through averages, percentages, or visual displays such as bar charts and histograms. Furthermore, they provide essential diagnostic functions. By summarising data, researchers can detect errors, anomalies, or patterns that may require further investigation before applying inferential analyses.

One of the most common forms of descriptive statistics is the measurement of central tendency. Central tendency describes the centre or typical value in a dataset. The three primary measures are mean, median, and mode. The mean, often referred to as the arithmetic average, is calculated by summing all values and dividing by the number of observations. It is widely used due to its mathematical properties and suitability for further statistical analysis. However, it is sensitive to extreme values, or outliers, which can distort representation. For example, in income data, a small number of very high earners can raise the mean substantially, making it an inaccurate representation of the typical income. The median, which identifies the middle value when data are ordered, is more robust against outliers. It is particularly useful for skewed distributions. For instance, median household income is often reported in government statistics because it provides a clearer picture of the "typical" household than the mean. The mode, representing the most frequently occurring value, is useful for categorical data or identifying the most common category. For example, in analysing favourite transport modes among students, the mode indicates which form of transport is most popular. While the mode may be less informative in continuous data, it remains important in understanding categorical or nominal datasets [25].

Furthermore, central tendency alone cannot fully describe a dataset. Two distributions may share the same mean but differ significantly in how values are spread. Measures of dispersion address this by describing variability within data. The range, the difference between the highest and lowest values, provides a simple measure of spread but is highly sensitive to outliers. The interquartile range (IQR), which captures the middle 50% of data, is more robust, making it useful for skewed distributions. Variance and standard deviation are the most commonly used measures of dispersion in qualitative research. Variance represents the average squared deviation from the mean, while the standard deviation, the square root of variance, expresses the spread in the same units as the data. A low standard deviation indicates that values cluster closely around the mean, while a high standard deviation suggests wide variability. These measures are particularly important because they form the basis for inferential statistics, such as confidence intervals and hypothesis testing.

Beyond central tendency and dispersion, descriptive statistics also involve understanding the distribution and shape of data. Distributions may be symmetrical, skewed, normal, or irregular. The normal distribution, often represented as a bell curve, is of particular importance in statistics due to its mathematical properties and prevalence in natural and social phenomena. Skewness measures the extent to which data deviates from symmetry. Positive skew indicates a long tail on the right, while negative skew reflects a long tail on the left. For example, income distributions often exhibit positive skew due to a small number of very high earners. Kurtosis, another descriptor, measures the "peakedness" of a distribution. High kurtosis indicates heavy tails and a greater probability of extreme values, while low kurtosis reflects flatter distributions. Understanding these characteristics is essential because they inform the choice of statistical techniques and influence interpretation. Graphs and charts are powerful tools for descriptive statistics, as they provide visual summaries that are often more intuitive than numerical measures. Common graphical methods include bar charts, histograms, pie charts, boxplots, and scatterplots. Bar charts and pie charts are particularly effective for categorical data, illustrating the proportion of responses in each category. Histograms display frequency distributions of continuous variables, revealing patterns such as normality, skewness, or bimodality. Boxplots highlight medians, quartiles, and potential outliers, providing a compact visual of dispersion. Scatterplots, meanwhile, enable researchers to visualise relationships between two variables, suggesting potential correlations for further investigation. Visualisation enhances accessibility for diverse audiences and supports pattern recognition. However, ethical presentation is vital. Misleading graphs, inappropriate scales, or selective emphasis can distort interpretation. Thus, researchers must present visual descriptive statistics accurately and transparently [26].

Applications of Descriptive Statistics

Descriptive statistics have wide-ranging applications across disciplines. In education, they are used to summarise student performance, attendance, or demographics. In healthcare, descriptive statistics track patient outcomes, disease prevalence, and service utilisation. In business, they inform decision-making by summarising sales, customer satisfaction, or financial performance. In social research, descriptive statistics often serve as the foundation for policy recommendations. For example, summarising unemployment rates by age, gender, or region provides crucial insights for policymakers. Even in experimental science, descriptive statistics are indispensable for summarising experimental results and providing initial insights before inferential testing.

While descriptive statistics simplify data, they also carry risks of misrepresentation. Averages can obscure variability, and selective use of measures can produce biased interpretations. For example, reporting the mean without acknowledging skew may provide misleading impressions of "typical" values. Graphs can also be manipulated through distorted scales or the omission of relevant categories. Ethical use of descriptive statistics requires transparency in reporting, careful choice of appropriate measures, and reflexivity about limitations. Researchers must resist the temptation to oversimplify complex realities, acknowledging when descriptive summaries mask inequalities or important variations. Reflexivity is particularly important in social research, where descriptive statistics may influence public perception and policy decisions. Although indispensable, descriptive statistics have limitations. They describe but do not explain. For instance, reporting average test scores does not reveal why differences exist between groups. Nor can descriptive statistics generalise to populations beyond the data at hand. Such tasks require inferential statistics. Descriptive measures also risk oversimplification, reducing complex human phenomena to numerical summaries. While this is often necessary, researchers must interpret descriptive statistics within broader methodological and theoretical frameworks.

Descriptive statistics are the cornerstone of data analysis, enabling researchers to simplify, summarise, and communicate complex datasets. Through measures of central tendency, dispersion, and distribution, as well as graphical representation, descriptive statistics provide the foundation for both understanding data and preparing it for inferential analysis. Their applications span education, healthcare, business, and social science, demonstrating their versatility and importance. However, descriptive statistics are not without risks. They can mislead if applied uncritically, conceal important variability, or be manipulated for persuasive ends. Ethical responsibility and methodological reflexivity are therefore essential. While they cannot explain causality or infer beyond observed data, descriptive statistics remain indispensable tools for organising information, guiding further analysis, and communicating findings to diverse audiences [27].

6.2.1 Measures of Central Tendency: Mean, Median, and Mode

Statistical analysis is essential in modern research, enabling scholars and practitioners to understand patterns within datasets and draw meaningful insights. One of the most fundamental aspects of descriptive statistics is the use of measures of central tendency, which provide simple yet powerful summaries of data by indicating where the "centre" of the distribution lies. The three most widely used measures of central tendency are the mean, median, and mode. Each measure offers unique insights, and their proper application is crucial for accurate interpretation. This section explores the concepts of mean, median, and mode, explaining how they are calculated, their strengths and weaknesses, and their applications across diverse fields. It also discusses how these measures interact with distributional characteristics, the impact of outliers, and the ethical implications of data reporting. A practical example with tabular data is provided, alongside real-world applications in education, healthcare, business, and social policy. Central tendency refers to the statistical concept of identifying a single value that best represents a dataset. In essence, it answers the question: "What is the typical value?" This is crucial in research, policymaking, and applied contexts, as stakeholders often require concise summaries rather than complex datasets. While variability measures such as range and standard deviation provide information about spread, central tendency distils a dataset into a representative figure [28, 29].

The Mean

The mean, commonly known as the average, is the most widely recognised measure of central tendency. It is calculated by dividing the sum of all values by the number of observations.

$$\mathbf{Mean} = \Sigma \mathbf{x}/\mathbf{n}$$

where x represents each data point, and n is the number of observations.

The Median

The median is the middle value when observations are arranged in ascending or descending order. For datasets with an odd number of values, it is the central observation, while for even-numbered datasets, it is the average of the two central observations.

$$\mathbf{Median\ position} = (\mathbf{n} + \mathbf{1})/\mathbf{2}$$

The Mode

The mode is the most frequently occurring value in a dataset. Unlike the mean and median, it can be used with nominal data, such as identifying the most common preference in a survey about favourite food, transport, or political party.

Illustrative Example

To illustrate the calculation and interpretation of these measures, consider a dataset representing the test scores of 10 students in a mathematics class:

Student	Score
1	50
2	55
3	60
4	65
5	70
6	75
7	80
8	85
9	90
10	95

- Mean: $(50 + 55 + 60 + 65 + 70 + 75 + 80 + 85 + 90 + 95) \div 10 = 72.5$
- Median: With 10 values, the median is the average of the 5th and 6th scores (70 and 75). Median $= 72.5$

- Mode: No value is repeated; therefore, there is no mode.

Real-World Applications

1. **Education**: In education, measures of central tendency summarise student performance. Teachers may calculate the mean score of a class test to gauge overall performance. However, if a few students perform exceptionally well or poorly, the mean may not accurately reflect typical performance, and the median becomes more informative.
2. **Healthcare**: Healthcare research relies on central tendency to summarise patient outcomes. The mean may be used to report average recovery times after surgery, while the median is often reported for survival times in clinical trials because it is less affected by extreme cases.
3. **Business and Economics**: Businesses use central tendency to analyse consumer behaviour. The mean expenditure per customer may indicate financial performance, but in skewed markets, the median provides a clearer picture of typical spending. Retailers often use the mode to identify the most frequently purchased product size or colour.
4. **Social Policy**: Governments use measures of central tendency to understand population trends. For example, the median age of a population informs healthcare and pension planning.

Measures of central tendency, mean, median, and mode, are fundamental tools for summarising data. Each measure offers unique strengths and limitations. Real-world applications in education, healthcare, business, and social policy demonstrate their practical significance. Ultimately, mean, median, and mode are not just statistical abstractions but tools with profound implications for research, decision-making, and the representation of social realities.

6.2.2 Measures of Dispersion

In statistics, it is not sufficient to describe data only by its central tendency, such as the mean, median, or mode. Two or more datasets may share the same average but differ widely in how individual values are spread around that average. To capture this variability, statisticians rely on measures of dispersion, which describe the extent to which data points deviate from the centre. Dispersion, also referred to as variability or spread, is therefore an essential complement to central tendency. It provides critical insights into the nature of data, influencing decisions in research, policy, and practice. This section explores the concept of dispersion in statistics, examining its purpose, major measures, and applications. It discusses the range, interquartile range, variance, and standard deviation, as well as advanced measures such as the coefficient of variation. Real-world applications from education, healthcare, business, and social sciences are used to illustrate the importance of dispersion. Ethical considerations and limitations are also addressed, underscoring the role of dispersion in responsible data interpretation [30].

Understanding Dispersion

Dispersion refers to the degree to which values in a dataset differ from one another and from measures of central tendency. While central tendency provides a "typical" value, dispersion highlights how representative that value is. For example, two classrooms may have the same average test score of 70. If scores in one class range narrowly between 68 and 72, and in the other between 40 and 100, the mean conveys little about the differences. Measures of dispersion, therefore, add nuance by describing consistency, reliability, and diversity within data.

1. **The Range**: The range is the simplest measure of dispersion, calculated as the difference between the maximum and minimum values in a dataset. For example, if the highest exam score in a class is 95 and the lowest is 55, the range is 40. It provides a quick sense of spread but is highly sensitive to outliers, as a single extreme value can distort interpretation.
2. **The Interquartile Range**: The interquartile range (IQR) addresses some of the weaknesses of the range by focusing on the middle 50% of the data. It is calculated as the difference between the third quartile (Q3) and the first quartile (Q1). By excluding extreme values, the IQR offers a more robust measure of variability.
3. **Variance**: Variance is one of the most important statistical measures of dispersion. It calculates the average squared deviation of values from the mean, providing a sense of how widely observations are distributed. While variance gives a mathematically precise account of variability, its unit of measurement becomes squared, making it less intuitive for interpretation.
4. **Standard Deviation**: The standard deviation, the square root of variance, addresses the interpretability issue by expressing dispersion in the same unit as the data. It indicates, on average, how far individual values lie from the mean. A small standard deviation suggests data points are clustered closely, while a large one indicates wide variability.
5. **Coefficient of Variation**: The coefficient of variation (CV) is a relative measure of dispersion, calculated as the standard deviation divided by the mean and expressed as a percentage. It enables comparison of variability across datasets with different units or scales.

Real-World Applications of Dispersion

1. Education: In educational assessment, measures of dispersion help evaluate fairness and consistency.
2. Healthcare: In clinical research, variability in patient outcomes is crucial for assessing reliability.
3. Business and Economics: In economics, dispersion measures illuminate income inequality.
4. Social Sciences: Social scientists use dispersion to understand diversity in attitudes and behaviours.

Reporting dispersion carries ethical implications. Selective presentation of averages without measures of spread can mislead audiences. Ethical reporting requires transparency, including both central tendency and dispersion, to avoid oversimplification and distortion. While essential, measures of dispersion are not without limitations. The range is overly sensitive to outliers, variance can be difficult to interpret due to squared units, and standard deviation assumes data are normally distributed. Measures of dispersion are indispensable in statistics, complementing central tendency by describing the variability within data. From the simple range to the widely applied standard deviation and coefficient of variation, these measures enhance understanding, improve decision-making, and safeguard against misleading interpretations.

Types of Measures of Dispersion

In statistics, it is rarely sufficient to describe data solely by central tendency. While measures such as the mean, median, and mode identify the "typical" value, they do not indicate how much the values vary around that centre. Two datasets may have the same mean but differ substantially in spread, leading to different interpretations. To capture this variability, statisticians employ measures of dispersion. Dispersion refers to the extent to which data points deviate from each other and from the central value. It is critical in understanding consistency, reliability, and diversity within datasets. This section examines the types of measures of dispersion, dividing them into two broad categories: absolute measures and relative measures. Absolute measures describe variability in original units of the dataset, while relative measures express variability as a proportion or percentage of a central value. Each type has distinct strengths, weaknesses, and applications across fields such as education, business, economics, and healthcare.

Absolute Measures of Dispersion

Absolute measures are expressed in the same units as the data. They provide direct estimates of spread but do not facilitate comparisons between datasets with different scales.

1. **Range**: The range is the simplest absolute measure, defined as the difference between the maximum and minimum values in a dataset. For example, if student test scores range from 40 to 90, the range is 50. Its simplicity makes it useful in exploratory analysis, but it is highly sensitive to outliers. A single extreme value can disproportionately inflate the range, making it an unreliable measure in many real-world datasets.
2. **Quartile Deviation (Semi-Interquartile Range)**: The quartile deviation, or semi-interquartile range, is half the difference between the third quartile (Q3) and the first quartile (Q1). It focuses on the middle 50% of the dataset, reducing the influence of extreme values. For example, if Q3 = 80 and Q1 = 60, the quartile deviation is (80 – 60)/2 = 10. This measure is particularly useful when distributions are skewed, such as in income or housing price data.

3. **Mean Deviation**: Mean deviation measures the average of the absolute deviations of each value from a central tendency (usually the mean or median). It reflects overall variability more accurately than the range. For instance, in a dataset with values close to each other, the mean deviation will be small, indicating consistency. However, because it relies on absolute deviations, it is less mathematically tractable in advanced statistical analyses.
4. **Variance**: Variance is the average of squared deviations from the mean. By squaring differences, variance ensures that positive and negative deviations do not cancel each other. It highlights variability more precisely but results in squared units, which are less intuitive. For example, if salaries are measured in dollars, variance is measured in dollars squared. Despite this, variance is essential in inferential statistics, probability theory, and hypothesis testing.
5. **Standard Deviation**: The standard deviation, the square root of variance, expresses variability in the same unit as the dataset, making it more interpretable. It is widely considered the most useful measure of dispersion because it provides a balance between precision and interpretability. In practical terms, a smaller standard deviation indicates greater consistency, while a larger one indicates greater spread.

Relative Measures of Dispersion

Relative measures express dispersion as a proportion of a central value, typically the mean. They are dimensionless, allowing comparisons between datasets with different scales or units.

1. **Coefficient of Range**: The coefficient of range is calculated as (Max – Min) / (Max + Min). It normalises the range, making it useful for comparing spreads across datasets with different scales. However, like the absolute range, it is still affected by extreme values.
2. **Coefficient of Quartile Deviation**: The coefficient of quartile deviation is (Q3 – Q1) / (Q3 + Q1). This ratio standardises the quartile deviation, making it possible to compare datasets of different magnitudes while reducing sensitivity to outliers.
3. **Coefficient of Mean Deviation**: The coefficient of mean deviation is the ratio of mean deviation to the mean or median. It expresses variability as a fraction of the central value, aiding cross-dataset comparisons.
4. **Coefficient of Variation (CV)**: The coefficient of variation is the most widely used relative measure. It is defined as (Standard Deviation / Mean) × 100. The CV indicates variability relative to the mean, making it ideal for comparing risk and stability across datasets. For example, in finance, two investments may have the same expected return, but the one with a higher CV is considered riskier.

Real-World Applications of Different Types

1. In education, absolute measures such as standard deviation are used to evaluate variability in student performance, while relative measures like CV help compare consistency across subjects.

2. In healthcare, variance and standard deviation summarise variability in clinical outcomes. Relative measures allow comparisons of variability across different treatments or dosage levels.
3. In business and economics, the CV is critical for comparing profitability and risk across sectors. Policymakers use quartile deviation and IQR to measure inequality, as they better represent the middle majority of populations.
4. In social sciences, mean deviation and quartile deviation are often applied to survey data, where robustness against outliers is essential.

Choosing between absolute and relative measures has ethical implications. For instance, reporting only absolute measures in income inequality studies may conceal disparities that relative measures would expose. Similarly, in healthcare, reporting only mean outcomes without dispersion may misrepresent the consistency of treatment effects. Researchers and policymakers must use both absolute and relative measures responsibly to ensure transparency and fairness. Measures of dispersion are indispensable for understanding variability in data. Absolute measures, such as range, variance, and standard deviation, describe spread in original units, while relative measures, such as the coefficient of variation, facilitate comparisons across datasets with different scales. Together, these measures provide a holistic picture of data variability. In practice, the choice between absolute and relative measures depends on the nature of the data and the purpose of the analysis. Standard deviation and CV are the most commonly used because they combine interpretability with analytical utility. Ultimately, measures of dispersion not only enrich statistical analysis but also safeguard ethical reporting, ensuring that data is represented with accuracy, fairness, and depth.

6.3 Techniques for Presenting Summary Statistics in Qualitative Studies

Qualitative research is traditionally associated with rich descriptions, thematic exploration, and in-depth accounts of participants' perspectives. However, contemporary trends in social sciences, education, healthcare, and applied fields have witnessed an increasing integration of qualitative elements into qualitative studies. Among these elements, the use of summary statistics plays an important role in enhancing clarity, improving rigour, and offering readers a clearer overview of complex qualitative datasets. While qualitative studies primarily focus on meaning, context, and interpretation, the incorporation of descriptive statistical techniques allows researchers to summarise frequencies, distributions, and trends without compromising the interpretive depth of the inquiry. This section critically examines techniques for presenting summary statistics in qualitative research. It explores the rationale for using summary statistics in non-numerical studies, identifies the most common techniques, evaluates their advantages and limitations, and discusses ethical and methodological considerations. Real-world applications from healthcare, education, sociology, and interdisciplinary research are included to illustrate how these techniques contribute to validity, transparency, and accessibility.

The Role of Summary Statistics in Qualitative Studies

The primary purpose of qualitative research is to generate insights into human experiences, behaviours, and social phenomena. Unlike qualitative research, which focuses on measurement and hypothesis testing, qualitative research privileges meaning, context, and narrative depth. Yet, qualitative datasets—such as interview transcripts, focus group discussions, or open-ended survey responses—often contain elements that can be numerically summarised. For example, researchers may wish to report how many participants mentioned a particular theme, how frequently a code was applied, or what proportion of respondents belonged to specific demographic categories.

Summary statistics serve several important purposes in this context. They:

1. Provide overview information, helping readers understand the scope and characteristics of the sample.
2. Enhance credibility by demonstrating systematic analysis rather than anecdotal reporting.
3. Facilitate comparison across cases, groups, or themes.
4. Support mixed-methods designs, where qualitative and qualitative findings are integrated.

Thus, while qualitative research resists reduction to numbers, summary statistics complement narrative accounts by offering structured snapshots that highlight patterns within data.

Common Techniques for Presenting Summary Statistics

1. **Frequency Counts**: One of the simplest and most widely used techniques is frequency counting. This involves tallying the number of times a particular code, theme, or response appears in the dataset. For example, in an interview study on barriers to healthcare access, researchers may report that "lack of transportation" was mentioned by 18 out of 25 participants. Frequency counts offer transparency, showing that themes are grounded in data rather than selectively highlighted. However, they should not be mistaken for indicators of importance. A theme mentioned by only a few participants may still be theoretically significant, particularly if it reflects marginalised perspectives.
2. **Percentages and Proportions**: Researchers often complement frequency counts with percentages, expressing the proportion of participants who expressed a particular view. For instance, in a study of teaching practices, reporting that "40% of teachers incorporated digital tools into classroom activities" conveys a clearer sense of distribution than raw counts. Percentages allow comparisons across subgroups and help contextualise findings within the broader sample. However, percentages must be interpreted cautiously in small samples typical

of qualitative research. Reporting "50%" in a study with 10 participants means only 5 people, which should be made explicit to avoid misinterpretation.

3. **Measures of Central Tendency**: Although less common, researchers sometimes use measures such as the mean, median, or mode to summarise numerical attributes within qualitative studies. For example, in narrative interviews about household caregiving, participants might report the number of hours they spend on care per week. Presenting the median provides a concise statistical summary without overshadowing qualitative insights. The mean and median can also be used when qualitative surveys include scaled responses (e.g., Likert-type ratings embedded in a predominantly open-ended questionnaire). However, central tendency measures must be interpreted in relation to qualitative context, not as standalone indicators.
4. **Cross-tabulations: Cross-tabulation** is a technique where frequencies are distributed across two or more variables. In qualitative research, this may involve showing how themes vary across demographic groups. For instance, in a study on work-life balance, researchers might report that "childcare concerns" were raised by 70% of female participants compared with 25% of male participants. Cross-tabulations are useful for highlighting subgroup differences and can inform more nuanced interpretations. However, given small qualitative samples, researchers must avoid overgeneralisation.
5. **Visual Summaries**: Beyond tables of counts and percentages, qualitative studies often employ visual summaries to present statistics. These include bar charts, pie charts, and histograms, which can be particularly effective for audiences less comfortable with dense textual explanations. For example, in a participatory action research project, a pie chart showing the proportion of participants emphasising different community needs (housing, education, healthcare) makes findings accessible to both academic and non-academic audiences. Visualisation techniques should be simple and should not distract from the interpretive narrative.
6. **Matrices and Data Displays**: Qualitative data analysis software (such as NVivo, Atlas.ti, or MAXQDA) enables the creation of matrices, where themes are mapped against participant characteristics. These displays often include both numbers and excerpts, integrating statistical summaries with rich qualitative evidence. For instance, a matrix may show that "safety concerns" were raised in 12 interviews, with exemplar quotes listed alongside. Matrices combine the strengths of qualitative depth and statistical overview, making them powerful tools for mixed representation.

Advantages of Using Summary Statistics in Qualitative Studies

The integration of summary statistics into qualitative research provides several methodological and communicative benefits.

1. First, they enhance rigour and transparency. By showing how often themes appear, researchers demonstrate that findings are systematically grounded. This

helps to counter criticisms that qualitative research is anecdotal or overly subjective.
2. Second, summary statistics improve clarity and accessibility. Readers who may be unfamiliar with qualitative traditions often find numerical summaries easier to digest. Presenting frequencies alongside narratives can thus bridge gaps between different scholarly audiences.
3. Third, they facilitate comparisons and triangulation. Summary statistics allow researchers to examine variations across participants, contexts, or time points, and they provide a basis for integrating qualitative and qualitative findings in mixed-methods designs.
4. Finally, they support policy and practice relevance. Policymakers and practitioners often seek clear numerical evidence to inform decisions. While qualitative narratives provide depth, summary statistics offer concise entry points that resonate with applied audiences.

Limitations and Risks

Despite these benefits, using summary statistics in qualitative studies poses challenges.

1. One limitation is the risk of qualitative reductionism, where rich qualitative insights are overshadowed by numerical summaries. Overemphasis on counts may encourage readers to equate frequency with importance, neglecting rare but significant themes.
2. Another challenge is sample size sensitivity. Qualitative research typically involves small, purposive samples. Presenting percentages may therefore create an illusion of generalisability that the study cannot support. Researchers must clearly contextualise numbers within the interpretive framework of qualitative inquiry.
3. A further risk is misinterpretation by audiences. Policymakers or readers unfamiliar with qualitative traditions may mistakenly treat summary statistics as representative of broader populations, rather than as descriptive of a specific sample.
4. Finally, ethical considerations must be addressed. When reporting summary statistics, especially in small samples, researchers must ensure that participants' confidentiality is preserved. Overly detailed tabulations may risk inadvertently disclosing identities.

Ethical and Methodological Considerations

Using summary statistics in qualitative research requires careful methodological justification. Researchers must:

1. Clarify that numerical summaries complement, rather than replace, interpretive analysis.
2. Contextualise percentages and frequencies within sample size and design.
3. Preserve participants' anonymity when reporting subgroup differences.

4. Ensure that rare or minority perspectives are not marginalised by frequency counts.

Ethical reflexivity is crucial. Researchers should consider whether the inclusion of numbers enhances or distorts the story being told, and they must avoid presenting statistics in ways that mislead audiences about generalisability.

Real-World Applications

1. In **healthcare research**, qualitative studies often report frequencies of themes related to patient experiences, such as barriers to accessing treatment. For example, a study of cancer patients may indicate that "financial costs" were reported as barriers by 65% of participants. This information helps policymakers prioritise interventions while narrative accounts provide depth.
2. In **education**, qualitative surveys exploring teaching strategies may combine narrative data with statistical summaries, such as the proportion of teachers using digital learning tools. This dual approach enriches understanding of pedagogical diversity.
3. In **social sciences**, cross-tabulations have been used in studies of migration to show how experiences differ across ethnic or gender groups. This provides both a statistical overview and nuanced interpretations of marginalised voices.
4. In **community research**, participatory projects often use visual summaries (charts or matrices) to share findings back with local stakeholders. Numerical summaries help communities see patterns in their collective experiences, while narratives capture lived realities.

Techniques for presenting summary statistics in qualitative studies occupy an important space in modern research practice. They provide clarity, enhance rigour, and support broader communication, particularly in applied settings where concise evidence is valued. However, they must be used judiciously to avoid qualitative reductionism, misinterpretation, or ethical risks. By balancing narrative depth with structured numerical summaries, qualitative researchers can ensure that their work remains accessible, credible, and impactful. The integration of summary statistics does not undermine the essence of qualitative inquiry; rather, it strengthens its communicative power, bridging traditions of meaning-making and measurement.

References

1. Boon, J., Yap, H., Leong, B., Skitmore, M.: Capitalising knowledge management (KM) for improving project delivery in construction. Ain Shams Eng. J. **13**(6), 101790 (2022). https://doi.org/10.1016/j.asej.2022.101790
2. Fadele, A.A., Diala, S.O., Najeem, O.: Evaluating artificial intelligence (Ai) tools for automating administrative tasks for colleges of education in Kaduna State. Adv. J. Educ. Soc. Sci. **10**(5), 27–46 (2025)
3. Almaiah, M.A., Al-Khasawneh, A., Althunibat, A.: Exploring the critical challenges and factors influencing the E-learning system usage during COVID-19 pandemic. Educ. Inf. Technol. (2020). https://doi.org/10.1007/s10639-020-10219-y

4. Purnomo, Y.J.: Digital marketing strategy to increase sales conversion on e-commerce platforms. J. Contemp. Adm. Manag., **1**(2), 54–62 (2023). https://doi.org/10.61100/adman.v1i2.23
5. Huang, T., Kou, S., Liu, D., Li, D., Xing, F.: A BIM-GIS-IoT-based system for excavated soil recycling. Buildings **12**(4), 1–17 (2022). https://doi.org/10.3390/buildings12040457
6. Kong, J., Zhang, L., Jiang, M., Liu, T.: Incorporating multi-level CNN and attention mechanism for Chinese clinical named entity recognition. J. Biomed. Inform **116**(March), 103737 (2021). https://doi.org/10.1016/j.jbi.2021.103737
7. Almustapha, J., Sani Dahiru, A., Bungudu, M.I., Hamisu Lawal, A.: A correlational analysis of school inventories management and teacher efficiency among primary schools of Zamfara State, Nigeria. J. Adv. Humanit. Res. **2**(1), 1–12 (2023). https://doi.org/10.56868/jadhur.v2i1.38
8. Mumtaz, H., et al.: Risk factors & perinatal outcomes of major depression during pregnancy: a population-based study during 2010–2020 in two major cities of Pakistan. Ann. Med. Surg. 84(October):104941 (2022). https://doi.org/10.1016/j.amsu.2022.104941
9. Mbona, I., Eloff, J.H.P.: Detecting zero-day intrusion attacks using semi-supervised machine learning approaches. IEEE Access **10**(July), 69822–69838 (2022). https://doi.org/10.1109/ACCESS.2022.3187116
10. Arvidsson, S., Dumay, J.: Corporate ESG reporting quantity, quality and performance: where to now for environmental policy and practice? Bus. Strateg. Environ. **31**(3), 1091–1110 (2022). https://doi.org/10.1002/bse.2937
11. Kognisi, P.K., et al.: The effect of athlete brand personality, brand image attributes and brand loyalty on athlete brand advocacy: the mediating effect of deservedness. Ind. High. Educ. **3**(1), 1689–1699 (2021). http://journal.unilak.ac.id/index.php/JIEB/article/view/3845
12. Achunguh, D.A.: The impact of leadership style on employee's performance in an organization. Sustainability **4**(1), 1–9 (2020). https://pesquisa.bvsalud.org/portal/resource/en/mdl-20203177951
13. Obi, K.C., Iwuanyanwu, O.: Innovation in the assessment of primary school mathematics: an approach to revamping Nigeria education. J. Teach. Perspect. **18**(1), 51–66 (2023). http://repositorio.unan.edu.ni/2986/1/5624.pdf%0A, http://fiskal.kemenkeu.go.id/ejournal%0A. https://doi.org/10.1016/j.cirp.2016.06.001%0A10.1016/j.powtec.2016.12.055%0A10.1016/j.ijfatigue.2019.02.006%0A10.1
14. Mokhtarzadeh, N.G., Mahdiraji, H.A., Jafarpanah, I., Cao, D.: Examining the influence of environmental turbulence on firm innovation performance in emerging markets: using an environment-strategy-performance framework. Int. J. Innov. Manag. **26**(4), 1–29 (2022). https://doi.org/10.1142/S1363919622500281
15. Ahsan, M.A.M., Ali, I., Imran, M., Bin Idris, M.Y.I., Khan, S., Khan, A.: A fog-centric secure cloud storage scheme. IEEE Trans. Sustain. Comput. **7**(2), 250–262 (2022). https://doi.org/10.1109/TSUSC.2019.2914954
16. Ltifi, M., Hichri, A.: The effects of corporate governance on the customer's recommendations: a study of the banking sector at the time of COVID-19. J. Knowl. Manag. **26**(1), 165–191 (2022). https://doi.org/10.1108/JKM-06-2020-0471
17. Gbadamosi, A.-Q., Oyedele, L., Mahamadu, A.-M., Kusimo, H., Olawale, O.: The role of internet of things in delivering smart construction. CIB World Build. Congr. **3**(6), 17–21 (2019)
18. Hu, Q., Kapucu, N., O'Byrne, L.: Strategic planning for community-based small nonprofit organizations: implementation, benefits, and challenges. J. Appl. Manag. Entrep. **19**(1), 83–101 (2016). https://doi.org/10.9774/gleaf.3709.2014.ja.00007
19. Iyanda, M.O., et al.: Enhancing the occupational safety and well-being of mining and construction workers: an application of emotional intelligence training (EIT). Saf. Extrem. Environ. **6**(2), 79–93 (2024). https://doi.org/10.1007/s42797-023-00098-0
20. Hu, J., Zheng, K.: A novel support vector regression for data set with outliers. Appl. Soft Comput. J. **31**, 405–411 (2015). https://doi.org/10.1016/j.asoc.2015.03.033
21. Wang, Y., et al.: Unsupervised anomalous sound detection for machine condition monitoring using classification-based methods. Appl. Sci. **11**(23) (2021). https://doi.org/10.3390/app112311128

22. Jayanthi, M.K., Srivatsa, S.K., Ramesh, T.: Object oriented analysis and design of e-learning system (2017). https://doi.org/10.3923/itj.2007.818.826
23. Oyedele, O.K., Fagbamigbe, A.F., Akinyemi, O.J., Adebowale, A.S.: Coverage-level and predictors of maternity continuum of care in Nigeria: implications for maternal, newborn and child health programming. BMC Pregnancy Childbirth **23**(1), 1–19 (2023). https://doi.org/10.1186/s12884-023-05372-4
24. Owoyemi, O., Ekwoaba, J.: Organisational culture: a tool for management for management control, motivate and performance enhancement. Am. J. Bus. Manag. **3**(3), 168–177 (2014). https://doi.org/10.11634/216796061403514
25. Ogbaji, D.I.: Teachers' perception of the utilization of instructional materials in teaching social studies in junior secondary schools in Calabar municipality, Cross river state, Nigeria. Glob. J. Educ. Res. **16**(2), 95 (2017). https://doi.org/10.4314/gjedr.v16i2.3
26. Uchendu, P., Cargo handling equipment and efficient Cargo delivery in Nigerian Ports. **13**(2), 26–42 (2020). www.arcnjournals.org
27. Odunayo Braimah, J., Comparative study of 2011 and 2015 presidential elections in Nigeria statistics. Glob. J. Human-Social Sci. Polit. Sci. **2**(November) (2015). https://www.researchgate.net/publication/328698172
28. Hauptmann, M., et al.: Brain cancer after radiation exposure from CT examinations of children and young adults: results from the EPI-CT cohort study. Lancet Oncol. **2045**(22) (2022). https://doi.org/10.1016/S1470-2045(22)00655-6
29. Alasan, A.I., Eyanuku, J.P.: Human resource management practices and Nigerian banks' performance. LASU J. Employ. Relations Hum. Resour. Manag. **2**(1), 124–140 (2020). https://doi.org/10.36108/ljerhrm/0202.02.0190
30. John-Olabode, S. O., Akintan, P., Okunade, K.S., Ajie, I.: Comparative assessment of serum selenium status in HIV-infected and non-infected children: a pilot study in a tertiary hospital in Nigeria. Cureus **15**(5) (2023). https://doi.org/10.7759/cureus.39626

Chapter 7
Conclusion and Future Research

7.1 Summary of Key Findings

Key conclusions drawn from each chapter are summarized below:

1. **Introduction to Qualitative Research** (Chap. 1): This chapter established the foundation of qualitative research by outlining its core principles, including objectivity, measurement, and statistical analysis. It emphasised how qualitative methods provide structured, replicable, and generalisable findings. The chapter also highlighted the importance of aligning research questions with methodological choices, setting the stage for the detailed techniques covered in subsequent chapters.
2. **Sampling Techniques** (Chap. 2): This chapter explored the critical role of sampling in ensuring validity and representativeness of research findings. Probability sampling methods (such as simple random, stratified, and cluster sampling) were shown to enhance generalisability, while non-probability methods (including purposive and convenience sampling) were highlighted for their practicality in certain contexts. The chapter concluded that sampling strategies must balance methodological rigour with real-world feasibility.
3. **Data Collection Methods** (Chap. 3): Chapter 3 examined structured approaches for gathering data, including surveys, questionnaires, and interviews. It highlighted the strengths of each method and the challenges of reliability, validity, and potential bias. The key finding was that method selection should be guided by research objectives, resource availability, and ethical considerations, ensuring the credibility of collected data.
4. **Observational Data Collection** (Chap. 4): This chapter discussed the systematic observation of behaviour and phenomena as a key method in qualitative research. Structured, unstructured, participant, and non-participant observation techniques were explained. The chapter underscored that observational data provide valuable insights but require careful planning to minimise observer bias and enhance objectivity.

F. A. Alaba et al., *Fundamentals of Qualitative Research*, Studies in Systems, Decision and Control 663, https://doi.org/10.1007/978-3-032-18427-6_7

5. **Measurements and Classifications** (Chap. 5): Chapter 5 analysed how data are categorised through scales of measurement—nominal, ordinal, interval, and ratio. It concluded that proper classification is vital to applying appropriate statistical tests and ensuring valid interpretation. This chapter stressed the importance of reliability and validity in measurement, linking theoretical constructs to empirical indicators.
6. **Data Preparation and Cleaning** (Chap. 6): The final methodological chapter addressed the processes necessary for ensuring accuracy before analysis. Techniques such as handling missing data, data transformation, and identifying outliers were presented as essential to maintaining data quality. The chapter highlighted that thorough preparation safeguards the validity of subsequent statistical procedures and strengthens research outcomes.

7.2 Contributions to Qualitative Research

Qualitative research has long been a cornerstone of scientific inquiry, providing the rigour and objectivity necessary to test hypotheses, establish causal links, and generate generalisable findings. This book contributes to the broader field of qualitative research in several meaningful ways, particularly by integrating classical methodological frameworks with contemporary perspectives. The following discussion outlines the main contributions this work makes to the study, practice, and pedagogy of qualitative methods.

Bridging Theory and Practice

One of the primary contributions of this text lies in its ability to bridge abstract theoretical concepts with practical application. Many textbooks either remain overly theoretical, limiting accessibility for novice researchers, or focus narrowly on applied techniques without sufficient attention to underlying principles. By carefully linking chapters on sampling, data collection, and measurement to their theoretical foundations, this book helps readers appreciate not only *how* to perform qualitative research but also *why* specific methodological choices matter. This balance ensures that students and practitioners can apply knowledge in diverse contexts, from academic projects to industry-based evaluations.

Emphasis on Methodological Rigor

The discussion of sampling strategies, measurement scales, and data cleaning contributes significantly to ongoing conversations about methodological rigour. In an era where reproducibility crises have challenged the credibility of research across fields, this book underscores the importance of careful preparation, systematic observation, and transparent reporting. By stressing practices such as handling missing data and ensuring reliability in measurement, it contributes to raising the standard of methodological soundness in future qualitative research.

Integration of Ethical Considerations

Another notable contribution is the integration of ethical principles throughout methodological discussions. While ethical concerns are often treated as separate from technical processes, this text shows how they intersect at every stage of the research cycle. For example, sampling decisions not only affect statistical validity but also raise questions of representation and fairness. Similarly, observational research requires balancing objectivity with respect for participants' privacy. By weaving ethical reflection into methodological instruction, the book contributes to a more holistic understanding of qualitative research practice.

Clarification of Complex Concepts

Complex ideas such as variance, standard deviation, or the implications of measurement scales are explained in accessible language while retaining technical precision. This pedagogical clarity contributes to the field by making the qualitative research approachable for readers with varying levels of statistical literacy. In doing so, the book broadens access to rigorous research methods, empowering a wider community of learners and practitioners.

Supporting Interdisciplinary Application

Qualitative methods are not confined to one academic discipline but are increasingly employed in fields ranging from health sciences to education, business, and social policy. By illustrating methodological principles with diverse examples, this book contributes to the interdisciplinary application of qualitative research. It equips researchers across domains with transferable skills while recognising discipline-specific challenges.

Contribution to Pedagogy

Finally, the structured progression from foundational principles to advanced techniques offers a pedagogical contribution. The inclusion of step-by-step discussions, summaries, and applied examples provides a roadmap for both independent learners and instructors. As such, the book not only contributes to scholarly literature but also serves as a teaching resource that supports capacity building in research education.

In sum, the contributions of this book to qualitative research are multifaceted. It clarifies foundational principles, strengthens methodological rigour, integrates ethics into practice, and enhances accessibility through clear explanations. By providing a resource that is at once theoretical, practical, and pedagogically sound, the book enriches the toolkit of students, educators, and practitioners. Its impact extends beyond academic discourse, enabling evidence-based decision-making in real-world contexts. In this way, the book makes a distinctive and lasting contribution to the evolving landscape of qualitative research.

7.3 Limitations of the Study

Every research endeavour, no matter how carefully planned or executed, is subject to certain limitations. These limitations do not necessarily diminish the value of the work but instead contextualise its scope and guide readers in interpreting findings appropriately. This book, while comprehensive in its coverage of qualitative research methodologies, is not without boundaries and constraints. Acknowledging these limitations is an essential part of scholarly integrity and transparency.

Scope of Coverage

The first limitation relates to the scope of topics covered. Qualitative research is a vast field encompassing an ever-expanding array of techniques, statistical models, and technological applications. While this book provides detailed explanations of foundational elements such as sampling, data collection, and measurement, it cannot cover the full breadth of advanced qualitative methods such as multilevel modelling, structural equation modelling, or machine learning algorithms. Readers seeking highly specialised techniques will therefore need to supplement this book with more advanced statistical texts.

Contextual Focus

A second limitation lies in the contextual focus of the book. The examples and applications provided are drawn primarily from the social sciences, education, healthcare, and business studies. While these fields represent common arenas for qualitative research, the absence of domain-specific examples from other areas such as engineering or natural sciences may limit the immediate relevance for readers in those disciplines. Although the principles remain transferable, some disciplinary audiences may require additional contextualisation.

Simplification of Complex Concepts

In order to make complex topics accessible, the book necessarily simplifies certain statistical concepts. Explanations are tailored to an introductory or intermediate audience, which means that nuanced debates within the statistical community—such as controversies over null hypothesis significance testing, Bayesian versus frequentist approaches, or the philosophical underpinnings of measurement—are not explored in full. While simplification enhances clarity, it may leave advanced readers seeking deeper theoretical engagement.

Emphasis on Classical Approaches

This text emphasises classical qualitative research approaches, including probability sampling, structured data collection, and traditional measures of dispersion. Although references are made to contemporary practices, such as mixed-methods designs and the use of computer-assisted data analysis software, these areas are not treated in depth. As such, the book may appear somewhat traditional when compared to cutting-edge methodological handbooks. This limitation is deliberate, as the aim was to provide a strong foundation rather than an exhaustive exploration of every modern innovation.

Resource Constraints in Examples

Another limitation concerns the illustrative examples used throughout the chapters. While hypothetical datasets and simplified scenarios were chosen for pedagogical clarity, real-world research often involves more complex, messy, and imperfect data. Thus, the examples may not fully capture the challenges faced by researchers in practice, such as negotiating access to sensitive populations, managing large-scale datasets, or addressing interdisciplinary constraints. Readers should therefore be cautious in assuming that the examples fully represent the realities of fieldwork.

Generalisability

Finally, because this book is intended as a general guide rather than a discipline-specific manual, some recommendations may not generalise seamlessly across all fields of research. Disciplines such as clinical medicine or economics may impose specific methodological standards that extend beyond the coverage offered here. Consequently, the applicability of some techniques may depend on disciplinary norms, regulatory environments, and contextual considerations.

In summary, the limitations of this study revolve around scope, contextual focus, simplification, and the balance between classical and contemporary approaches. These boundaries reflect deliberate choices made to ensure clarity, accessibility, and pedagogical value. While the text may not meet the needs of highly specialised or advanced researchers, it provides a robust foundation upon which such expertise can be built. Acknowledging these limitations does not weaken the book's contribution; rather, it strengthens its credibility by situating its claims within a realistic and transparent framework.

7.4 Recommendations for Future Research

The field of qualitative research continues to evolve as new technologies, analytical methods, and interdisciplinary applications emerge. While this book provides a strong foundation in the fundamentals of qualitative inquiry, there remain many opportunities for future research to expand, refine, and innovate upon these principles. The following recommendations highlight key directions that researchers, educators, and students might pursue to advance both the theory and practice of qualitative research.

Integration of Advanced Statistical Methods

One important recommendation for future research is the incorporation of advanced statistical techniques that go beyond the foundational methods covered in this text. Increasingly, fields such as psychology, healthcare, and economics employ complex models like multilevel analysis, structural equation modelling, and Bayesian inference to capture the intricacies of real-world data. Future research should focus on making these advanced techniques more accessible, providing clearer explanations and applications that bridge the gap between basic statistics and cutting-edge modelling.

Exploration of Emerging Data Sources

Another area of growth lies in the exploration of emerging data sources. The rise of digital technologies, social media platforms, and big data has transformed the landscape of research. Unlike traditional surveys and observational studies, these sources provide vast amounts of information that are dynamic, unstructured, and multidimensional. Future research must develop methods for integrating these new data streams into rigorous qualitative frameworks while addressing issues of validity, reliability, and ethical responsibility.

Expansion of Mixed-Methods Approaches

The boundary between qualitative and qualitative research is increasingly porous, with mixed-methods studies gaining prominence. Future research should continue exploring strategies for integrating numerical summaries with rich qualitative narratives. Doing so not only strengthens triangulation but also offers a more holistic understanding of complex phenomena. Recommendations include refining analytical frameworks for combining qualitative coding with qualitative frequency counts and developing software tools that support seamless integration of both traditions.

Attention to Ethical Challenges in a Digital Age

As research practices evolve, so too do ethical challenges. The growing use of online surveys, wearable technologies, and administrative databases raises important questions about privacy, consent, and data security. Future research must prioritise the development of ethical guidelines and frameworks tailored to these new realities. In addition, scholars should investigate how transparency and open-science practices can be embedded in qualitative research to improve trust and accountability.

Cross-Disciplinary Applications

Future research should also place greater emphasis on cross-disciplinary applications. While this book has drawn examples from education, healthcare, and social sciences, other fields such as environmental science, engineering, and artificial intelligence increasingly rely on qualitative methods. By fostering collaborations across disciplines, researchers can generate new insights and adapt methodologies to address global challenges such as climate change, public health crises, and technological innovation.

Pedagogical Innovations

Finally, recommendations extend to the teaching and learning of qualitative research. Many students struggle with statistics due to abstract explanations or fear of mathematics. Future research should investigate innovative pedagogical strategies, such as simulation-based learning, gamification, and interactive visualisation tools, to make qualitative methods more engaging and accessible. Research into how learners from diverse backgrounds engage with qualitative concepts will help create inclusive educational practices.

In summary, future research should build upon the foundational principles presented in this book by incorporating advanced methods, embracing emerging data sources, strengthening mixed-methods approaches, and addressing new ethical challenges. Cross-disciplinary collaboration and pedagogical innovation will also be essential for keeping qualitative research relevant and impactful in a rapidly changing world. By pursuing these recommendations, researchers can ensure that qualitative methods remain both rigorous and adaptable, capable of meeting the evolving needs of scholarship, policy, and practice.

7.5 Practical Implications

Qualitative research does not exist solely in the realm of theory. Its value lies in how it informs decisions, shapes policy, and guides practice across disciplines. The principles and methods explored in this book carry significant practical implications for researchers, educators, policymakers, and practitioners. By translating methodological rigour into real-world impact, qualitative research bridges the gap between knowledge and application.

Implications for Researchers

For academic and professional researchers, the techniques outlined in this book highlight the importance of systematic design and data integrity. Careful sampling ensures that results are representative, while thorough data cleaning safeguards the validity of analysis. These practices enhance the credibility of findings, allowing researchers to contribute meaningfully to scholarly debates. The emphasis on ethical considerations also reinforces researchers' responsibility to uphold transparency, respect for participants, and reproducibility. Practically, this means that researchers who adopt these principles are better equipped to produce studies that withstand critical scrutiny and inform evidence-based practice.

Implications for Educators and Students

Qualitative research skills are increasingly essential in higher education and professional training. The practical implication for educators is the need to embed methodological rigour into curricula, using clear examples, interactive exercises, and accessible language to demystify statistical concepts. For students, the lessons in this book emphasise that qualitative research is not just about mastering formulas but about cultivating a mindset of critical inquiry. Practical applications include the ability to evaluate published research, design projects with appropriate methods, and engage confidently with statistical tools in both academic and professional contexts.

Implications for Policy and Decision-Making

In the policy arena, qualitative research provides a foundation for evidence-based decision-making. Governments, NGOs, and institutions rely on reliable data to allocate resources, evaluate programs, and assess social needs. The practical implication is that rigorous research methods—such as representative sampling and valid measurement—ensure that policies are grounded in accurate evidence rather than assumptions. For instance, in public health, statistical findings on disease prevalence inform vaccination strategies, while in education, performance data guide curriculum reforms. By following the principles described in this book, policymakers can make decisions that are equitable, efficient, and responsive to real-world challenges.

Implications for Industry and Practice

In business, healthcare, and other professional sectors, qualitative research underpins operational efficiency, product development, and service delivery. For example, businesses use surveys and market analysis to identify consumer trends, while hospitals employ statistical data to monitor patient outcomes and improve the quality of care. The practical implication here is that professionals equipped with qualitative research skills can harness data to optimise performance, reduce risk, and innovate responsibly. Furthermore, the ability to interpret summary statistics allows practitioners to communicate findings effectively to stakeholders, fostering data-driven cultures within organisations.

Implications for Technology and Innovation

The rise of digital technologies and big data has transformed the practical relevance of qualitative research. Data analytics, artificial intelligence, and machine learning all rely on the foundational principles of measurement, sampling, and statistical inference. The implication is that researchers and practitioners who understand these foundations are better prepared to engage with technological innovations, ensuring that emerging tools are applied ethically and effectively. This book provides the groundwork for navigating such technological frontiers while maintaining a critical awareness of their limitations.

The practical implications of this book are wide-ranging. For researchers, the message is the necessity of methodological rigour. For educators and students, it underscores the importance of cultivating accessible and critical statistical literacy. For policymakers and practitioners, it highlights how reliable data can drive effective decisions. And for technologists, it provides the foundation for engaging responsibly with advanced analytic tools. Ultimately, the real-world implication is that strong qualitative research skills empower individuals and institutions to act with confidence, responsibility, and impact.

7.6 Final Thoughts

The journey through the fundamentals of qualitative research presented in this book has sought to provide readers with a clear, accessible, and comprehensive understanding of key principles and practices. From the foundational concepts of research design and sampling to the technical considerations of measurement, data preparation, and statistical summarisation, each chapter has built progressively toward a fuller appreciation of the role that qualitative methods play in generating reliable knowledge. These concluding reflections aim to draw the threads together, reaffirming the value of qualitative inquiry while looking forward to its evolving role in scholarship and practice.

The Enduring Value of Qualitative Research

Qualitative research continues to hold a central place in the sciences and applied disciplines because of its ability to produce findings that are generalisable, replicable, and systematically validated. Its structured methodologies provide tools for addressing questions of scale, prevalence, and causality that are essential for informed decision-making. Despite debates about the limitations of positivist approaches, the enduring value of qualitative research lies in its capacity to reduce complexity into meaningful patterns without abandoning rigour. This strength makes it indispensable in addressing contemporary challenges ranging from public health crises to climate change.

Reflection on the Learning Journey

For many readers, engaging with qualitative research can be an intimidating experience, particularly when statistical terminology and mathematical reasoning appear overwhelming. This book has sought to demystify the process by presenting concepts step by step, supplemented with illustrative examples and accessible explanations. A key final thought is that qualitative research should not be viewed as an exclusive domain reserved for mathematicians or statisticians. Rather, it is a practical and adaptable toolkit available to all researchers willing to engage critically and ethically with data.

Balancing Tradition and Innovation

Another important reflection concerns the balance between traditional methods and emerging innovations. While classical approaches such as sampling, survey research, and measures of central tendency remain vital, researchers must also remain open to new developments. The rise of digital data, big data analytics, and interdisciplinary collaboration is reshaping how qualitative research is conducted. Future generations of scholars will need to navigate this evolving landscape by blending foundational rigour with innovative adaptability. In this sense, the final thought is one of continuity and transformation—qualitative research remains grounded in its principles but must continually evolve to meet new demands.

Ethical and Human Dimensions

It is also important to recognise that research is never purely technical. At every stage, from sampling decisions to data analysis, human values, ethical considerations, and power dynamics shape the process. A final thought, therefore, is that qualitative research must always remain accountable not only to scientific standards but also to the communities it seeks to understand. Ethical reflexivity and inclusivity are not optional extras; they are integral to the legitimacy of research. As new data sources such as online platforms or wearable technologies emerge, these ethical questions become even more pressing.

Looking Forward

In concluding this book, it is worth emphasising that learning qualitative research is not an endpoint but a beginning. The skills acquired through studying the principles outlined here provide a foundation for lifelong learning and application. Readers are encouraged to continue building upon these basics by exploring advanced methods, engaging in interdisciplinary projects, and applying their knowledge to real-world problems. The ultimate goal is not simply to master techniques but to cultivate a mindset of curiosity, rigour, and responsibility in the pursuit of knowledge.

In sum, the final thoughts of this book highlight the enduring relevance, accessibility, and transformative potential of qualitative research. While grounded in tradition, the field continues to evolve in response to technological, ethical, and societal changes. The hope is that this book has provided readers not only with tools but also with confidence—the confidence to design, conduct, and critically evaluate qualitative studies with integrity. As the boundaries of research expand, so too does the opportunity for qualitative methods to contribute meaningfully to science, policy, and practice. The final message, therefore, is one of empowerment: with sound methods and ethical commitment, researchers can generate insights that make a real difference in the world.

MIX
Papier aus verantwortungsvollen Quellen
Paper from responsible sources
FSC® C105338

If you have any concerns about our products,
you can contact us on
ProductSafety@springernature.com

In case Publisher is established outside the EU,
the EU authorized representative is:
Springer Nature Customer Service Center GmbH
Europaplatz 3, 69115 Heidelberg, Germany

Printed by Libri Plureos GmbH
in Hamburg, Germany